RESEARCH METHODS FOR THE BIOSCIENCES

This edition is dedicated to the authors' families for their love and support.

RESEARCH METHODS
FOR THE **BIOSCIENCES**

Third Edition

D. Holmes

P. Moody

D. Dine

L. Trueman

Great Clarendon Street, Oxford, OX2 6DP,
United Kingdom

Oxford University Press is a department of the University of Oxford.
It furthers the University's objective of excellence in research, scholarship,
and education by publishing worldwide. Oxford is a registered trade mark of
Oxford University Press in the UK and in certain other countries

Published in the United States of America by Oxford University Press
198 Madison Avenue, New York, NY 10016, United States of America

British Library Cataloguing in Publication Data
Data available

Library of Congress Control Number: 2016940757

ISBN 978-0-19-872849-8

Printed in Great Britain by
CPI Group (UK) Ltd, Croydon, CR0 4YY

Contents

Detailed contents

SECTION 2 Handling your data

List of boxes

Preface

Getting the most out of this book

We write this section with some uncertainty since we rarely read a preface in a book and wonder if anyone is going to read what we've written here. Nonetheless, there are several important things we need to explain to you to enable you to get the best out of this book and its Online Resource Centre, so we'd encourage you to take a few moments to read on.

Who should read this book?

This book is primarily written for undergraduates who wish to develop their understanding of designing, carrying out, and reporting research. We anticipate that graduates, although familiar with most of this material, will find this book and its Online Resource Centre to be a useful reference tool. We have therefore used examples almost entirely drawn from real undergraduate and some graduate research projects. We are indebted to all our students who have allowed us to use their ideas and data in this book.

Content

The content is designed to take you through all the steps you need to follow when choosing, planning, evaluating, and reporting undergraduate research. The content is therefore laid out in a number of sections:

How to choose a suitable topic for undergraduate research (Appendix A)

Section 1: Planning an experiment (Chapters 1–4)

Section 2: Handling your data (Chapters 5–11)

Section 3: Reporting your results (Chapter 12)

In addition, in the appendices, we have included an explanation of the mathematical processes used in the chapters (Appendix B) and a quick guide to the correct statistical test to test hypotheses (Appendix C).

There is a close and essential link between planning research and understanding how statistics fit into this process. We have demonstrated these links in Section 1 by cross references to content in Section 2. In Section 2 we have highlighted the design elements that affect the data analysis and discuss how to overcome shortcomings in experimental designs.

Getting started

You may be coming to this book with very little in the way of training in mathematics. If you do not know how to calculate this sum $(4 \times 3)^2/2$, or if you do not recognize the symbols $<$ or $>$, then we suggest you first look at Appendix B and the additional examples on the Online Resource Centre.

You may wish to analyse data you have gathered from an experiment carried out in your course. For this we suggest you start with Chapter 7, which will then direct you to the correct sections in Chapters 8 to 11.

You may wish to prepare a critique of published research. The chapters that primarily consider this are Chapters 2 and 7.

If you wish to design a research project and you are familiar with terms such as variable, parametric, aim, hypothesis, etc., then you can start with Appendix A and Chapter 2. If you are not familiar with these terms still refer to Appendix A, and then continue from Chapter 1.

Learning features

Sign posts

There are a number of ways we have tried to help you quickly find the information you need. We have included an overview ('In a nutshell') and summary in each chapter and we have included a number of cross references providing links between related content.

In a nutshell

In this chapter, we consider statistical tests and experimental designs that will b suitable for testing the hypothesis 'Do my data fit an expected ratio?' The statistic test most often used to test this hypothesis is the chi-squared goodness-of-fit tes We explain more about this type of hypothesis, how to carry out the necessary ca culations, and how to resolve some of the problems you may encounter.

The statistics tests covered in this chapter are:

8.3 Chi-squared goodness-of-fit test: one sample

8.4 How to check whether your data have a normal distribution using the ch squared goodness-of-fit test

Key terms

We know that most people will dip into this book and so we have included a glossary of most terms that you need to be familiar with towards the end. Definitions are placed in the margins in the chapters in which the terms are first used.

Boxes

In our experience we have found that students best understand the statistics element of this book if they first use a calculator to work out the calculation. Therefore we have arranged the statistical information in boxes with general details and a worked example for you to follow.

In the calculations we've included in this book we have rounded all values, usually to five decimal places. However, we carried out all the calculations using all decimal places (as you should). This means that some of our sums do not appear to quite add up. Any minor differences in the calculations as presented here should be the result of this rounding of values.

In the boxes some of the mathematical steps have not been included. We have therefore included the full calculations in the Online Resource Centre. We have also included an explanation of how to use SPSS, Excel, Minitab, and R to carry out the same calculations in the Online Resource Centre.

BOX 10.15 How to carry out a three-way factorial parametric replicates

This calculation is given in full in the Online Resource Centre. For presentation purposes all values have been rounded to five decimal places. This calculation is illustrated using data from Example 10.11.

1. **General hypotheses to be tested**

As there are six pairs of hypotheses, we have not included general hypotheses but only those from our example.

H_{0A}: There is no difference between the mean lead concentration in the soil samples ($\mu g/g$) due to distance from the smelter (km) (factor A).

H_{1A}: There is a difference between the mean lead concentration in the soil samples ($\mu g/g$) due to distance from the smelter (km) (factor A).

H_{0B}: There is no difference between the mean lead concentration in the soil samples ($\mu g/g$) due to the bearing from the smelter (factor B).

H_{1B}: There is a difference between the mean lead concentration in the soil samples ($\mu g/g$) due to the bearing from the smelter (factor B).

H_{0C}: There is no difference between the mean lead concentration in the soil samples ($\mu g/g$) due to the depth at which the soil was sampled (cm) (factor C).

H_{1C}: There is a difference between the mean lead concentration in the soil samples ($\mu g/g$) due to the

(cm) (factor C) in thei concentration ($\mu g/g$)

$H_{0(B \times C)}$: There is no i (factor B) and depth in their effects on the ($\mu g/g$) in the soil.

$H_{1(B \times C)}$: There is an i (factor B) and depth in their effects on the ($\mu g/g$) in the soil.

2. **Have the criteria for**

As far as we can tell, t (10.17.2ii).

3. **How to work out F_{ca}**

A. Calculate gener

1. Calculate the grand t together all the obser the total number of

$$\sum x_T = 155 + 96 + 365$$

$$N = 60$$

2. Square each and ever together: $\sum(x^2)$

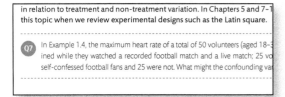

in relation to treatment and non-treatment variation. In Chapters 5 and 7–1
this topic when we review experimental designs such as the Latin square.

Q7   In Example 1.4, the maximum heart rate of a total of 50 volunteers (aged 18–3
     ined while they watched a recorded football match and a live match; 25 vo
     self-confessed football fans and 25 were not. What might the confounding va

Questions

To help you check your understanding of the topics covered by this book we have included a number of questions. The answers are provided at the end of each chapter. More questions are included in the Online Resource Centre.

Online Resource Centre

Research Methods for the Biosciences is more than just this printed book. The *Research Methods for the Biosciences* Online Resource Centre features extensive online materials to help you really get to grips with the skills you need to carry out research work. This can be found at: www.oxfordtextbooks.co.uk/orc/holmes3e/

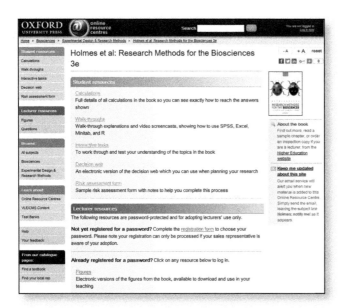

The student area of the Online Resource Centre includes:

- full details of all calculations in this book: every step in each calculation is shown so you can see exactly how we reach the answers shown in the book;

- walk-through explanations, in addition to video screencasts, showing how to use SPSS, Excel, Minitab, and R to carry out these calculations

- interactive tasks for you to work through to test your understanding of the topics in this book, and hone your research methods skills;

- an electronic version of the decision web which you can use when planning your research; and

- an electronic risk assessment which may be used to help you complete this process.

You'll see the Online Resource Centre mentioned throughout the book.

The lecturer area of the Online Resource Centre provides additional materials to make the book easier to teach from. These materials include:

- figures from the book, available to download for use in lecture slides; and
- a test bank of questions.

Simply go to www.oxfordtextbooks.co.uk/orc/holmes3e/ and register as a user of the book to gain free access to these materials.

And finally...

Just in case OUP ask us to produce a fourth edition, we would like to hear about any errors (we hope there are none) and any suggestions you have for improvements. You can contact us by using the 'Send us your feedback' option in the Online Resource Centre.

Acknowledgements

We have used many of the bright ideas and results that our undergraduates and graduates have produced to ensure that this book is relevant to you. We especially wish to thank Helen Bagley, Alys Black, Roz Chalk, Elaine Chape, Suzanne Charlton, Ruxandra Ciobotaaru, Shirley Coolican, Jeremy Cox, Stephanie Ellis, Angela Jones, Sam Fulgoni, Shaan Gabriel, Emma Georgiou, Hannah Griffin, Mikaela Haslem, Robin Holbrook, Becky Lee, Sean Macauley, Holly Merrifield, Sarah Nicholas, Dan Price, Anita Rattu, Olivia Renshaw, Michelle Robertson, Angie Roxburgh, Nicola Shale, Stephanie Stallard, Aron Stubbins, Gemma Sykes, Viv Tolley, Robert Vernon, and Gaby Wright. We also wish to thank Jack Huffer for his insights into the content of this book and to Lauren Blackburn and Catherine Woodhead for their help in the final proof reading stages of production.

We are grateful for the contributions to this third edition made by Dr Christopher Butler, who has written a new online section on US law. We have also benefited considerably from the excellent, thorough comments provided by our reviewers: Dr Stephanie Schroeder, Dr Christopher Butler, Professor Nancy E. Todd, Dr Miriam Dwek, Dr Michael Hornsey, Dr Ian Hardy, and Dr Charlotte Chalmers. We are most grateful to Minitab.com for providing us with the Minitab software. Most of the figures have been produced by Rosemary Holmes and the tables of critical values by Louise Heath. The wonderful illustrations are by Jenny Joseph.

And finally our continuing thanks go to Jonathan Crowe, Alice Roberts, and OUP for giving us the opportunity to develop this book, to review our thinking about research methods, and to explain our understanding of this topic to you.

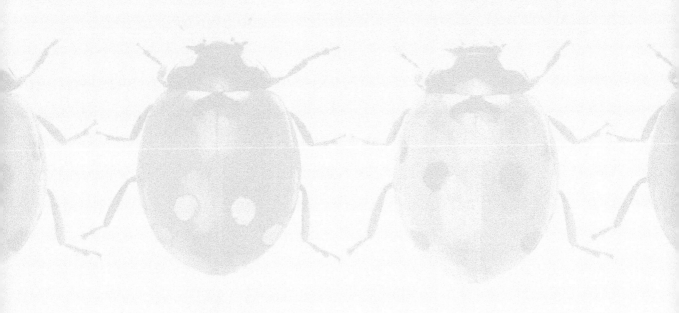

Section 1

Planning your experiment

Where do I begin?

 In a nutshell

This chapter gets you started with your research and explains some of the terms you will need to be familiar with when designing or reviewing experimental work. First you need to define an aim and at least one objective; then you will need to decide what you are measuring and whether you are collecting these measurements (observations) from the statistical population or a sample. Finally, you need to decide if your data can be analysed by testing hypotheses. Having introduced these ideas here, these points are developed in subsequent chapters along with further demonstrations of how these terms are used in practice.

You may be using this book because you are starting to plan a piece of independent research, you have been carrying out an experiment designed by someone else, or you are reading about published research. This book is written for those of you who need to have an understanding of experimental design and, as part of this, need to develop your statistical skills. We begin in this chapter by introducing you to some of the basic principles and terms that you will come across in your research. If you are about to embark on a piece of research and need help choosing a research topic, we have included some suggestions as to the approach to take in Appendix A. If you are working through this chapter, it should take about 1 hour to complete all the questions. The answers are at the end of the chapter. Nearly all our examples are based on undergraduate research projects. If these examples are not in your subject area, you will find more in the Online Resource Centre.

1.1 Aims and objectives

When designing research, you start by choosing an aim and usually then subdivide the work you are doing under the heading 'objectives'. It is a common failing to confuse these two terms. This can lead to a lack of direction in the research and research report. All investigations require at least one aim and at least one objective. You therefore need to understand the distinction between these terms and why this difference is important.

1.1.1 The aim

Aim
A generalized statement that describes the area you are investigating.

Most research is developed through a series of investigations. All the investigations are carried out within a specific area of interest. This area of interest is the aim.

Example 1.1 The effect of the introduction of new individuals on behaviour in an established herd of wildebeest

An undergraduate had links with a local safari park and was interested in the changes in behaviour of animals when new individuals were introduced into an established herd.

In Example 1.1, the student's aim was therefore to 'investigate the effect of the introduction of new individuals on behaviour in an established herd of wildebeest'. Most students are very good at writing aims even if they are not aware of it. If you are in any doubt as to how to write an aim imagine yourself writing a report based on the investigation; the title you would use for this report is usually very similar to the aim as it has the same role providing an overarching statement that explains what you are investigating. So if in doubt write a title and use this as your aim.

Example 1.2 The uptake of heavy metals by *Calluna vulgaris* near electricity pylons

An undergraduate wanted to study the capacity of *Calluna vulgaris* (heather) to take up heavy metals from soils contaminated by wash from electricity pylons. They examined the soils around seven pylons that were present on a heath and compared the heavy metal concentration in the soil samples to samples taken 1km away.

- -

 How might the aim be phrased?

- -

1.1.2 The objective

To thoroughly investigate the topic described by an aim can require one or more experiments, each of which will examine one aspect of the overall topic. The experiments will have a particular design and record particular data. It is therefore important to have objectives that provide additional information about each step of the overall investigation. This enables the researcher to keep track of their work, and anyone reading their reports can see how the separate experiments relate to each other and to the aim. One way to imagine the difference between an aim and objective is to think of yourself looking at the sky; this broad perspective is the visual equivalent of an aim. If you now take a telescope and look at parts of the sky in detail these separate views that are part of the whole are the equivalent of objectives.

An objective
Provides succinct and specific details about an experiment that is being carried out in order to examine all or part of an aim.

Example 1.3 The effect of cell culture volume on parameters of growth of tobacco cells in liquid culture

Researchers were interested to see whether they could effectively culture tobacco cells in microfuge tubes (0.5ml) rather than as larger cultures (>5.0ml). The effectiveness of culturing in a small volume of 0.5ml was tested under a range of temperatures and sub-culturing methods. To assess the effects of these treatments the cell viability and cell number were recorded.

The aim for Example 1.3 was therefore to examine the effect of culture volume on the growth of tobacco cells. The objectives were:

1) To examine the effect of temperature (20°C, 25°C, 30°C) after 24 hours on the viability of tobacco cells in 5ml and 0.5ml (microfuge) cultures.

2) To examine the effect of temperature (20°C, 25°C, 30°C) after 24 hours on the number of tobacco cells in 5ml and 0.5ml (microfuge) cultures.

3) To examine the effect of the frequency of sub-culturing on the viability of tobacco cells in 5ml and 0.5ml (microfuge) cultures.

4) To examine the effect of the frequency of sub-culturing on the number of tobacco cells in 5ml and 0.5ml (microfuge) cultures.

i. **Clumpers or splitters**

Objectives can be 'clumped' or 'split'. In Example 1.3, it would be possible to rephrase the four objectives outlined so we only have two:

5) To examine the effect of temperature after 24 hours on tobacco cell viability and cell number in 5ml and 0.5ml (microfuge) cultures.

6) To examine the effect of the frequency of sub-culturing on tobacco cell viability and cell number in 5ml and 0.5ml (microfuge) cultures.

When reading, these two objectives are much neater and less repetitive than the previous four and can therefore be a more efficient way of reporting your investigation. However, you may wish to use your objectives as a basis for reporting your results. In our first version when our objectives are split into four statements, this works well as each objective has only one data set (either cell viability or cell number). In the second 'clumped' version, there are two data sets for each objective (cell viability and cell number). Using only two objectives as a basis on which to present your results may therefore be less suitable. In this book, we will therefore be 'splitters' and have more rather than fewer objectives. You may like to discuss this aspect of any report you prepare with a member of staff and either be a 'splitter' or a 'clumper' according to their advice.

ii. Only one objective

In undergraduate work, it can be difficult to appreciate the need for objectives, as many investigations of a topic only involve a single experiment. This is particularly true of under-graduate practicals. You may not therefore have many opportunities to see the real value of having objectives. Where there is only one experiment carried out to examine an aim, the aim looks almost identical to any objective. Hence the confusion that can arise in understanding what is an aim and what is an objective.

iii. Personal and experimental objectives

Objectives can be personal or experimental. For example, in the experiment on the tobacco cells in culture (Example 1.3) there are four objectives. These all describe experiments. How-ever, you could add two more objectives to this list:

1) To analyse the data.
2) To write a report.

These are personal rather than experimental objectives. In most undergraduate reports and any publications where objectives are included, you will only see experimental objectives. However, when you are asked to include objectives, you may need to check that only experi-mental objectives are wanted.

1.2 Data, items, and observations

Data
A number of observations or measurements on the subject you are investigating.

Observation
A single measurement taken from one item.

Item
One representative of the subject you wish to measure.

When carrying out an investigation, you will generate data (singular—datum). Data consist of a number of measurements or observations on the topic you are investigating.

The individual from which the measurement is taken is usually called the item. We use these terms throughout the book; however, in our Online Resource Centre, we also show you how to use the statistical software SPSS to carry out calculations. In this software, the term 'case' is used in place of the word 'item'. Throughout the examples in the chapters on statistics tests (Chapters 7–11) we show you which is the item and which the observation.

Example 1.4 The effect on heart rate of watching live and recorded football matches in self-declared football fans and non-football fans

One undergraduate wanted to assess the effect of watching live and recorded football matches on the heart rate of self-confessed football fans and those who were not football fans. Having identified a number of volunteers, their heart rate was recorded first during a recorded football match and then during a live match.

In Example 1.4, each person taking part is the item and each recorded heart rate is an observation. All the results together are the data.

Q2 Examine Example 1.3. What are the items and observations?

1.3 Populations

There are several different definitions for the word population. In ecology, the term usually means the individuals that can be observed within an identifiably discrete group; for example, the number of individuals in a wood, river, or city. In genetics, a population usually means the grouping in which all the individuals of a particular species mate at random. When you are using statistics to analyse your data, the term population is used in yet another way. A statistical population is all the individual items that are the subject of your research.

In Example 1.1, the effect of the introduction of new individuals on behaviour in an established herd of wildebeest was examined. Here the statistical population would be all the wildebeest in the population at the safari park. You might argue that safari park conditions are sufficiently similar that the statistical population might be extended to all safari park populations of wildebeest. However, it would be difficult to argue a case for extending the statistical population to all wildebeest including those in the wild.

The statistical population reflects your aim. Sometimes a statistical population will be the same as an ecological population. For example, if your aim was to investigate the characteristics of individuals of one species within a wood, then the ecological population and the statistical population are the same. However, if your aim was to examine the characteristics of a particular species worldwide, then the statistical population will be all individuals of that species on the planet. The ecological population will still be the group of individuals of that species in each wood.

Identifying the statistical population is of great importance to your work. Any hypothesis testing revolves around the statistical population. Similarly, when you come to discuss your results you will automatically apply your findings to the statistical population. For example, having studied wildebeest behavior at a safari park (Example 1.1), the student presented some suggestions on the management of such populations. She needed to make it clear in her report whether her recommendations were just for the safari park she studied, for safari parks in general, or for the management of all wildebeest. Her previous consideration of the 'statistical population' meant she was sure where her results could be justly applied.

Statistical population
All the individual items that are the subject of your research. A statistical population may be the same as an ecological or genetic population but not necessarily.

 Q3 In Example 1.4 where the effect of watching live and recorded football matches on the heart rate of self-confessed football and non-football fans was being examined, the student decided to limit his study to undergraduates in an English university with an age range of 18–30 years and no known health problems. What is an appropriate statistical population?

1.4 Sample

It is often not possible to study the whole statistical population because of practical and economic constraints. The testing procedure may also be destructive and you may therefore wish to limit your investigation. For example, if you wanted to carry out taste testing on cheese and onion crisps, it would not help the factory managers if you tasted all the crisps. Usually a subset of the population is examined. This is the sample.

For example, in the investigation into the effect on heart rate of football and non-football fans when watching live and recorded matches (Example 1.4), the student might study only five Manchester United football supporters and one of their non-football-fan friends. But is this a representative sample?

1.4.1 Representative samples

If it is not possible to measure the whole statistical population, then a representative sample is required. A representative sample should be obtained by a clearly defined and consistently applied method. The observations recorded by this method should closely match the result that would be obtained if every item in the population was measured.

In Example 1.4, if only five Manchester United fans and one non-football fan were studied, this sample is not representative of the number and range of football fans and non-fans. To resolve this, the student might decide to test the non-fan five times as often to obtain a similar number of observations as the observations of the five football fans. The measurements on the non-fan will be more similar to each other than would be expected to be obtained from five different people. This is an example of a lack of independence. The five measures on the non-fan are not independent of each other, whereas the single measures on each of the five fans are independent of each other.

A mix of independent and non-independent measures will also not lead to a representative sample. What should be done in the case of Example 1.4?

1.4.2 How do you obtain a representative sample?

There are a number of sampling methods that may be used. Most of these are designed for particular circumstances and to produce a representative sample. We describe the common methods here. The implications of your choice are explained alongside the specific statistical tests and should be considered before you finalize your method.

i. Random sampling

Here each item must have an equal chance of being sampled each time. This method avoids conscious and unconscious bias. On average, as long as the sample size is not small, random sampling does produce a representative sample and is therefore in common usage.

How do you randomly sample? In ecology, a common error is to believe that throwing quadrats over your shoulder is going to provide you with a random sample. This is untrue: you will only sample within a small area around yourself and will tend to throw forwards or backwards because this is how arms work. A better approach is to use a random number table to identify an item within your population.

For example, there are a number of practical approaches the undergraduate might use to sample football and non-football fans (Example 1.4). One would be to survey as many people as possible to identify those willing to take part, who are within the age range 18–30 and where there are no known health problems. If these are then grouped into self-confessed football fans and non-football fans, it would be possible to allocate each person a number

Sample
A number of observations recorded from a subset of items from the statistical population.

Representative sample of items
A sample that generates sample statistics that closely match the population parameters through the use of a clearly defined and consistently applied method.

Independent observations
Where only one observation is measured from each item for one treatment variable.

Random sampling
Each item must have an equal chance of being sampled each time.

and use a random number table to select individuals from this pool of volunteers. The number of volunteers (i.e. sample size) is something we will come back to (2.2.6, 3.5, and 6.2).

ii. Systematic/periodic sampling

Systematic/periodic sampling is a regularized sampling method, where an item is chosen for observation at regular intervals. This method can be suitable, for example, for examining the general distribution of plants across a footpath using a line transect where data are collected at regular intervals. This method is also commonly used when studying the distribution of flora and fauna along a shore from cliff to sea.

Systematic/periodic sampling
A regularized sampling method where observations are recorded from items chosen at regular intervals.

iii. Stratified random sampling

This is a combination of the two methods outlined above, where random sampling occurs at regular intervals. Thus, an area may be divided into sections, and random sampling occurs within each section, the numbers sampled reflecting the relative size of the section. An example of this would be where soil samples are collected from a contaminated site and where it is important that all parts of the site are represented within the sample. The site can be divided into a number of discrete areas and random samples are taken within each area.

Stratified random sampling
Where random sampling occurs at regular intervals.

iv. Homogeneous stands

In some work, you may need to be selective and to identify areas for your investigation that are homogeneous, i.e. similar. In the UK, this approach is most common when collecting data about a plant community and assigning a National Vegetation Classification (e.g. Rodwell, 1991). Full details about this approach, including suitable quadrat sizes for the canopy, ground, and field layer, are given within this series of books.

Homogeneous stands
Where a number of areas that are similar to each other are selected for investigation.

Q4 Anthills are dispersed throughout a grassland. You wish to study the percentage coverage of specific plant species on the anthills. What sampling strategy would you use?

1.5 Population parameters and sample statistics

Later in this book we introduce you to mathematical methods that you may apply to your data: both the methods used to summarize data (Chapter 5) and those used to test hypotheses (Chapters 7–11). Where the answers from these calculations relate to a population, they are usually called parameters. Where they relate to a sample, they are called statistics. Most research, of necessity, collects observations from samples, and therefore all our chapters focus on designing investigations and evaluating data from samples. Even so, there are frequent occasions when you will come across terms that relate specifically to a statistical population or to a sample.

1.5.1 Mathematical notation for populations and samples

In statistics, we use mathematical notation. For those not familiar with this notation, we have included further explanation in Appendix B. Different symbols are used if we are thinking about populations or samples. The most common symbols are shown in Table 1.1. Different symbols are used because there are important differences in the meanings and methods involved when calculating parameters of populations or statistics of samples. More information, including an explanation of the terms mean, variance, and standard deviation, is given in Chapter 5.

1.5.2 Calculations

Some mathematical tests exist in two forms, one suitable when all population values are known and one when a sample is being studied. One value you may often use is called a variance. We look at these equations in Chapter 5. The variance you calculate may be a population parameter and in this instance the formula for the variance from a normal distribution can be written as:

$$\sigma^2 = \frac{\sum (x - \mu)^2}{N}$$

If the variance you are working out is for a sample, then the equation you use is different and is written using notation relating to the sample:

$$s^2 = \frac{\sum (x - \bar{x})^2}{n - 1}$$

Table 1.1 Symbols used to indicate population parameters and sample statistics

Term	Population	Sample
Mean	μ	$\bar{x}$
Variance	σ^2	s^2
Standard deviation	Σ	s

(Two of these symbols are lower case Greek letters pronounced 'mu' (μ) and 'sigma' (σ).)

(Where Σ means 'sum'; $\bar{x}$ is the sample mean; x is the observation; N in this context is the size of the population; and n is the sample size.)

Calculators and computer software invariably have the facility to calculate both population parameters and sample statistics, and it is therefore important to ensure that you are using the correct version. Calculators and computer software often indicate the version being used either by the symbols shown in Table 1.1 or by the symbols n (population) or $n-1$ (sample). This is examined again in the Online Resource Centre.

1.6 Treatments

Treatment is a word that has two valid meanings and both are used in this book. The examples we use here appear in Chapters 12 and 9, respectively. In the first definition the word treatment is used to mean the particular factor you are manipulating in your experiment so that you can see what effect it has. In this instance you are indeed treating your samples. However, when data are collected they can be arranged into tables and sometimes there are cells or blocks of one or more numbers in these tables. These are also called treatments.

> **Treatment (definition 1)**
> When carrying out an experiment, your items are exposed to a particular environment that is manipulated by you, the investigator.

> **Example 12.1** The response of tobacco explants to auxin
>
> In an undergraduate investigation into the effect of different concentrations of auxin on the growth of tobacco in tissue culture, the relative increase in diameter (mm) of leaf explants was recorded after 2 weeks.

> **Treatment (definition 2)**
> A statistical term in which any samples or groups of observations that are being compared are called treatments.

In this example from Chapter 12, the tobacco explants were *treated* by exposure to different concentrations of auxin.

> **Example 9.1** Shell colour in *Cepaea nemoralis* in coastal and hedgerow habitats
>
> An investigation was carried out into the frequency of banding and colour patterns in snail shells in two different habitats (a coastal region and a hedgerow).

In this example from Chapter 9, although the investigator has not imposed these habitats on the snails, when the data are analysed each sample is called a treatment. Thus, the coastal region is one 'treatment' and the hedgerow is a second 'treatment'.

The word treatment is often found coupled with other terms; hence, you will come across treatment variable, non-treatment variable, and treatment effects. In all these contexts, it is the broader definition of 'treatment' (definition 2) that is used.

 In Example 10.2, representative samples of *Littorina obtusata* and *Littorina mariae* were collected from Porthcawl in 2002 and their shell height (mm) recorded. The investigators wished to test the proposal that there is no difference between shell height of the two groups of periwinkles (*L. obtusata* and *L. mariae*). What is/are the treatment(s)?

1.7 Variation and variables

1.7.1 Variation

Treatment variable
The factor(s) under investigation where you may examine your data for a difference, or the effect of the treatment variables, or you are looking to see if there is an association.

In any investigation, you will record observations. Usually these observations do not all have the same value—they vary. A set of observations is said to show variation. The factors that bring about this variation are called variables. They can be divided into a number of types of variables. The first division is between the factors that are part of your experiment and those that are not.

1.7.2 Variables that are designed to be part of your experiment

The variables that are part of your experiment fall into two groups. In the first group there are treatment variables. You may look for some type of difference or the effect of the treatment variables, or for an association.

Dependent variable
Variables that you measure to assess whether they respond to the treatment variables. Not all investigations include dependent variables. This term does not automatically imply that there is a significant biological relationship between the dependent variables and treatment variables.

In addition to treatment variables there are sometimes, though not always, other variables that you measure. These are used to assess the effect of the treatment and are known as respondent or dependent variables. The term 'dependent' variable unfortunately implies that you already know that this variable is affected by the factors you are investigating, which is most definitely not the case. We have decided, however, to use this term rather than 'respondent variable' as this is the phrase that is most often used in published research and other texts.

To illustrate how the terms treatment variable and dependent variable are used we can re-examine the examples in this chapter and identify the treatment and dependent variables in each.

In Example 1.1 the factor that is being investigated is the introduction of new individuals into the wildebeest herd, so this is the treatment variable. The investigation also involves measurements of behaviour. These are dependent variables.

In Example 1.2 the treatment variable is the presence/absence of the electricity pylon. There is a second variable that is being investigated which is the concentration of heavy metals. This is the dependent variable.

In Example 1.3 the investigation is all about the culture volume (0.5ml compared to 5ml). This then is the treatment variable. The responses of two other variables (cell viability and cell number) were recorded. These are dependent variables.

In Example 1.4, 50 volunteers took part: 25 were self-confessed football fans and 25 were not football fans. All volunteers were fitted with a heart rate monitor and as a group were shown a recorded football match. Then all the volunteers attended a live football match. Maximum heart rate was recorded for each volunteer at each event. Identify the treatment variables(s) and dependent variable(s).

There are three more terms relating to variables that you need to become familiar with. These are 'fixed', 'random', and 'independent'.

All the treatment variables we have looked at so far are under the control of the investigator. For example, when the investigator was looking at the effect of culture volume on cell growth they would be certain to ensure that every replicate with 0.5ml culture volume would indeed have a volume of 0.5ml. Investigations where the conditions are exact like this are called 'fixed'. In some experiments this is not the case. For example, if we wished to examine the association between height and weight by measuring these characteristics in 100 females then both height and weight are factors that we are investigating and these are our 'treatment' variables. However, clearly neither is under the investigator's control, they are both 'random', and both are subject to sampling error. We consider this again in Chapter 10 when you are more familiar with the concept of sampling error and when it becomes relevant to your selection of particular statistics tests.

The final term you need to be familiar with in relation to variables is 'independent'. In some situations, usually where you are plotting data on a scatter plot and have one treatment variable (x axis) and one dependent variable (y axis), the treatment variable is then referred to as the independent variable. Again this can be confusing as this term also implies that the variation in the dependent variable is directly caused by the changes in the independent variable—you do not know this until your research is complete, and only then you may then show this to be the case.

Having this number of terms that all describe the variables in your research is not helpful, so authors select which to use so that they use clear and consistent language to convey their meaning. In this book we have decided to use the terms treatment variable and dependent variable. To choose the correct statistics test you have to be able to identify the treatment variables and whether these are fixed or random, and the dependent variables. To help you become proficient we have explicitly annotated all the examples in Chapters 7–11 with these details so that you can see how these various terms are used.

Fixed variable

When each treatment is consistent every time, there is no sampling error. Most fixed variables are under the control of the investigator.

Random variable

When the variable is not under the control of the investigator and is subject to sampling error.

Independent variable

Is a term usually used when the data are plotted on a scatter plot and describes the treatment variable plotted on the x axis.

1.7.3 Confounding variation: confounding variables

Variation between observations can be due to many factors. For example, the variation in human height is due to genetics, diet, and disease. If you wanted to investigate the effect of diet on human height, how can you decide how much of the variation is due to diet and how much variation is due to other factors? As an investigator you want to be able to 'partition' the variation in your data:

In our example this would be:

Variation in human height = variation in height due to diet
+ variation in height due to all other factors

There are three terms that are used in the context of variation due to 'other factors': confounding variables, confounding variation, and sampling error (1.7.4). The distinction between their meanings is subtle and as a result the terms are often used interchangeably. We have elected, however, to use these as three separate terms.

Confounding variation

Variation in your data that is caused by the combined effect of confounding variables and sampling error. You need to explicitly work to minimize the confounding variation in your experimental design.

Confounding variables

Variables that are outside the scope of the investigation but may affect the results.

Confounding variation arises from two elements: confounding variables and sampling error, both of which must be explicitly managed in your experimental design. The confounding variables are variables that are outside the scope of the investigation but may affect the results. You are very familiar with this form of non-treatment variation as you have been taught to control factors such as temperature or pH when these are not part of your experiment. In our example on the effect of diet there are many other variables that may affect the observations and these need to be explicitly managed. The most obvious of these factors are genetics, age, and gender.

It is important for you to reduce the variation in your data that results from the confounding variables, otherwise the effect you are trying to understand and identify may be masked by all this experimental background noise. You therefore need to identify the confounding variables and then consider how to manage both these and the effect of sampling error. Two sources of information can be used to help you identify the confounding variables: background reading and pilot studies. Reading about the techniques you are planning to use and other studies similar to your own can enable you to quickly identify many of the factors that can impinge on your experiment.

1.7.4 Confounding variation: sampling error

Sampling error is the term used to describe the chance effects that can occur in an experiment and cause your samples to differ from the statistical population from which they are derived. We can illustrate this best with an example.

Example 1.5 Sampling error in a study of *Cepaea nemoralis* in Jack's Wood

As part of a study of the ecological genetics of *Cepaea nemoralis*, the shell patterns of the snails in a wood were recorded. Every snail in the area (i.e. the whole population) was examined and their shell patterns recorded. Two days later, three samples were collected at random and their shell patterns were also recorded (Table 1.2).

Table 1.2 Percentage of *Cepaea nemoralis* with particular shell patterns in Jack's Wood

	Cepaea nemoralis with a particular shell pattern and colour (%)			
	Banded Yellow	No bands Yellow	Banded Pink	No bands Pink
Sample 1	90	0	5	5
Sample 2	20	5	30	45
Sample 3	0	50	50	0
Population	30	10	30	30

Clearly the samples differ from the population results and from each other. This variation is said to be due to sampling error. By chance, the samples collected did not reflect the actual population frequencies of shell pattern in this species. Two important effects of sampling error are firstly that the samples may not reflect the statistical population values, and secondly that the sampling error can lead to variation between samples by chance. A further graphic example which you can easily try for yourselves is to record sock colour in students in a lecture theatre, comparing those in the back rows with those in the front rows. These two samples are never identical. Assuming that there is no reason for those with a particular sock colour to sit in a particular position this then this difference is entirely the result of sampling error.

Sampling error Variation between samples collected from a single population that has occurred by chance and the variation between the statistical population value and sample value that has arisen by chance.

1.7.5 Minimizing the effect of confounding variation

To be able to quantify any effect of your experimental system you need to minimize the confounding variation. This can be achieved in four ways: 'Reduction', 'Equalize the effect', 'Randomization', and 'Partition'.

i. Reducing the confounding variation

Having identified the confounding variables it is often possible to develop a method which will remove as many of the causes of confounding variation from the experiment. This is often one of the central reasons for developing laboratory rather than field-based investigations.

ii. Equalize the effect

When within your experiment you wish to compare one treatment to another or one sample to another you can reduce the effect of the confounding variation by ensuring that the confounding variation is similar or kept constant in either treatments or samples. This is a well-known approach and you probably take for granted the need to control temperature, pH etc. for a 'fair test'. This type of control is not what is meant by the term 'control' which we encounter in the next chapter. These are important but different aspects of designing an experiment.

iii. *Randomization*

Randomizing your items or samples or blocks of items in relation to your treatments is a common practice that, on average, will act to reduce the impact of confounding variation. It is another way of seeking to evenly distribute the effect across your experiment. What you randomize will depend on your investigation and the confounding variables. For example, items may be randomized to treatments, the treatments may be randomized in space, or the items in a sample may be randomly collected from the population.

iv. Mathematically separate out (partition) the variation

There is a group of statistics tests called Analysis of Variance (ANOVAs) that will mathematically partition the variation in your experiment due to the factors you are investigating and the confounding variation. Therefore, when you are planning your experimental design you should investigate whether you can use the designs suitable for analysis using an ANOVA (10.6). To partition variation in this way requires replication within the experimental design.

In the rest of this book, we develop these ideas and show how experimental design and data analysis can be used as powerful tools to allow you to examine the effect of the treatment(s) within your research. In Chapters 2–4, we begin by considering good practice in experimental design, and in Chapter 6 we focus on hypothesis testing and data analysis in relation to treatment and non-treatment variation. In Chapters 5 and 7–11, we return to this topic when we review experimental designs such as the Latin square.

 In Example 1.4, the maximum heart rate of a total of 50 volunteers (aged 18–30) was examined while they watched a recorded football match and a live match; 25 volunteers were
self-confessed football fans and 25 were not. What might the confounding variables be?

1.8 Hypotheses

As part of the process of designing your experiment, you need to give some thought as to how you will analyse your data. This is really important as it can help you decide how many samples or observations you need and ensures your experiments are likely to produce data that can be analysed. There are two main approaches to analysing data. Either you choose a null hypothesis to test your data against or you analyse your data to see which of a number of alternative models best fits it (information-theoretic approach).

1.8.1 Null hypotheses

Hypothesis
The formal phrasing of each experimental objective; includes details relating to the experiment and the way in which the data will be tested statistically.

Experiments set out to test a specific question. This question may be written in a format called a hypothesis (plural hypotheses). Chapter 6 introduces you to hypothesis testing and explains in more detail how to write hypotheses, why this format is used, and why this is the basis of one approach to statistical analysis.

One aspect of the experiment outlined in Example 1.3 was to assess the effect of culture temperature on tobacco cell viability. The hypothesis would be:

H_0: There is no significant difference in the median cell viability of tobacco cells in 5ml and 0.5ml (microfuge) cultures in response to temperature (°C).

H_1: There is a significant difference in median cell viability of tobacco cells in 5ml and 0.5ml (microfuge) cultures in response to temperature (°C).

From the example we include here, you can see that hypotheses come in pairs. These two hypotheses are called the null hypothesis (H0) and the alternate hypothesis (H1). The word 'significant' has a specific meaning in statistics that we will discuss later (Chapter 6). For the moment, you can think of it as being used in the 'everyday' sense of 'big enough to be important'.

In many experiments, you may choose to identify a null hypothesis and then test your data to see if this proposed model appears to probably explain the data. Most often, this null hypothesis is a general statement that there is no difference between sample A and sample B that cannot be accounted for by chance. For example, an undergraduate wished to examine the differences in seed quality between seed sold by a number of different companies. After carrying out a seed purity experiment, the student tested her data against a 'general' null hypothesis that there was *no difference* in the seed purity.

You may wish to devise a null hypothesis other than 'there is no difference'. We introduce you to some 'specific' hypotheses in Chapters 8–11. However, most statistical tests included in this book are designed to test 'general' null hypotheses. In Chapters 2 and 3 we discuss the design of investigations. There are more examples of developing hypotheses in these chapters. Exercises for you to test your understanding are included in the Online Resource Centre.

1.8.2 Information-theoretic models

In this alternative approach to hypothesis testing you collect data and then assess it to see which are the mathematically distinctive groupings within the data. This alternative approach to analysing data is becoming increasing popular. One occasion when you may encounter this is in studies where little is known about the topic you are investigating. For example, you may be surveying a particular location such as woodland to see which species are present and the relative abundance of each species. There are mathematical techniques that will allow the data to be grouped into observations that are similar. For example, a forensic science undergraduate carried out a survey of diatom species at two rivers and a number of sites along these rivers during the year. She used cluster analysis to see whether there were distinct groupings within the data. She was then able to see whether these groupings related to location, season, or other factors. This is a more open approach to evaluating your data. You do not compare your data to the one model you devised before you analysed your data. Instead, you *allow* the data to indicate a suitable model. This is known as an information-theoretic model.

There is often a concern that all experiments must be tested against a null hypothesis. This is not the case for many types of investigation. There is also an argument that, by only testing against one null hypothesis, the data analysis is constrained and less information can be discovered by this approach than by taking an information-theoretic model approach. At present, the use of an information-theoretic approach is rare for undergraduate research other than when surveys of species are carried out or if you have large samples and many variables in behavioural studies. We have not therefore covered this approach, but refer you to texts such as Fielding (2007) and Quinn & Keough (2002). Even if you do take this approach, you may also wish to summarize your data (Chapter 5) and use tables and/or figures to communicate your findings (Chapter 12).

Information-theoretic models
When data are collected and analysed to see whether there are significant groupings. Only then may hypotheses about the factors causing these groupings or the nature of these groupings be tested, i.e. models are constructed from the observations.

Summary of Chapter 1

- To provide a focus for your research, you will have an aim and objectives. These are not the same thing, and when designing your investigation you will need to distinguish between them (1.1 and developed in Chapter 2).

- Within your investigation, you will record observations for particular items. You may collect observations for every item in a population or you may sample (1.3, 1.4, and 1.5, and developed in Chapter 2).

- When sampling, your intention is usually to obtain a representative sample that reflects what is happening in the population. Statistical tests, including those programmed in calculators and computer software, use different symbols when referring to populations and samples (1.3, 1.4, and 1.5, and developed in Chapters 2 and 6).

- The term 'treatment' is used when designing and evaluating experiments. There are two definitions and both are used in this book (1.6 and developed in Chapter 2).

- In most investigations, the observations do not all have the same value—they vary. This overall variation can be portioned into treatment variation and confounding variation.

  - The factors that you are studying are called the treatment variable(s) and the dependent variables(s) (1.7 and Chapters 2, 6–11).

  - Confounding variation can arise from confounding variables and/or sampling error (1.7 and applied in Chapters 2, 6–11).

- Some investigations start out with a question (hypothesis). When you use statistics to test the hypothesis, then the hypothesis is phrased in a specific way (1.8 and developed in Chapters 7 and 8).

- The Online Resource Centre includes interactive exercises that test your understanding of this chapter with other topics, particularly those considered in Chapters 2, 6, and 7.

Answers to chapter questions

A1 Aim: The capacity of *Calluna vulgaris* to take up heavy metals from soils contaminated by wash from electricity pylons.

A2 The items are the tobacco cells in their particular culture vessels. For example, if the researcher used ten 0.5ml cultures, each of these ten cultures is an item. The same applies to the 5ml cultures. The observations are the number of viable cells and the total number of cells in samples taken from each culture.

A3 There are two statistical populations here. Self-confessed football fans, aged 18–30 with no known health problems, and self-confessed non-football fans aged 18–30 with no known health problems.

A4 There may be several practical and effective answers. Here is ours. Take numbered plant pot labels with you and place one in each anthill. Number a corresponding set of squares of paper and place in an opaque container. In planning your investigation you will already have determined how many anthills need to be sampled. Draw out this number of squares and these will give you the number of the anthills you then include in your random sample.

A5 For this example you would use definition 2 to explain your use of the term treatment. The treatment here is 'species of *Littorina*'.

A6 In this example the investigator was examining two factors. The first factor was the 'live' versus 'recorded' football matches. The second factor was the 'football fans' versus 'non-football fans'. These are both treatment variables. There is another variable: the heart rate. This is the dependent variable.

A7   There are a number of potentially confounding variables. You may identify others. Confounding variables might include: gender; an assumption that volunteers correctly identified themselves as fans or not fans; social interaction between volunteers; personal preference or dislike for particular teams; and resting heart rate.

2 Planning your experiment

 In a nutshell

In this chapter, we take you through the process of planning your experiments. First we consider how to critically review published experimental designs. The first step in this process is to identify the aim and objectives and then to consider the strengths and weaknesses of the work. In doing this, you will develop your ability to be a critical reader and extend your understanding of what makes a competent experimental design. This topic complements Chapter 7 where we consider the role of the critical reader in relation to published data.

In the second part of this chapter, we take you through the process of designing your own experiments using the decision web that is available in this book and online. If you are unsure about a topic to investigate, refer to Appendix A. As we take you through the process of planning an experiment, we apply the principles introduced in Chapter 1 and in this way develop your understanding.

The best way to learn about the design of investigations is to look at what other people have done and to have a go yourself. In this chapter, we take you through both these approaches. The aim is to show you that you are already very able to understand the strengths and weaknesses of an experimental design and identify the ways in which the design might be improved. We also provide information about the process of designing experiments and the things you need to consider.

Designing an experiment requires an understanding of certain design principles. One of the most common problems in designing research is that many people ignore the importance of considering how their data will be analysed before they begin their investigation: they only come to this when the data are collected and they then find out that the data cannot be analysed. This means that, although in this chapter we cover all the steps you need to consider when designing an investigation, you may not fully appreciate the importance of some of our comments until you have become more familiar with the content of other chapters relating to statistical analysis.

If you work through this chapter, it should take about two hours to complete all the exercises. The answers for these exercises are at the end of the chapter.

2.1 Evaluating published research

Evaluating published research requires you to be a critical reader. In this context 'critical' does not mean 'negative' but more that you do not take anything for granted. Just because research is published does not always mean it is good work and it is important that you learn how to judge the merits of scientific studies. To do so you need to examine a number of aspects including the experimental design, data analysis, and general communication of the work. In this chapter we focus predominantly on the experimental design aspects of a study. This complements 7.3 where we consider how to critically review the data analysis and reporting of the analysis.

Example 2.1 is based on a short research paper that appeared some time ago in a prestigious journal. Read it through and, if possible, discuss this example of an experimental design with your friends and combine your ideas. Think about the following three questions:

1) What are the aim and objective(s)?

2) What are the strengths of this experimental design?

3) What are the weaknesses of this experimental design?

Example 2.1 Social drinking and morning-after breath alcohol levels

The investigator wished to study the level of morning-after breath alcohol concentrations after an evening of social drinking to see how these levels were affected by the levels of habitual intake. An important element of this investigation was that real social drinking behaviour was examined rather than effects observed in a laboratory trial.

(continued)

> Fifty-eight men aged 20–50 years took part in the study. Habitual alcohol intake was estimated by a 4-week prospective drinking diary or accounts of the frequency of drinking and quantity per session in a typical month. The participants were asked to count the number of drinks they consumed at an evening social event where they anticipated drinking heavily. Other people present were asked to verify the amount drunk and the time that drinking stopped. All the men ate shortly before or during drinking. The morning-after alcohol concentrations were then recorded in each person's home 7–8 hours after drinking had stopped (Wright, 1997).

2.1.1 What are the aim and objective(s)?

The aim describes the broad area being examined by the researcher, and the objectives break this down to reflect the various investigations that were carried out. We identify the aim and objectives for the study described in Example 2.1 as follows.

Aim: The influence of habitual intake on morning-after breath alcohol concentrations in men.

Objectives:

1) To quantify the habitual intake of alcohol in 58 men.
2) To assess morning-after breath alcohol levels in 58 men.

Compare our aim and objectives with your own. How, and in what way, did they differ?

2.1.2 Strengths of the experimental design

It is always easy to be critical, and sometimes overcritical, of research. It is therefore best if you start by considering what is good about the experimental design before considering the experiment's limitations. What strengths have you listed for this experiment? We identified the following strengths:

- The experiment was carried out in the environment in which this activity usually takes place.
- Attempts were made to keep the effect of the experiment on people's behaviour to a minimum.
- Fifty-eight is a reasonable sample size (1.4, 2.2.5, 2.2.6).

2.1.3 Weaknesses of the experimental design

There are three types of weakness that you may identify in any report of an investigation.

i. Faults

When you are considering published research, the first type of weakness is a fault that could be avoided through a change in the design of the investigation without changing the aim. In any publication or report, these faults should be acknowledged and should be taken into account when evaluating the results (Chapter 12). If you design and carry out a piece of research, then careful preparation and planning should ensure that there are few, if any, faults in your work.

ii. Limitations

The second type of weakness, known as a limitation, can arise in an investigation, but cannot be overcome without changing the aim. These limitations should also be taken into account when the results are discussed (Chapter 12). If you are evaluating published research, then your task is to identify these limitations, as you clearly cannot change the aim. If you were designing an investigation and were aware before you started of significant limitations in your design, then you could consider changing your aim to enable you to carry out an investigation without these weaknesses. You should also show that you are aware of any limitations in your design, when discussing your research.

iii. Communication

The third weakness is one of communication. Omissions and lack of clarity when reporting research can make it difficult to fully comprehend the design of an investigation. The format of the paper required by a journal can sometimes exacerbate this.

If you look at Example 2.1 you will quickly appreciate that there is plenty of scope for discussion here about what is a fault in the experimental design, what is a limitation, and what is due to omissions in the journal article. We think the following fit into these three categories. What did you list?

Fault in the design

The age of the men varied considerably and it is not known how this may affect the results. There is a need to demonstrate that this sample is truly representative of the average male drinking population, and if it were, then this range in age is a strength more than a weakness. Alternatively, using statistics to confirm that 'age' as a factor did not influence the results would resolve this concern.

It seems unlikely that the drinker and friends will remember accurately how much was drunk and over what time period. This may be a factor that affects the heavy drinkers more. The author does acknowledge this weakness in the article. Could a bartender not be recruited as an independent recorder?

Two methods for establishing habitual intake were used. Will the drinkers be honest about how much they habitually drink? There is a need to test these methods of recording in advance to see how reliable these forms of reporting are and which is the better method, and then to be consistent in the use of just one method.

Limitation

The times of eating varied and may have affected alcohol metabolism. If eating was controlled so that certain items were eaten at certain times during the drinking session, this may resolve the concern over the impact of eating on alcohol metabolism, but would almost certainly interfere with the normal behaviour of the participants.

The rate at which the alcohol was consumed was not recorded. Quickly drinking three pints and then nursing the fourth may result in a different rate of alcohol metabolism than a steady drinking pattern. It is difficult to see how this can be either controlled or recorded without affecting the participants' behaviour.

Communication

The published report did not indicate the length of the drinking period or what was eaten by the participants during their drinking session. It was not clear how the two methods for establishing habitual intake were used. Were they alternative methods or used in conjunction with each other?

What weaknesses have you identified? How do they compare with those we have included? Do you disagree with any parts of our evaluation? Why? By thinking about this investigation, we have introduced you to some of the important elements you need to consider when designing an investigation.

2.2 Have a go

In our Online Resource Centre, we have a range of different topics for you to have a go at planning an experiment. They cover a broad range of subject areas, so there should be one that is closely related to your area of study. There are many acceptable experiments that could be devised. We have provided you with one version on the 'effect of wind strength on maximum seed dispersal in the common creeping thistle (*Cirsium arvense*)' to illustrate each of the design steps. We explain how we develop this experiment in each section and this example is in a border so it is easy to identify within the chapter. An interactive version of the decision web (Fig. 2.1) is also provided in the Online Resource Centre.

When designing an investigation you will find that you need to be prepared to go through the process several times, first drafting your design and then redrafting as all the elements fall into place. So many factors need to be considered that, as you adjust one, you will need to go back and rethink all the others again. We have organized this process under eleven subheadings.

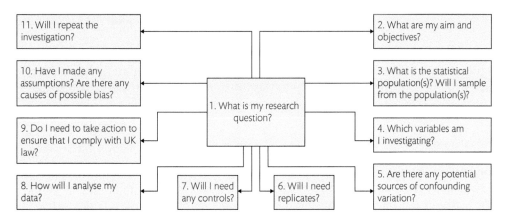

Fig. 2.1 Overview of the steps to be taken when planning an experiment

2.2.1 Identification of a research topic

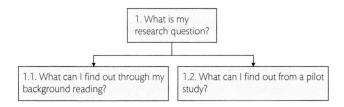

The starting point for any research is to know your research question. You may find that you are given a topic to research or you may need to identify your own research question. (If you are not sure where to begin in choosing a topic for research, we make some suggestions in Appendix A.) In either case having found an area that interests you, you then need to read about the subject.

You cannot gather too much background information. It provides the following critical elements for the researcher: a biological and/or environmental context, the justification, practical information, potential methods, and an indication of possible approaches to the data analysis. Background information can come from a variety of sources, most often from your review of the literature and a pilot study.

i. Literature review

Reading about other published studies in the same area as your research proposal is invaluable, yet often the time given to this stage in planning is seriously underestimated. You need to carry out a sufficient review of the literature to ensure that you have found all key references (usually those most frequently cited). Abstracts will not provide you with sufficient detail to be useful when planning an experiment.

a. The context

Having an understanding of what is known about the biological system you are investigating is critical. It helps you to see what has already been discovered and what is unknown or uncertain. When you report your results you will also need to demonstrate that you understand how your work fits into the existing body of research.

b. The justification

Just because you can record or measure something does not make it appropriate research. For example, you may be able to record the effect of adding whisky to soil on the growth rate of daisies. But why investigate this? Do you have a sensible reason? Your experience to date has been to consider the rationale when writing the introduction to a practical report. However, you now need to consider this as part of the planning stage as an invalid rationale completely negates the value of the research.

c. The practicalities

Many topics may be affected by seasonal factors, such as climate, timing of growth in plants, or behaviour in animals. If you were intending to examine the effect of temperature on the behaviour of *Oniscus asellus* (woodlice), what temperatures do they normally experience in their habitat? What temperature is likely to kill them?

d. Previous designs

Background reading can be really helpful in showing you how similar previous studies have been designed and how the data were then analysed, and can indicate suitable sample sizes. If you use other studies in this way, the authors' work must be cited in reports. You also need to be a critical reader and see if there are ways in which their design and/or analysis can be improved for your own work.

ii. Pilot studies

One of the most effective steps to take in any research planning is to carry out a pilot study. Not only can this allow you to identify the concentrations you need to use in a laboratory experiment or allow you to develop identification skills in a field experiment, but it also gives you a clear sense of how long every task takes to complete.

You may carry out power calculations which enable you to determine a suitable sample size. However, to do so you need estimates of population parameters such as the standard deviation which can come from the values obtained during a pilot study. Even with the best planning in the world the reality of executing your project can throw up some unexpected issues. Far better to identify this during a pilot study than when you have committed yourself to your full experiment, when it may be too late to resolve any problems. Our advice is that you always run a pilot study if possible.

The effect of wind strength on maximum seed dispersal in the common creeping thistle (*Cirsium arvense*)

What background information would you need if you were to develop an investigation into the effects of wind speed on maximum seed dispersal in *Cirsium arvense?* We suggest the following:

The context and justification An understanding of the reproductive biology of this species and its ecology would enable you to put your work into a broad context. For this investigation, a key reference is: Klinkhamer & de Jong (1993). Biological flora of the British Isles: *Cirsium vulgare. Journal of Ecology 81*: 177–91. This reviews the ecology of the species. Other studies on dispersal of seed in other species would demonstrate the areas of investigation where wind dispersal is relevant and would provide a justification for the investigation; for example, niche breadth (e.g. Thompson *et al.*, 1999); the effect of grazing (e.g. Bullock *et al.*, 1994) and conservation (e.g. Westbury *et al.*, 2008).

Practical implications This investigation requires 'live' material; consequently there are many practicalities to consider. These include: when the seeds are produced; the normal range of wind speeds experienced by the plants; the equipment we would need to examine wind speed; whether *Cirsium arvense* can be grown in pots and how long it might take to flower; whether it is known from other studies what the natural dispersal of thistle seed is; and the factors that are known to affect seed dispersal in general and *Cirsium arvense* in particular. In the review by Klinkhamer & de Jong (1993), it is clear that seed production per plant varies considerably but is reported usually as hundreds to thousands of seeds per plant. These seeds are large enough to be handled individually (each seed weighs between 1.3mg and 1.8mg) and can travel between 3.7m and 32m when dispersed naturally. This dispersal is affected by the height of the inflorescence. The first ripe seeds are known to be found by the end of July.

For our investigation into wind dispersal of seed a pilot study would be invaluable. The pilot study would enable us to check that the plants can be translocated into pots successfully and without losing their seed. We can collect the data we need to determine the wind speeds to use and to confirm that our choice of wind speed is sensible. We can collect sufficient data to allow us to carry out a power calculation and check that our sample size will be adequate to detect a reasonable minimum difference between each wind speed treatment. We will also be able to see how long it takes us to carry out the investigation and to respond to any previously unknown factors such as the equipment overheating through prolonged use or the researcher overheating in the enclosed space of the wind tunnel.

Previous designs and data analysis Background research can help determine how other researchers have carried out similar studies in this or related wind-dispersed species. For example, in a study by Fresnillo & Ehlers (2008), dispersability was studied by recording the time taken for more than 900 *Cirsium arvense* seeds to fall 1m. These data were analysed using a nested ANOVA to allow the comparison of seed from mainland and island populations. In a study by Skarpaas *et al.* (2004), the authors manually released about 50 seeds taken from ten plants of *Crepis praemorsa* in two locations and recorded wind speed, seed characteristics, and distance travelled. In addition, these researchers also collected 101 plants with mature seeds, placed them in a mown field, and recorded the distance the seeds travelled. These results were transformed and analysed using a linear regression. (For details on these statistics tests see Chapters 7–11.)

2.2.2 Aim and objectives

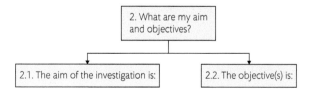

We explained in 1.1 that an aim is the general overarching statement about the topic you are investigating whilst the objectives are focused and reflect specific parts of your investigation. You can realistically only write an aim and objectives after quite a lot of preliminary planning and discussions. Having collected your background information and prepared a preliminary method you should be able to write a draft aim and objective(s). This draft aim and objective will be finalized towards the end of the planning process.

The effect of wind strength on maximum seed dispersal in the common creeping thistle (*Cirsium arvense*)

Based on the background reading we propose the following for this experiment:
   Draft Aim: To investigate the effect of wind speed on maximum seed dispersal in *Cirsium arvense*.
   Draft Objective: To examine the maximum distance travelled by the seeds of *Cirsium arvense* when placed in a wind tunnel and exposed to wind speeds reflecting the range and average wind speed in the natural population.

2.2.3 **What is/are the statistical population(s)? Will I sample from the population(s)?**

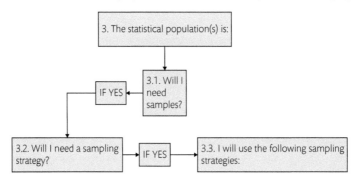

It is rarely possible or even of any value to measure every individual in your population. You will usually sample and you will therefore need to define the statistical population that the sample is representing (1.4) and develop a suitable sampling strategy. You need to consider each treatment variable in turn. For example if you were examining the pathogenicity of *Phytopthora parasitica* on *Arabidopsis thaliana* knockout mutants you are sampling both the *A. thaliana* individuals and the *P. parasitica* as you will not use every potentially available plant nor every *P. parasitica*. The first will come from a particular accession and the second from a particular batch. These are samples from the statistical populations of '*A. thaliana* with a particular knockout mutation' and '*P. parasitica*'.

The effect of wind strength on maximum seed dispersal in the common creeping thistle (*Cirsium arvense*)

Our aim indicates our interest in maximum seed dispersal in the species *Cirsium arvense*. We can propose that all individuals of this species are our statistical population.

(continued)

In an investigation on seed dispersal, you cannot realistically examine every plant in the statistical population; therefore, you will need a representative seed sample. For our sampling strategy, we have decided to use whole plants, as this will then mean that the seeds are held at the same height as they would be in the field and within an inflorescence.

Therefore, in our design one ecological population of *Cirsium arvense* was chosen at random and all the apparently mature flowering plants with an apparently undisturbed inflorescence were given a number. Fifteen numbers were chosen at random using random number tables and these plants were dug up. Care was taken not to dislodge any seeds.

See Q1.

Q1 Do you think that our design so far in the study on the effect of wind strength on maximum seed dispersal in the common creeping thistle (*Cirsium arvense*) will provide us with a representative sample?

A1 It is difficult to say with any confidence that this will be the case, as a sample of plants from a single ecological population may not reflect this species across its range. We must consider revising our experiment, and there are two ways forward at this stage. We can either extend the sampling into a number of ecological populations over a broad range of the species distribution, or we can revise the aim and objective so that they reflect the limited nature of this study. We will take the second course. Therefore, the revised aim and objective are:

Aim: To investigate the effect of wind speed on the maximum distance seed *Cirsium arvense* taken from Windmill Hill, Worcester is dispersed.

Objective: To examine the maximum distance travelled by the seeds of *Cirsium arvense* sampled from an ecological population in Worcester when placed in a wind tunnel and exposed to wind speeds reflecting the range and average wind speed in the natural population.

The statistical population is now all the individuals of *Cirsium arvense* at Windmill Hill.

2.2.4 Which variables am I investigating?

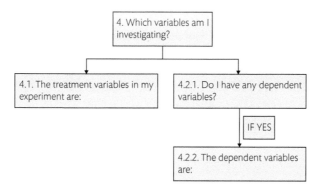

This may seem to be an unnecessary step in our planning process. However, there are so many terms that relate to the factors we investigate, such as 'variable', 'treatment', and 'sample', that it can become confusing as to which is which. The identification of the variables that we are interested in also allows us to determine with more confidence the factors that we are *not* investigating in our experiment (1.7.3). You need at this stage to determine which are your treatment variables and which (if any) are your dependent variables (1.7.2). If you find the distinction difficult to understand you can look at Chapters 7–11 where we have highlighted which are the treatment variables and any dependent variables in speech bubbles on every example.

> ## The effect of wind strength on maximum seed dispersal in the common creeping thistle (*Cirsium arvense*)
>
> The factors we are investigating in this experiment are the distance travelled by the seeds and the wind speed. The wind speed is the treatment variable and the distance travelled by the seed is the dependent variable.

2.2.5 Are there any potential sources of confounding variation?

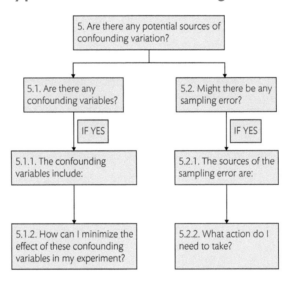

When you record a set of measurements in an experiment, you usually see variation in these measurements. This variation may be due to the factors you are investigating, i.e. the treatments (1.6 and 1.7) or other causes, i.e. confounding variation. Confounding variation can be due to confounding variables (1.7.3) and/or sampling error (1.7.4). If the confounding variation in your data is considerable, it can mask the effect of the factors you are investigating. Therefore, it is important to be clear about what you are investigating, identify possible confounding variables, minimize the effect of these confounding variables, and be aware of sampling error.

i. Identifying confounding variables

Identifying the confounding variables has to be your next step as you cannot minimize the effect of something if you are not aware that it might affect your experimental system.

Background or preliminary investigations, common sense, and reading help in identifying causes of this non-treatment variation.

Example 2.2   The antibacterial properties of triclosan and tea-tree oil

An undergraduate took swabs from the hands of volunteers working in the medical profession. These were the 'before' measures. She then washed the volunteers' hands. One hand was washed with triclosan and one with tea-tree oil. Having allowed these washed hands to air dry, the student took a second swab from each hand. These gave the 'after' results. At the laboratory, the swabs were used to inoculate plates. As the bacterial count could be high, serial dilutions were also made from the swabs and plates were then inoculated. After two days' incubation, the numbers of bacterial colonies on the plates were counted.

In Example 2.2, there are a number of potential confounding variables. These include the method used to wash the hands, the method used to swab the hands, the consistency of the poured agar plates, the method used to inoculate the plates, the temperature of the incubator, the time of incubation, and the method used to count the resulting bacterial colonies.

ii. Minimizing the effect of non-treatment variation

In your investigation, you will be collecting observations from a number of items. Often your items are grouped in some way, such as different samples. If you wish to gain an understanding of the degree to which the variation in your data is due to the factors you are investigating, you need to ensure that all other non-treatment effects are equalized in some way across all the items in all the groups. This can be achieved by equalizing the effect of the confounding variables across the experiment and through randomization. Where possible, both approaches should be incorporated into your design.

In Example 2.2, the student ensured that she was consistent in her use of all methods throughout the experiment. In addition, she made up a single batch of agar and poured all the plates at same time. The inoculated plates were then arranged at random within the same incubator. In this way all replicates had an equal chance of exposure to the confounding variables.

iii Sampling error

In most biological investigations there is a level of sampling that occurs and chance alone can lead to variation between a sample and a statistical population. It is impossible to remove sampling error from an investigation. However, the effect of sampling error is lessened if you have a large representative sample.

The effect of wind strength on maximum seed dispersal in the common creeping thistle (*Cirsium arvense*)

In the investigation into seed dispersal in *Cirsium arvense*, there are many possible causes of confounding variation. These include: the effects of digging up the plants, the maturity of the seeds

(continued)

---

and inflorescence, the height of the inflorescence, the number of seeds in the inflorescence, the temperature and humidity of the wind tunnel, and the location of the plants during testing. To minimize the non-treatment effects in the experiment on seed dispersal, the temperature and humidity of the test area were consistent throughout the experiment. Each plant was placed at the same point in the wind tunnel, the inflorescence being placed above a mark on the floor. Three wind speeds (low, medium, and high) were selected to reflect the range of wind speeds experienced by the thistles in their natural population. The 15 plants were assigned at random, five to each of the three speed treatments. In this way, any variation between the plants in terms of their maturity, height of inflorescence, and size of inflorescence should be distributed equally across each of the three treatments.

2.2.6 **Will I need replicates?**

Replication

Where you have more than one item, group of items, or sample in an experiment.

A replicate is when you have more than one of something. It can be more than one item being exposed to a treatment or more than one group of items exposed to one treatment or more than one item in a sample. Replication should be used whenever possible for two reasons: firstly, to increase the reliability of your estimate of the population parameters; and secondly, to allow mathematical estimates of variation due to confounding factors.

i. Replicates can increase the reliability of your estimate of the population parameters

In Example 2.2, the student could carry out her investigation on one person, where one hand is washed in triclosan and one in tea-tree oil. If the aim is to investigate the effectiveness of these two hand washes on people working in a medical profession, then this is clearly a very small sample and any values from a single individual are not likely to closely match the population parameters. In fact, 25 people took part in this investigation so that each hand washed in triclosan was a replicate within the treatment 'triclosan' and each hand washed with tea-tree oil was a replicate within the treatment 'tea-tree oil'. The results showed that for some people tea-tree oil reduced bacterial counts, in some people the counts were much higher, and in some people counts were too low or too high to be measured effectively by the method used. In a practical setting such as this, the variation demonstrated between people (replicates) was crucial to gauging the real effectiveness of not only the investigation, but also the hand washes.

ii Replication and mathematical estimates of non-treatment variation

A final approach to dealing with all the confounding variation in your experiment regardless of its source is to design your investigation in such a way that you can mathematically partition the total variation in your data into that due to the factors you are investigating and the confounding variation. One requirement to enable you to partition variation in this way is the need for replication within your experiment. There are specific designs that allow for the simple mathematical partitioning of the variation due to confounding variables. These designs are discussed in Chapter 10.

iii. Features of replicates

Replicates must be an integral part of the experimental design and must all be established and examined within the single experiment and not subsequently. Any work carried out later is a new experiment. Replicates must also be 'independent' of one another. This means that you should not use the same animals, plots, etc. as replicates. For example, if you were investigating the response of plants to a watering regime, you might measure the growth of several leaves on the same plant exposed to one treatment and several leaves on another plant exposed to a different watering regime. This is an example of pseudo-replication. Clearly the leaves on any one plant are not independent of each other: they share the same genotype and microclimate. A more appropriate method for replication would be to expose many plants assigned at random to each treatment and, either randomly or using a stratified method (1.4.2), sample one leaf from each plant.

Another example of pseudo-replication frequently arises when items are grouped in the same Petri dish, pot, etc. For example, in an investigation into the effect of fertilizer on the growth of marigolds, there were ten plant pots each containing four marigolds. Five of these pots were treated with a new fertilizer and the other five pots were treated with the standard plant food. After six weeks, the dry mass of the shoots and roots from each plant was measured. A common mistake is to treat each plant as a replicate and therefore believe that there are 20 replicates for each treatment, but the four plants in a pot are not independent of each other. (If you do design an experiment like this, you should refer to 10.6, which discusses blocking in experiments, and 10.16 as the four plants are 'nested' within the five pots.)

iv. How many replicates?

One of the hardest things to establish is how many replicates are needed in any particular investigation. An indication of the number of replicates comes from the requirements of the particular type of statistical test to be used and the extent of the confounding variation. It may be necessary to carry out a preliminary investigation to identify the degree of confounding variation before choosing the number of replicates for the main experiment. We look at this thorny issue again in relation to sample sizes in 2.2.6 and review this fully in 6.2.

The effect of wind strength on maximum seed dispersal in the common creeping thistle (*Cirsium arvense*)

In the current design five plants were exposed to each of the three wind speeds. Each plant in any one group was a replicate. The 15 plants were each given a number. A random number table was used to determine which plant was to be tested first. The plant had already been assigned to a treatment (a particular wind speed) so this was the first wind speed tested. This approach was used for each plant. As a result, the order in which the plants were tested (and therefore each replicate and each treatment) was randomized. Each plant was exposed to the given wind regime for 10min. After this time the distance (mm) travelled from the mark on the floor to the seed furthest from the plant was recorded.

2.2.7 **Will I need any controls?**

Control

An experimental baseline against which any effects of the treatment(s) may be compared or to check that aspects of your experiment are working correctly. It must be an integral part of an experiment.

In 2.2.5ii we discussed the practice of keeping confounding variables the same across all your experiment as a way of minimizing the effect of confounding variation on your ability to detect any real biological differences. The word often used to describe this keeping non-treatment factors constant is 'control'. However, this word is used in another sense in experimental design which is where you add additional elements that can either provide a baseline against which you compare the treatments or to check that aspects of your experiment are working correctly. When researchers refer to a 'control' it is this meaning they are using. For example if you are carrying out a polymerase chain reaction (PCR) you will include a number of additional microfuge tubes which have only some of the PCR reactants in. For example in one tube it is common practice to leave out the primers, in another to leave out the DNA. This checks that the PCR only works when all the correct reactants are present and you are not detecting false negatives or positives due to contamination.

In laboratory experiments it is usually relatively easy to construct a control. In field experiments, if a control is necessary then you will need to select an area that is as similar as possible to the one used in the rest of your experiment. In complex experiments where you are investigating the effects of more than one variable, you may require more than one control.

One of the most important aspects of a control is that it must be an integral part of the investigation and must be carried out at the same time as the rest of the investigation. If we ran the PCR controls in a later batch we would not have checked that the actual test batch was working as it should.

 In the experiment described in Example 2.2 on the antibacterial properties of triclosan and tea-tree oil, are any controls needed? If so what?

A control may not always be necessary. For example, in an investigation into the effect of temperature on the behaviour of *Oniscus asellus* (woodlice), you might select temperatures of –5, 0, 5, 10, 15, and 20°C as your treatment range. There is no one temperature that could be called a control.

> ### The effect of wind strength on maximum seed dispersal in the common creeping thistle (*Cirsium arvense*)
>
> The investigation into the effects of wind speed on seed dispersal in *Cirsium arvense* is similar to the *Oniscus asellus* example in that there are no controls.

2.2.8 **How will I analyse my data?**

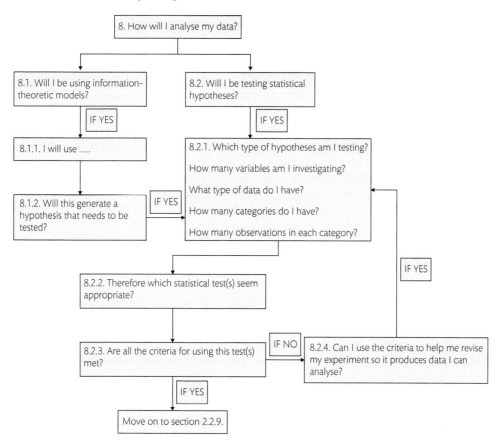

This critical stage of experimental design is the one most frequently left out of the planning stages, and in our experience is the most frequent cause of disappointing research. This step is very much a forwards and backwards process. Your draft design is used to determine which is the most appropriate statistical test to use, but this choice of test may then influence factors such as how many observations you should have in each sample.

In the first step you need to decide if you are testing a hypothesis or not (1.8.2). The next step is to decide which type of hypothesis you are testing and the correct statistical test to use. These decisions are determined by the type of data you have, how many variables you wish to test, and how many observations you have. We illustrate this process for our thistle example here but also take you through this particularly in Chapter 7, and illustrate this process in relation to each of the statistical tests we have included in Chapters 8–11. There are also more examples in the Online Resource Centre.

When selecting statistical tests there are a number of criteria that your design and data need to meet. For example if you wished to use the two-sample z-test for unmatched data you should have more than 30 observations in your samples (10.1.2) At this planning stage you will find you cannot check all the criteria as some require you to examine your data.

Therefore you do as much as you are able to at this stage but should check the criteria again when you have collected your data to ensure they have all been met. If they have not you will then have to revise your choice of statistical test.

Many statistical tests of hypotheses such as the z-tests have minimum and maximum guides to the numbers of observations in the sample. These may be determined by the range over which the test is most effective. Alternatively, with large samples, it may be that you will obtain no additional useful information from having more observations. Clearly, by checking your design in the manner we have described you can select the most appropriate statistical test and, as part of this process, you will also have checked on details such as how many observations you need in each category. If your draft design does not meet the criteria, you still have time to amend it before you carry out the experiment.

This and previous steps in the planning process can lead you to change your objective(s) and sometimes your aim. Therefore, you should now finalize them, taking into account the planning stages you have completed.

The effect of wind strength on maximum seed dispersal in the common creeping thistle (*Cirsium arvense*)

In our experiment, we have no reason at the outset to expect the data to match a particular mathematical or biological model and we are examining the effect of only one treatment variable (wind speed). However, we do want to know whether all the samples behave in the same way, i.e. is the maximum distance travelled by a seed the same for all wind speeds? The generic hypothesis for our experiment is therefore: do samples come from the same or different populations?

In this experiment, only three wind speeds (low, medium, and high) are being tested so the observations will be organized in three columns. This treatment variable is therefore described as a discrete ordinal variable with three categories. The distances travelled by the seed (in mm) are measured on an interval scale and are quantitative, continuous, and can be ranked. There are five plants (replicates) in each category. The distance travelled by any one seed has only been recorded once, so the data are independent and not matched. Ideally at this stage we would also wish to determine whether the distribution of the observations would be normal if every seed from our statistical population was measured. To estimate this we consider the sample we are measuring but at this point we have no observations to check. However, background reading about seed dispersal indicates that the data are unlikely to be normally distributed (i.e. this distribution is 'non-parametric'). This then is what we will assume in the planning stage, but we must confirm this after the experiment has been carried out and we have our observations to check.

By using the key in Chapter 7 we can determine that a Kruskal–Wallis test will probably be a suitable test to use if we make some modifications to our design. In our current draft design, the sample size we had chosen arbitrarily is not going to be adequate. Therefore, we will increase the sample size to 60 and assign 20 plants at random to each treatment. The minimum number of observations for our experiment has therefore been determined by the criteria for using a Kruskal–Wallis test and the maximum is determined by practicalities and by ensuring that we do not devastate the ecological population at Windmill Hill by our sampling. We can now finalize our aim and objective.

Aim: To investigate the effect of wind speed on the maximum distance *Cirsium arvense* seeds from Windmill Hill are dispersed in a wind tunnel.

(continued)

Objective: To examine the maximum distance travelled by seeds from 60 *Cirsium arvense* sampled from one ecological population in Worcester when placed in a wind tunnel and exposed to low, medium, or high wind speeds that reflect the range and average wind speed in the natural population.

2.2.9 Do I need to take action to ensure that I comply with UK law?

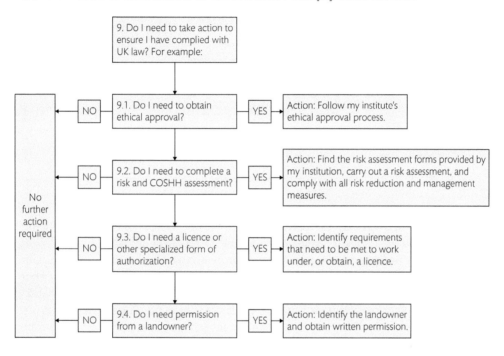

A critical aspect of research planning is ensuring compliance with relevant legislation and codes of conduct whether in the UK and/or elsewhere. It is not possible to consider all UK legislation within the scope of this book so we have focused on the four areas that our students most often encounter: health and safety, sampling and access, animal welfare, and working with humans. (Research and the law is covered in detail in Chapter 4.) It can take a considerable amount of time to obtain a license or ethical approval so forward planning will help you avoid having to delay the start of your research.

The effect of wind strength on maximum seed dispersal in the common creeping thistle (*Cirsium arvense*)

Cirsium arvense is not a protected species. The sampling will be carried out on private land managed by the Worcestershire Wildlife Trust and therefore permission to sample the plants must be sought. As we are using electrical equipment and handling soil, we will complete a risk assessment.

2.2.10 Are there any causes of possible bias? Have I made any assumptions?

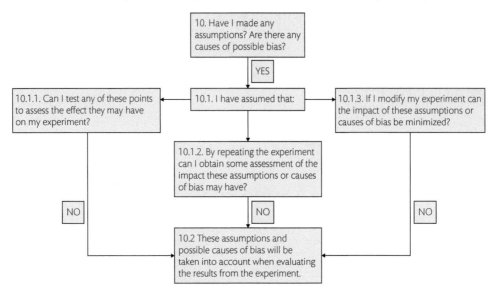

Some particular methods may themselves influence the outcome of an investigation. If this is likely, then you should try to estimate the extent of this effect and take it into account when planning your experiment and when you interpret the results. One of the most obvious examples of this can occur when studying animals, where your own presence and behaviour and the presence of experimental apparatus may influence the outcome. The contamination of samples from your own body or equipment is another example. The problems that can arise from not wishing to influence the investigation are clearly seen in Example 2.2 where the undergraduate needed to ensure that her presence did not alter her volunteers' behaviour.

Whilst you cannot plan an investigation without background reading, one problem that can arise is that you may then believe you know what the outcome of your research should be. We call this a 'prediction'. In our experience, if you have a prediction in mind at the outset, you are likely to carry out poor research as a result of conscious or unconscious bias. Therefore, although you need to know the background information to plan an experiment, you still need to approach your work with an open mind.

The most common assumption which is inherent in most research is that the samples truly reflect the attributes of the statistical population. You should always show that you are aware of these assumptions and possible causes of bias when you interpret your results and communicate your findings.

The effect of wind strength on maximum seed dispersal in the common creeping thistle (*Cirsium arvense*)

Look through the information you have noted down when 'having a go'. How many assumptions and possible causes of bias can you identify? For our design these include the following:

- We assume that the random method employed when collecting plants from the wild did result in a sample that is representative of the statistical population.

(continued)

- No bias was introduced by the loss of any seeds from the inflorescence during the sampling.
- We assume that the random allocation of plants to the treatments minimized the effect of confounding variables.
- We assume that the wind speeds used in the experiment were representative of those experienced by the species at Windmill Hill, Worcestershire.
- We assume that the order of testing the plants within the wind tunnel had no effect on the distance travelled by the seeds.
- We assume that the seeds were blown along in a manner that is similar to that in the field, including seeds that were blown along the ground.

2.2.11 Will I repeat the investigation?

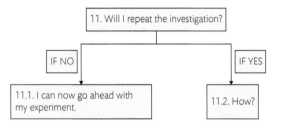

If an effect is real, then it is reasonable to expect that a repeated investigation, carried out using the same method, will identify the same trend. A scientific finding that has been confirmed in this manner is considered to be 'sound'. Clearly there may be circumstances that do not allow for a repeated experiment. For example, an investigation into the effects of hurricane Ivan on the distribution of manatees along the Florida coast could not be repeated.

One common failing is to confuse a repeated experiment with a replicate. If an experiment is repeated at a different time, this is not a replicate. The data obtained in this way cannot be combined with the data from the earlier experiment without using certain mathematical steps to prove that it can be pooled. We give one example of how this might be done in 8.5.

The effect of wind strength on maximum seed dispersal in the common creeping thistle (*Cirsium arvense*)

In our example, the experiment could be repeated with another 60 plants taken from the same population in a similar manner.

2.2.12 Back to the beginning

i. Review your planning

As we explained at the beginning of this chapter, when designing an investigation you usually have to develop a draft and then refine it, because at each step any change you make may affect steps you have already been through. So at this point you need to look through your design again using our decision web and make sure your design is as good as it can be.

ii. Experimental design and methods

Our decision web focuses on the design of our experiment, but this does not necessarily mean that you will have a record at this point of all the technical details in your method. As part of your final review of your plan, you can check that all technical details such as volumes or concentrations are now included in your design.

The experimental design we came up with is summarized below. How does it compare with your design?

The effect of wind strength on maximum seed dispersal in the common creeping thistle *Cirsium arvense*

Aim: To investigate the effect of wind speed on the maximum distance *Cirsium arvense* seeds from Windmill Hill, Worcestershire are dispersed.

Objective: To examine the maximum distance travelled by the seeds of 60 *Cirsium arvense* sampled from one ecological population in Worcestershire when placed in a wind tunnel and exposed to low, medium, and high wind speeds that reflect the range and average wind speed in the natural population.

Method

All the apparently mature flowering plants with an apparently undisturbed inflorescence within a specific ecological population at Windmill Hill in Worcestershire were given a number. Sixty numbers were chosen at random using a random number table and these plants were dug up. Care was taken not to dislodge any seeds. Each plant was potted up in 12cm pots in John Innes No. 1 compost.

Three wind speeds (low, medium, and high) that reflected the range of wind speeds experienced by these plants in the natural population were selected. Each plant was given a number. The first 20 numbers chosen at random were assigned to the low wind speed treatment, the second 20 numbers chosen at random were assigned to the medium wind speed treatment, and the remainder were assigned to the high wind speed treatment. In this way, any variation between the plants in terms of their maturity, inflorescence size, and height should be distributed equally across each of the three treatments.

This numbering system was also used to determine the order in which the plants were tested. Again, using a random number table, a plant was selected and then exposed to its predetermined treatment. A second plant and treatment were selected in the same way until all plants had been tested. This ensured that the order in which the three treatments were tested was also randomized. To minimize the non-treatment effects in the experiment, the temperature and humidity of the test area were consistent throughout the experiment. Each plant was placed at the same point in the wind tunnel. Each plant was exposed to the given wind regimen for ten minutes and the distance from the centre of the pot to the seed that had travelled the furthest was measured (mm). The hypotheses for this experiment will be tested using a Kruskal–Wallis test. The experiment will be repeated with another 60 plants.

2.3 Managing research

To be effective as a researcher, either as an undergraduate or a graduate, you must develop good management skills. In research, management falls into at least three areas: managing time, managing space, and managing data.

2.3.1 Time management

Time management is critical. You may have steps in a method that need to be carried out for certain periods of time, or you may need to collect samples or complete preparations all on one day. In addition, most research is carried out within a time frame determined either by your course or by funding. It is therefore important to (*i*) find out which are the time-sensitive aspects of your project and (*ii*) draw up a timetable.

i. Time-sensitive aspects of research

The following lists the time-sensitive aspects of research that you may encounter and are roughly in the order that they affect your work:

- Obtaining permissions, approval, or a license can take several months.
- Seasonality of investigations can determine at what time of year the work can be carried out.
- Ordering and obtaining chemicals or equipment can take several weeks or longer.
- Preparation of solutions, sterile items, agar plates etc. can take several days.
- Maintaining stocks. Failure to do so, especially if stocks need to be reordered, can hold up your work.
- Access to equipment can be restricted and equipment may need to be booked in advance.
- Institution closed days such as Christmas or New Year need to be factored in to your planning.
- How long do you have to complete all work on the project? If you have to analyse data and write a report within a set time this needs to be allowed for when planning how much time you have to complete the practical aspects of your research.

Careful preparation will mean that you are aware of these potential sources of delay and build them into your planning.

ii. Timetable

A timetable should help you identify critical points in the execution of your investigation or flag up experimental designs that just cannot be carried out in the time available. The time constraints imposed by seasonality, for example, must also be recognized in the planning, as should external factors such as assessment points for other courses. In Example 2.2, the student taking and examining swabs from volunteers' hands collected the samples on one morning, but it then took all afternoon and into the very late evening to inoculate the microbial plates. This student's planning was exemplary and she knew from her experience with a pilot experiment how long the work would take. She was therefore able to plan her time effectively and carry out the project, having made proper arrangements in advance to ensure she could complete this part of her experiment in one day.

iii. The inexperienced researcher

From our experience, there are two common errors that arise in relation to time management because of inexperience. The first of these is developing complex experimental designs. It is very tempting to try to investigate lots of factors all within one experiment. This can be counter-productive for two reasons. The factors may interact and you may not be able to untangle which factors had which effect. Trying to do so can be very time-consuming and is usually not very successful. Complex designs often need complex statistics. This is fine as long as you have thought this all through before you start and then allowed in your planning for the time it will take to analyse the data correctly. It is usually better to have more objectives to examine an aim than one complex objective.

   The second aspect of research that only comes with experience is knowing that everything will take longer than you think. A general rule of thumb is to plan your investigation and then double the time you have allowed for each part. This should then give you time to make mistakes and solve unexpected problems.

2.3.2 Space management

In research, you are often working in a confined space, such as a laboratory, and in the company of other scientists. This means that your management of your space is very important, allowing you to complete your research without interfering with other people's work. You must be considerate of other users of the areas around where you are working. We consider the health and safety aspects of shared working environments in 4.2 and 4.4.

2.3.3 Data management

When designing and carrying out research, your ability to keep and manage comprehensive records is paramount. For most of us, records are kept in two ways: in a transportable hard-copy format, such as a laboratory or field notebook, and as electronic files.

i. Laboratory or field notebooks

The best approach to recording information in such notebooks is to record more detail rather than less. You should date each day's work, give full details of the method used that day, which may for fieldwork include a site description and details of the weather, and a record of the results you obtained. You may also need to record contacts' details, but you should be careful when making any records that may compromise volunteers' confidentiality (4.2). Finally, add notes about your ideas, why you are taking the research in a particular direction, what the data appear to show you, and how this may relate to other people's studies. All this information will be invaluable, both in tutorials and when communicating your results in a more formal way. Having more detail rather than less is particularly important for graduates, who may not write formal reports for several months and therefore will be less able to rely on memory to compensate for unclear notes.

Example 2.3 The growth rate of *Secale cereale* (rye) seedlings

Researchers at an agricultural research station were interested in the natural variation in the rate of growth of rye. The first measurement was made when the rye was at the seedling stage and the leaf length (mm) was recorded for 15 seedlings. In her notebook, the researcher noted down the following:
    12.0, 12.0, 11.5, 18.0, 14.0, 11.0, 14.5, 11.5, 10.0, 10.0, 19.5, 19.0, 21.0, 15.5, 14.5.

 Q3 Look at Example 2.3. What additional details should the researcher have also recorded?

In Chapter 4, we consider health and safety legislation. One outcome from this is that you will need to produce a risk and/or a COSHH assessment. This should also be pasted firmly into your field or laboratory notebook. There is no point in finding out how to deal with an emergency if you do not have the relevant information with you when it is needed!

ii. Electronic records

There are two useful practices here that can save enormous heartache from lost or apparently missing files. The first of these is keeping track and the second is backing-up. As your research project develops and you alter electronic files, you need to know which version of a file you are working with. The simplest method to achieve this is to include a table at the top of each file as illustrated in Table 2.1. Using a table like this will prompt you to make sure you keep track of the date the file was last altered and your thoughts as you develop your work. When saved, files do indicate the date last altered, but we have found this 'table' system to be more effective as a file-management tool.

Backing-up electronic files is essential and you should get into the habit of doing this **at least** every day, so a file-management system (e.g. using a table as Table 2.1) is invaluable. Backing-up means transferring files to a second location that is independent from the first. One system you could consider for backing-up files as it is not dependent on a local server or local hard drive is a system such as the Cloud. Most academic institutions give advice on their preferred method for members of the institution to back-up material.

Table 2.1 Information that helps you keep track of electronic files

Title of project
Last revised
Aim
Objectives
Check for accuracy
Check for inadvertent plagiarism
Checked supervisor's feedback
COMMENTS

Data storage and collection are subject to several laws, including the Human Rights Act 1998 (as amended) and the Data Protection Act 1998 (4.2 and 4.8): you must therefore make sure that you are careful about how all records are stored, where, and for what period of time.

Summary of Chapter 2

- By working through the first section in this chapter, you will realize that you already know about experimental design and are able to identify the strengths and weaknesses of a design (2.1). We consider how some of the weaknesses may be avoidable and some not, without changing the aim.

- We then consider the eleven steps that take you through designing an experiment including: identifying an aim and objective(s) (2.2.2), determining the statistical population (2.2.3), identifying the treatment and confounding variables (2.2.4 and 2.2.5), replication (2.2.6), the use of controls (2.2.7), the role of statistics when planning experiments (2.2.8), the importance of compliance with UK law (2.2.9), causes of bias and making assumptions (2.2.10), and repeating your investigation (2.2.11). We introduce you to a decision web that may be used when planning an experiment.

- Tools for managing research in relation to the planning stages, time management, space considerations, and data management are also discussed in this chapter (2.3) and developed in Chapters 4 and 12.

- The Online Resource Centre includes interactive exercises that test your understanding of this chapter along with other topics, particularly those considered in Chapters 7–12.

Answers to chapter questions

A2 Yes, controls are needed to check that the bacteria are coming only from the participants' hands. The controls in this case need to be a series of plates inoculated with: (i) nothing; (ii) a sample from the triclosan only; (iii) a sample from the tea-tree oil only; (iv) an unused swab; and (v) the solution used for serial dilution. These are all to check that the experiment is working correctly and there are no sources of contamination.

A3 Clearly not enough information has been recorded. In a few weeks' time, the data will be useless, as not enough will be reliably known about them. Other information that should be recorded includes the units of measurement, date, time of recording, details about the experiment, and the name of the investigator (if more than one is involved in the project).

3 Questionnaires, focus groups, and interviews

 In a nutshell

Here we give a brief introduction to the use of questionnaires, focus groups, and interviews to gather scientific information. We discuss the design of the questions in relation to the sample size and how the data may be analysed quantitatively.

We have considered the essential steps needed to design most investigations in Chapter 2. However, some information may be gathered using methods such as unstructured and structured interviews, questionnaires, and focus groups. If you are planning to use one of these methods, there are additional points that need to be considered.

Questionnaires, focus groups, and interviews may allow you to collect both quantitative (numerical) data and qualitative (descriptive) information. Qualitative approaches are based on the notion that reality varies for different people in different contexts; therefore, you cannot use a single scale against which you make measurements. An example of this is people's perception of pain. Qualitative information can be summarized but the tools used to do this are outside the scope of this book. In this section, we therefore focus on how these methods may be used effectively to obtain numerical or quantitative data and how these data may be analysed. We have, of necessity, had to assume some prior understanding of the use of hypothesis testing (Chapter 6) and chi-squared statistical tests (Chapter 8).

Further examples relating to the analysis of quantitative data from questionnaires are included in the Online Resource Centre. It will take about 1 hour to work through this chapter. The answers for these exercises are at the end of the chapter.

3.1 What is a questionnaire, interview, or focus group?

These three methods for obtaining information differ in terms of the degree of interaction between participants and the interaction between the researcher and the participant.

3.1.1 Questionnaires

Questionnaire
A collection of questions given to all participants in an investigation, for completion without the involvement of the researcher.

Questionnaires are, as the term implies, collections of questions given to all participants. The participant writes down their answer and returns the questionnaire to the researcher. Questionnaires are usually anonymous and confidential (4.2 and 4.8.1). They require very little contact between the researcher and volunteers and usually there is no interaction between participants. A questionnaire is most useful if you wish to generate quantitative data.

3.1.2 Focus groups

Focus group
A discussion-based interview involving more than two people, managed by the researcher

A focus group is a discussion-based interview involving more than two people. A theme or focus is provided by the researcher, who also directs the discussion. These discussions are usually recorded as audio- or videotapes and evaluated later. Most of the output from a focus group will be a record of who said what when. Therefore, less-quantitative data are usually generated by this method.

3.1.3 Interviews

Interview
A directed discussion between the researcher and a participant.

An interview is a meeting between the researcher and a participant. Interviews can be structured where the interviewer asks the same series of questions to all the interviewees, or unstructured where the interviewee is left to make their own comments about a topic. Recording the results from an interview can be done at the time if the researcher is able to take notes. More often, the interviews are recorded as audio or video images to be evaluated later. Interviews can be used in a similar manner to questionnaires and may also be useful if you wish to generate numerical data.

3.2 Closed and open questions

Closed question
Where you give the participant a limited number of answers and they choose one or more of these options.

Open question
Where the answers are not prescribed

There are two types of questions: open and closed. A closed question is one where you give the participant a limited number of answers and they have to choose one or more of these options. An open question is where the answers are not prescribed.

Example 3.1 Dietary analysis and comparison of coeliac and non-coeliac university students in the UK

An undergraduate examined the diets of a sample of university students nationally who have coeliac disease and a sample of students who do not have coeliac disease who were matched by age and gender. As part of the study the students were asked to complete a questionnaire. The questionnaire is reproduced (with permission) below and has been annotated so you can see why the various questions were included.

Please provide some basic information.

(Tick box that most applies to you)

Gender   Male   ☐

          Female   ☐

          Prefer not to say   ☐

> Published research has indicated that these are all known to be potential confounding variables. By collecting this information these factors can be considered during the analysis of the data. These are closed questions with opt-out answers where deemed ethically appropriate.

Age   18-21   ☐

          22-25   ☐

          25-30   ☐

Which course are you studying?

....

Year of study at University (Optional)

....

Ethnicity (please tick box or write in your ethnicity)

White

• British   ☐

• Irish   ☐

> Ethnicity categories follow UK government guidelines

• Any other White background (please write in)

....

Mixed

• White and Black Caribbean   ☐

• White and Black African   ☐

(continued)

- White and Asian   ☐
- Any other mixed background (please write in)

....

Asian or Asian British

- Indian   ☐

Asian or Asian British

- Indian   ☐
- Pakistani   ☐
- Bangladeshi   ☐
- Any other Asian background (please write in)

....

Black or Black British

- Caribbean   ☐
- African   ☐
- Any other Black background (please write in)

Chinese or other ethnic group

- Chinese   ☐
- Any other (please write in)

....

Prefer not to Say   ☐

(continued)

You need to strike a balance between the number of questions and the participants likelihood of completing these honestly. These are all closed questions.

Questions	Everyday	Most days	Occasionally	Never
1. Do you buy breakfast from catering facilities at the university campus?				
2. Do you buy lunch from catering facilities at the university campus?				
3. Do you buy your main evening meal from catering facilities at the university campus?				
4. Do you purchase snacks from the catering facilities at the university campus?				
5. Have you ever had difficulties finding suitable meal choices whilst on the university campus?				
6. Do you prepare your own breakfast to eat whilst at university?				
7. Do you prepare your own lunch to eat whilst at university?				
8. Do you prepare your evening meal at home to eat whilst at university?				
9. Do you have difficulties finding food you will enjoy whilst at university?				
10. Have you ever skipped meals whilst at university?				
11. Do you eat ready-prepared convenience foods (e.g. ready meals, frozen pizzas)?				
12. Do you eat savoury snacks (e.g. crisps, tortilla chips, breadsticks)?				

(continued)

Questions	All of the time	Most of the time	Sometimes	Never
13. Do you eat sweet snacks (e.g. chocolate, soft sweets, boiled sweets, cakes, biscuits, pastries)?				
14. How often do you eat fruit (e.g. apples, oranges, grapes)?				
15. How often do you eat vegetables (e.g. carrots, broccoli, sweet potato)?				
16. How often do you go out for meals with friends/family/colleagues?				
17. Do you find difficulty finding suitable restaurant meal choices?				
18. Have you ever avoided meals out?				
19. Do you feel that your diet is limited because of food availability?				
20. Do you ever feel your diet is limited because of food prices?				
21. Do you have difficulty finding suitable choices for snacks whilst out and about (e.g. gluten-free alternatives)?				
22. Do you feel fulfilled and satisfied by your diet whilst at university?				

23. Do you wish to make any further comments on your diet at university?

24. Do you wish to make any further comments on factors that you feel impacts upon your diet whilst at university?

Thank you very much for taking the time to complete my questionnaire.

Questions 23 and 24 are open questions.

3.2.1 Closed questions

Closed questions may be used in questionnaires or structured interviews. They have an advantage over open questions in that the answers can be collated and the frequency of respondents giving a particular answer can be recorded. Table 3.1 summarizes the responses to question 1 in Example 3.1 and it appears that students nationally with coeliac disease tend not to buy breakfast at their university campus.

Table 3.1 Responses by a national sample of students with or without coeliac disease to question 1: how often do you purchase breakfast from catering facilities at the university campus?

Students with or without coeliac disease	Frequency of responses to question 1: How often do you purchase breakfast from catering facilities at the university campus?			
	Every day	Most days	Occasionally	Never
Coeliac disease	0	2	3	18
Non-coeliac disease	0	7	7	9

There are a number of different types of answers that can be used for closed questions. Of these three types of answers your choice will be determined in part by the question but you should also consider the implications this choice has for the sample size you will then need in your study and the difference in how the data will be analysed (3.5).

i. Nominal categories

Having carried out your survey and collated the data the results will fall into a number of categories determined by the answers to each question. In Example 3.1 the question relating to ethnicity has a choice of 18 answers so your data table will have 18 categories. The headings to these categories are all descriptive (e.g. Bangladeshi) and there is no reason for organizing these answers in any particular order. These are known as 'nominal' categories (5.1).

ii. Ordinal categories

Some answers can be ordered as is seen in questions 1–15 in Example 3.1. Even though the choice of answers here ('Every day', 'Most days', 'Occasionally', 'Never') are descriptive, nonetheless there are clearly understood reasons why these phrases have a specific order. The ordinal types of answers can be phrases such as 'Every day' or can be numbers but the midpoint between these answers may not be meaningful or obvious or the scale is not necessarily linear. For example, in Example 3.1 for questions 1–15 to try to envisage a mid-point between the categories 'Every day' and 'Most days' does not make sense.

The most commonly used ordinal scale is a Likert scale. In this scale attitudes are assessed usually using a choice of 5–9 possible answers for a number of different statements. The answers for the 5 answer set are usually: Strongly disagree; Disagree; Neither agree nor disagree; Agree: Strongly agree. The choice of answers frames the neutral state (Neither agree nor disagree) with an equal number of positive choice answers and negative choice answers. Using a Likert scale has the advantage that this scale is familiar to many people. However, there is a tendency for participants to avoid the extreme answers so their choice does not reflect the range of answers available.

Likert scale An ordinal scale commonly used to assess attitudes to a number of different statements.

iii. Interval data

The last type of answers is those demanding a numerical response where the values are given on a larger linear scale. In Example 3.1 the student may have asked her participants to grade out of 100 their answer to 'How much difficulty do you have finding suitable restaurant meal choices?' Each respondent will provide their own grading and are not restricted to a choice of a few possible answers. The interval scale differs from the ordinal scale in that the intervening points along the scale are meaningful. If one student graded their answer as 50/100 and a second gave a score of 70/100 we could sensibly consider the average between these two scores; this mid-point is meaningful.

3.2.2 Open questions

Open questions may be used in all the methods we are considering in this section, including structured interviews and questionnaires. Open questions lead to descriptive answers or comments. The chief advantage of open questions is that you do not restrict the information you gather. However, this information is much harder to summarize (3.5.2). In Example 3.1, questions 23 and 24 are open questions.

 Q1 In Example 3.1, are questions 16–22 open or closed questions?

3.3 Phrasing questions

Whether you use open or closed questions or a combination of both, all questions need to be phrased so that the meaning is clear. Table 3.2 provides a checklist with examples that you can use to check your own questions against.

Table 3.2 Potential issues with phrasing of questions in questionnaires

Fault in the phrasing of your question	Examples of poorly structured questions	Improved structure to questions
Vague	How did you get here?	Give the mode of transport used to travel here, e.g. on foot, bus, train, car, bike, motorbike, etc.
Too few options in a closed question	How regularly do you go swimming: every day/never?	How regularly do you go swimming? Every day/a few times a week/once a week/once a month/a few times a year/never.
Leading, likely to create a biased response	What is your opinion of the terrible and frightening developments in molecular biology?	What is your opinion of the developments made in molecular biology in the last 5 years?
Double negatives	Do you not believe that your degree course is not adequate in preparing you for employment?	Do you believe that your degree course is adequate in preparing you for employment?
Jargon	Are you a member of the HEA?	Are you a member of the Higher Education Academy?
Too many topics	Are you in favour of fox-hunting and the control of deer populations through shooting?	a. Are you in favour of fox-hunting? b. Are you in favour of the control of deer populations through shooting?

Table 3.2 (*Continued*)

Fault in the phrasing of your question	Examples of poorly structured questions	Improved structure to questions
Unrealistic	How old were you when you read your first word?	Delete such questions.
Status questions	Are you employed/unemployed? ('Employed' is the first answer, implying that this has a higher status.)	These questions are very difficult to improve but you need to be aware of the potential impact that the order of the answers may have on a volunteer. The participant may be more likely to lie and/or be offended.
Insufficient details for analysis	How old are you? 10–20, 21–70	How old are you? 10–19, 20–29, 30–39, 40–49, 50–59, 60–69, 70 or over
Overlap in answers	How old are you? 18–20, 20–22, 22–24	How old are you? 18–19, 20–21, 22–23, 24–25

If your method for collecting information is structured as in a structured interview or questionnaire, you should carry out a pilot study before you begin. This is an invaluable way of checking that the questions really do provide you with the answers you are expecting. In an unstructured interview or a focus group, you will have the opportunity to clarify any questions during the interview/discussion.

3.4 Your participants

For any investigation that involves people, you need to make sure your approach is suitable for your 'audience' (4.2 and 4.8). There are many elements in this and not all will apply to all the methods you might use. Here are some prompts:

- **Language.** Use appropriate language. If necessary define terms and provide explanations of terms or ideas without influencing responses.
- **Accessibility.** Consider the layout of a questionnaire. Is the format and font suitable, for example, for dyslexic participants or for children? Do you need to produce a Braille copy? If you are providing a questionnaire online, can it be read by the standard screen-reader software? Is the room you are using accessible to all participants? Do you need a translator?
- **Environment.** If the location for carrying out this research is important, as in the interviews and focus groups, is the room free from distractions?
- **Time.** How long will the interview, questionnaire, or discussion take? Do your participants have that much time? Do you?

3.5 Sample sizes and data analysis

For most investigations, you will not involve the whole population; instead, you will sample (1.4). Apart from deciding on your sampling strategy (1.4.2), you will also need to decide on the size of your sample. The size of your sample will be determined in part by you satisfying yourself that your sampling strategy will generate a representative sample. In addition, sample size is determined by the way in which the data will be analysed. We consider this last point for closed (3.5.1) and open (3.5.2) questions.

3.5.1 Closed questions

As we discuss in Chapters 1 and 2, if you are planning an experiment to test hypotheses, you need to identify the statistical test you will use before you carry out your investigation, as it is this, along with the notion of a representative sample (1.4.1) and practicalities, which determines the size of your sample.

i. Nominal and ordinal answers

When your choice of answers are nominal or ordinal the statistical tests used most often are the chi-squared goodness-of-fit test (8.3), the chi-squared test for association (9.3), or the Fisher's exact test (9.4). Which test you use will depend on the number of answers in the closed question, and the number of groups of people you intend to sample and the sample sizes. The point we wish to highlight in this chapter relates to sample size and the chi-squared tests as it has implications for the number of answers you include in your questionnaire.

We compare the responses from Example 3.1 to the two background questions about gender and ethnicity. The gender question has a choice of three answers: Male; Female; and 'Prefer not to say'. The responses of the 23 students in each sample will therefore be spread across these three possible answers. In contrast the question about ethnicity has 18 possible answers so the same 23 responses will be spread across far more possible answers. The chi-squared tests have a requirement for one of the values in the calculation (called an 'expected' value) to be ≥ 5. The more possible answers you have the less likely this criteria will be met as your data becomes spread more thinly.

As you cannot calculate expected values until you have carried out the investigation and collected your data, how then can you determine what sample size to use? A pilot study on a group of people similar to those who will take part in your study will enable you to judge whether you are likely to get an appropriate distribution of answers and the effect this has on the 'expected' chi-squared values when these are calculated. It is also possible to derive a very rough estimate of a sample size that on average should be expected to meet this requirement of a chi-squared goodness-of-fit test. A very rough estimate can be obtained as the minimum proposed sample size divided by the number of answers for any question should be more than or equal to 5.

ii. Interval answers

Where you have asked your participants to provide interval answers you will obtain a data table that has a number of columns of interval data. The number of columns will be determined by the groups you are comparing. In Example 3.1 the student compared individuals with coeliac disease and those without so would have two columns of interval data. Here you most likely to analyse these data using a t-test (10.2), a z-test (10.1) or a Mann–Whitney U test (11.1). If you are comparing the responses of more than two groups you are likely to use a parametric (10.7) or non-parametric (11.3) one-way ANOVA. Each of these tests has specific criteria that your data and experimental design need to meet and some of the requirements indicate a minimum or maximum sample size.

3.5.2 Open questions

In this book, we only consider how you may extract quantitative information from open questions and analyse this. If you wish to evaluate the qualitative elements of your information, you will need to look to other texts (e.g. Robson, 2011). To generate quantitative data from qualitative answers, one approach is to record the number of times a keyword appears within the answers. You could total the number of times a keyword or phrase is mentioned. But if one person mentions four keywords and others only one, then the person who includes four keywords is over-represented in your sample and these observations are said to be not independent of each other. An alternative approach would be to record the first keyword in each answer. Although this resolves the problem of independence, clearly this approach does not then reflect all the information you have. Trying to obtain quantitative data from qualitative answers can be indicative at best.

Having obtained a numerical estimate of the use of a keyword or phrase, these frequencies can then be evaluated in the same way as data from closed questions. For example, in response to question 24 in Example 3.1 it was clear that several keywords, such as cost, convenience, and preference, were used consistently. As few students added comments to this open question we have collated the responses of those that did (Table 3.3). One of the chi-squared tests may be appropriate to analyse this type of data. Clearly you again have a problem when trying to determine sample size, as you will not know in advance how many keywords or phrases you will identify from your respondents' answers. The solution is to run a pilot study on a similar group of people. The general relationship between numbers of possible answers and sample size applies here to the number of keywords and phrases you identify. The more keywords and phrases there are, then in general the larger the sample size required to meet the criteria for using these statistical tests.

3.5.3 Achieving the required sample size

One of the great drawbacks of using these methods to gather information is obtaining a large enough sample size simply because the members of your population are not sufficiently motivated or do not have enough time to take part. You need to bear this in mind when

Table 3.3 Responses to question 24: 'Do you wish to make any further comments on factors that you feel impacts upon your diet whilst at university?', categorized by one keyword in each answer

	Keywords used to answer question 24 'Do you wish to make any further comments on factors that you feel impacts upon your diet whilst at university?'		
	Cost	Convenience	Preference
Number of students	11	5	2

writing your aim and objectives. Will the population identified by the aim provide sufficient numbers of participants to allow you to test any hypotheses you have? You must work ethically in your research, so any investigation that includes an element of coercion is not acceptable. Forms of coercion can include asking your family and friends, asking everyone in a lecture or tutorial group, or offering a reward. We discuss this in 4.8.1.

Summary of Chapter 3

- There are additional points that need to be considered when designing investigations in which you may use questionnaires, focus groups, or interviews to obtain quantitative data, including the need to consider the type of question and how this may determine how the data are analysed, and the sample size.
- The Online Resource Centre includes interactive exercises that test your understanding of this chapter along with other topics, particularly those considered in Chapters 7–11.

Answers to chapter questions

A1   In Example 3.1 q 16–21 are closed questions.

4 Research, the law, and you

 In a nutshell

This chapter examines UK law primarily as it relates to undergraduate bioscience research. The way in which law is developed within the UK is outlined. In most respects at present UK law is similar across England, Northern Ireland, Scotland, and Wales, with the exception of the laws of trespass. This chapter examines the law as it relates to health and safety, access, wild species, domesticated animal species, and humans.

Research and the law are intertwined in numerous ways and at all levels from the individual to the institution in which the research is carried out. It is a huge area, and our aim in this chapter cannot be to cover all relevant legislation. Instead, we provide an introduction to the areas that appear to be most relevant to bioscience undergraduates and graduates. Although there are differences between the law in England, Wales, Scotland, and Northern Ireland, on the whole the law that impinges on bioscience research is sufficiently similar that we can cover all four countries at the same time. Where there are notable differences, these are highlighted. The Internet in general is an invaluable and usually reliable resource in relation to the law for any student requiring further details. We provide links to relevant websites in our Online Resource Centre and further details about US law.

Clearly, as an undergraduate or graduate student, your institution carries some responsibility for you and what you do. As a result, your institution will make available to you guidelines and information that it is your responsibility to read and respond to. Some of the legal issues we cover in this chapter may therefore be addressed by your institution centrally and through staff such as your supervisor and technical staff. You have responsibilities both to your institution and under the law to comply with legislation whether it is channelled through your institution or by some other means. For example, your institution will have health and safety guidelines in place that put into practice the health and safety legislation. You are required to comply with these guidelines and be aware of the relevant legislation. However, if you work with wild plants and animals, your institution is less likely to have a central policy; nonetheless, you are responsible for being aware of the relevant legislation and complying with it.

Unlike other chapters, there are not many exercises here. Instead, we have suggested a number of discussion topics at the end that will allow you to relate real undergraduate project proposals to the areas of law we cover in this chapter. It will take you about 1 hour to consider these discussion topics in detail.

4.1 About the law

In the UK, the body of law is derived from two sources: international law and national law. We use the term 'law' to mean any legislation or regulation derived from an Act of Parliament (or equivalent national body) or other ratification of international agreements that is potentially binding on the individual and where failure to comply may lead to sanctions.

4.1.1 International law

Many nations voluntarily negotiate agreements at an international level. These agreements, however, are not legally binding until ratified by that country. An example of this is the Council of Europe where the European Convention on Human Rights and Fundamental Freedoms was developed. This was later ratified within the UK as the Human Rights Act 1998. Another example is the Convention on International Trade in Endangered Species of Wild Flora and Fauna (CITES). Originally agreed in 1975, this was ratified in the UK in 1976 as the Endangered Species (Import and Export) Act.

EU Regulation
Legislation that is drawn up by members of the EU and takes immediate effect as law.

EU Directive
A binding EU agreement that must then be passed as national law by the national government.

Other international law derives from the European Union (EU). At present the United Kingdom (UK) is a member of the EU and this form of legislation is negotiated by UK representatives on behalf of all countries in the UK. Agreements usually take on one of five formats: *regulations, directives, decisions, recommendations,* and *opinions*. Regulations take immediate effect as law in all member states and may subsequently also be included in national law. A Directive is binding but it is not law in a particular country until the national authorities pass an appropriate Act. Decisions are legally binding consequences resulting from EU law and can be directed at individuals or institutions. Recommendations and opinions do not have a binding effect on nation states. An example of EU law is the Habitats Directive 1992, which became incorporated into legislation primarily in the Conservation (Natural Habitats, &c.) Regulations (as amended) 1994 and the Conservation (Natural Habitats, &c.) Regulations (Northern Ireland) 1995.

4.1.2 National law

The UK comprises four countries—England, Scotland, Northern Ireland, and Wales. Each has its own body of law developed over many centuries. The formation of the UK as a combined political unit at various times has resulted in points at which law for all four countries was made by the one parliament in London. In 1998, governance in relation to many aspects of specific law making was devolved to Scotland, Wales, and Northern Ireland, with the UK

Parliament retaining powers to adopt Acts of Parliament for the UK as a whole. This process of devolution is ongoing.

In Northern Ireland, parliament had powers to adopt legislation for Northern Ireland from 1921 to 1972. These powers were suspended in 1972, and until 1998, the Westminster Parliament used direct rule powers to legislate for Northern Ireland and legislation was then adopted by Order of Council rather than by Act of Parliament.

Law developed within the UK is, for the purposes of this book, called national law. In the UK, the body of national law in general comes from five sources: common (case) law, legislation, institutional writers, canon law, and custom. Where national law and the legislative processes differ significantly between countries within the UK, this has been indicated.

National law
Law developed in the UK from common law, legislation, institutional writers, canon law, and custom.

i. Common law (case law)

The legal system in England, Northern Ireland, and Wales is based on common law derived from legal decisions made during specific court cases, i.e. there is a doctrine of legal precedent. These cases make rules that must be followed subsequently by all courts of similar or lower standing in the court hierarchy. Common law has been particularly important in shaping the law relating to trespass.

In Scotland, whenever a judge or sheriff pronounces his or her findings, he or she is influencing the development of the law on the point at issue. Like other countries in the UK, this decision has to be followed by all lower courts if the same point is disputed at a future date. However, unlike common (case) law in other UK countries, the Scottish courts seek to discover the principle that justifies a law rather than searching for an example as a precedent.

ii. Legislation

These are laws enshrined in an Act of Parliament, e.g. the Health and Safety Act 1974. The Acts of Parliament are divided into chapters, sections, and parts. Detailed information is usually included in Schedules at the end of the Act. For example, the lists of species covered by specific parts of the Wildlife and Countryside Act 1981 are included in the Schedules.

Law introduced through an Act of the United Kingdom Parliament or of the parliaments/assemblies in Scotland, Wales, and Northern Ireland may empower other authorities to make further provisions or amendments. The provisions are often known as Orders or Regulations. Principal among those empowered to make secondary legislation are ministers in a particular government office, e.g. the Home Secretary. The law may also empower government departments to license certain otherwise prohibited activities, and in developing the conditions for the licence further statutory controls are introduced. For example, to obtain a licence under the terms of the Animals (Scientific Procedures) Act 1986, an applicant must adhere to the code of practice produced by the Home Office for the housing and care of animals used in scientific procedures.

iii. Local law

Parliament has identified certain national bodies as being able to make laws. These bodies include the Crown, local authorities, and public corporations, such as the Environment

Codes of conduct
Desired actions outlined by a public body where compliance is usually statutory.

Agency, Forestry Commission, Natural England, and Natural Resources Wales. These bodies may establish laws within certain remits prescribed by the Statutory Instruments Act 1946 and Local Government Act 1972 and often in response to other specific statutes such as the Wildlife and Countryside Act 1981. Where a public body develops these laws they are usually known as codes of conduct, and where local authorities do so they are known as bye-laws.

iv. Institutional writers

Particularly in Scotland, writers of legal text books may attain a reputation so high that their works acquire authority, and if there is no case law or statute relating to a particular issue, the courts will turn to the views of these writers. The principal of these institutional writers is Sir James Dalrymple, Viscount Stair (1619–95), whose *Institutions* was published in 1681. Other important writers are Barons Hume (1756–1838), Erskine (1695–1768), and Bell (1770–1843).

v. Canon law

This derives from the legal system established by the medieval Roman Church in Western Europe, in turn derived from Roman law. After the Reformation of 1560, church courts disappeared and their powers, particularly relating to the law on the family, were taken over by the secular courts. Modern law has now largely overshadowed the importance of canon law.

vi. Custom

Custom plays only a very small part as a separate source of law. Many of the older customs have been embedded in Acts of Parliament and judicial decisions in court cases.

vii. Quasi-legislation

Apart from the sources of law outlined above, there are many individuals, and organizations that also develop policies, codes of conduct, etc. These are voluntary codes and do not have the force of law. This can be confusing, especially where such terms as 'code of conduct' are used. You may therefore need to confirm whether a code of conduct is enforceable under the law or not. Failure to follow some recommendations and codes of practice may be used as evidence that a law has been broken, even though these guidelines may not themselves be part of statute law.

viii. Enforcement

The agency responsible for enforcement of legislation varies enormously and is outlined in the original legislation. The police are responsible for enforcing most bye-laws and most statute laws, but there are many other agencies with powers of enforcement. For example, Customs and Excise staff have a major role in enforcing laws relating to imports and exports whereas the Animals (Scientific Procedures) Act 1986 is enforced through an Inspectorate, which reports to the Animals Procedures Committee and to the Home Secretary.

4.2 Ethics

Research in the UK should be guided primarily by the two basic principles behind an ethical basis for research: 'to do no harm' (non-maleficence) and 'to do good' (beneficence). Both terms (non-maleficence and beneficence) are essential, although at first glance they appear to be two sides of the same coin. However, to 'do good' does not necessarily mean that we do no harm. The actions you are required to take as a response to an ethical approach to research will differ depending on the organism and system you are studying. Legislation reflects society's ethical standards in relation to different species. The highest expectations of ethical practice are for higher organisms including ourselves. Common practices that are deemed appropriate in relation to research on animals (4.5 and 4.6) and on humans (4.8) are covered in the relevant sections.

Non-maleficence To do no harm.

Beneficence To do good.

The overarching question we must first ask ourselves is whether the research being planned is worthwhile and necessary. For undergraduates this is a difficult question as study at this level frequently repeats aspects of previous research and on a small scale so that it may not necessarily add new knowledge. Therefore questions of 'do no harm' are harder to balance against the outcomes from the work.

Ethics requires a number of value judgements to be made. Legislation undertakes to set suitable frameworks for these decisions and guidance on what is considered ethical and what is not. Most institutions have some form of ethics scrutiny where a panel of experienced researchers consider the ethical issues in work proposed by their peers and students and consider this in relation to institutional and national guidelines and legislation. This scrutiny is essential in research for several reasons. Firstly, researchers themselves, especially if they are inexperienced, may not foresee circumstances which may cause harm. Secondly, the researchers have their own ethical and moral principles and these will influence the design of a project. Researchers may also have a vested interest in carrying out this particular research project. For most students, this will relate to their wish to complete and be successful on a degree or higher degree programme, but can also relate to career opportunities and obtaining further funding. Therefore, it is common practice to subject research proposals to this independent scrutiny.

4.3 Intellectual property rights

Intellectual property is protected by a raft of legislation including: the UK Patents Act 1977 (as amended); the Copyright, Designs and Patents Act 1988; and the Intellectual Property Act 2014. Intellectual property is something that you have created and the legislation both protects those who create novel intellectual product and may include those who have rights through ownership. For example a company may purchase a patent. The protection offered by the legislation limits the ways in which other people may exploit the work, copy it, or in any way deprive you, the 'creator', of reward and incentive. You do not usually own the intellectual property for something you created as part of your work while you were employed by someone else. However some universities do recognize the intellectual property rights of both their undergraduates and graduates.

There are four main areas where rights have been granted: patents, trademarks, designs, and copyright. Some rights have to be applied for, such as patents, whilst others, such as copyright, are automatic. You need to comply with this legislation and ensure that in carrying out your work you do not infringe someone else's rights. You also need to know that under this law the work you do may also have potential or automatic rights.

In general undergraduates are most familiar with issues relating to copyright and are used to advise on what may or may not be copied and the need in some circumstances to work within the terms of a licence that has been granted to the Institution. Copyright protects a range of types of work including original music, films, and written work. You will be advised by bodies such as the library about how to comply with copyright of material you wish to read or use during your studies. Correctly citing sources is also required to demonstrate this compliance. Your rights in relation to the work you do will be laid out in a policy on intellectual property rights provided by your institution.

If you are creating or using a database this may be subject both to copyright and also a 'database right'. For example if you have invested in obtaining, verifying, or presenting the contents of a database it is probably covered by a database right. If the database has a high level of originality in its content or arrangement of contents then it may also be covered by copyright. The database right like copyright is automatic.

Patents can be granted within a specific country to protect products and processes. Within biological sciences you may need to comply with patents relating to equipment, some methods, and some consumables. Your technical staff in particular should be able to advise you on this. Commercial software is generally covered by patents which may limit access or length of use. Your institution should be able to give you guidance on patented software and the terms of any licence granted for its use.

Most universities have a department dedicated to helping its staff and students develop commercial applications of any innovation. Therefore if you develop a product or process that you believe warrants a patent you will be able to seek support and advice within your institution. Where a patent is infringed the patent owner can undertake court proceeding to claim for damages.

4.4 Health and safety

You must work in such a way that your safety and that of others is paramount. To do this, you need to identify the hazards of each step of your research protocol and then incorporate practices that minimize the risk to health. At the end of this process, you need to make a judgement as to whether the risk to your health and that of others is sufficiently small that the research may proceed. There are a number of laws that cover health and safety, including the Health and Safety at Work Act 1974, the Environmental Protection Act 1990, and the Environmental Act 1995, and a number of detailed regulations including the Management of Health and Safety at Work Regulations 1992, the Control of Substances Hazardous to Health Regulations 1988 (as amended) (COSHH), the Chemicals (Hazard Information and Packaging for Supply) Regulations 2002 (CHIP), the Special Waste Regulations 1996, and the Reporting of Injuries, Diseases and Dangerous Occurrences Regulations 2013 (RIDDOR).

Health and safety is something that is usually considered in laboratory work where chemicals or micro-organisms are being handled. However, health and safety must be considered in all types of research from carrying out a questionnaire survey to sampling soil in the field. The method for considering health and safety is a risk assessment. We include an example risk assessment form on the Online Resource Centre.

This can be divided into seven steps:

1) **Hazard identification and rating.** What hazards may be present if you carry out your research as you have designed it? In what way may these hazards affect your own or other people's health?

2) **Activity.** How might you become exposed to the hazard and what is the possible scale of exposure?

3) **Probability of harm.** Given the nature of the hazard, the possible effect on health, and the possible scale of exposure, what is the probability of harm?

4) **Minimize the risk.** Decide what precautions are needed to control or reduce this risk.

5) **Risk evaluation.** Is this an acceptable level of risk?

6) **Emergencies.** Be aware of, or prepare procedures to deal with, accidents and emergencies.

7) **Action.** Ensure these precautions are in place and are followed.

Risk assessment
A process whereby risks associated with work are identified, minimized, and finally evaluated to determine if the work can proceed.

4.4.1 Hazard identification and rating

What hazards may be present if you carry out your research as you have designed it? In what way may these hazards affect your own or other people's health?

In most student research, the hazards will fall into three broad categories.

i. The environment

Your working environment may contain a number of hazards. This is true of both laboratory work and working in the field. In the laboratory, you should consider physical features including temperature extremes such as working in a cold room or in a greenhouse in the summer. There are often many electrical sources, equipment with moving mechanical parts, or dust in the atmosphere from, for example, grinding bones, sorting soil, or allergen studies.

In the field, you will almost certainly encounter particularly hot or cold days. The environment may have rough terrain and may be subject to flash flooding. Additional hazards can come from living things in the environment that are not strictly part of your research. For example, you may need to consider threats from strangers, encounters with wild or agricultural animals, bites that may cause skin damage and introduce infections, poisonous or allergenic plants or fungi, etc.

You must also be aware of activities that may be carried out in the environment of your research that you are not involved in but that may be a source of hazards. For example, areas that have been sprayed with pesticides, chemical spills, the use of microbes, the presence of a UV source, etc. Chemical hazards found in the environment are considered further in 4.4.1ii.

One example where the environment may have hazards that need to be considered is seen in Example 8.2: the genetics of flower colour in *Allium schoenoprasum*. In this experiment, the investigator is working in a greenhouse. Common hazards of a greenhouse include: working in a high temperature, which can bring about heat stroke; broken glass, which may cause cuts; electrical sources in contact with water, which may result in electrocution and death; soil-borne pathogens, which can bring about serious illnesses; and allergens from fungi and plants, which may cause asthma. Most universities and other similar institutions carry out general risk assessments for areas such as laboratories. You may therefore be able to make use of these when carrying out your own risk assessment.

ii. Chemicals

COSHH assessment
Where the use of hazardous substances is reviewed to ensure compliance with current legislation.

Obtaining, storing, using, and safe disposal of chemicals are covered primarily by the Dangerous Substances and Explosive Atmospheres Regulations 2002 and by the Control of Substances Hazardous to Health Regulations 2002 (COSHH). The first of these two Acts covers all flammable substances. COSHH covers 'hazardous substances' that are to be used during work, substances generated during work, naturally occurring substances, and biological agents such as bacteria and other micro-organisms. The term 'hazardous substance' is taken to mean a substance or mixture of substances classified as dangerous to health under the Chemicals (Hazard Information and Packaging for Supply) Regulations 2002 (CHIP). These regulations do not cover all substances, however, and specific regulations cover the safe handling of gases such as helium and the use of pesticides, medicines, cosmetics, lead, and radioactive substances. Under the terms of the Health and Safety at Work Act 1974, these substances must also be considered in hazard identification.

To identify the hazard relating to any substance that you will use in your work, you can refer to the label on the container in which the substance is supplied. Further details can easily be obtained by searching the Internet using the chemical name, or by referring to suppliers' catalogues, or through Health and Safety Executive (HSE) publications such as the *Approved Supply List*. Links to useful websites are included in the Online Resource Centre. When carrying out research you may not always know how substances you are using will react and therefore you may not be able to predict what substances will be produced during your work. If you are not sure for any particular protocol, you should refer this matter to your supervisor.

The potential effects of a chemical hazard are described using a number of specific terms. For example, 'caustic' is the chemical burning that occurs on exposed surfaces such as skin, eyes, or internal organs if they become exposed to the substance. The hazard is therefore potentially damaging if you eat/drink it or allow your skin or eyes to come into contact with it. Other common descriptive terms are flammable, explosive, toxic, harmful, mutagen, carcinogen, corrosive or strongly oxidizing, irritant, radioactive, and harmful to the environment. Links to definitions of these 'risk phrases' are included on many chemical and biochemical hazards websites, in suppliers' catalogues, and on the Health and Safety Executive (HSE) website.

Q1 Use one of the sources outlined above to identify the hazard(s) associated with concentrated ethanoic acid (glacial acetic acid).

---

iii. Biological agents

When biological agents are discussed in health and safety, most thought is given to micro-organisms. For example, in the field you may encounter *Leptospirosis* sp., which is a bacterium that can be found in water contaminated with the urine of infected animals. You also need to be aware of organisms that may present a potential hazard, including the common intestinal parasites of dogs and cats (*Toxocara* spp. (ascarids) and *Ancylostoma* spp. (hookworms)), which can be distributed in the environment from animal faeces.

Micro-organisms (bacteria and fungi) present a different challenge from most chemical and environmental hazards. If you are working with microbes in the laboratory, you may not know which of the possible thousands of species you are incubating and the potential virulence and toxicity of a species may vary among strains. It is therefore sensible to consider all micro-organisms as potential pathogens unless the specific nature of the hazard is known. Information about pathogenic micro-organisms can be obtained from a number of sources such as the Advisory Committee on Dangerous Pathogens (ACDP). In general, the hazards associated with micro-organisms are: infection (e.g. botulism, meticillin-resistant *Staphylococcus aureus* (MRSA)), the production of toxic substances by the micro-organism (e.g. *Aspergillus flavus* and *A. parasiticus* are both known to produce aflatoxin), and the potential allergenicity of live or dead airborne micro-organisms leading to the onset of asthma or dermatitis.

Other species including animals, birds, reptiles, and large fish can also be potential biological hazards and should be taken seriously if present or likely to be present in the area you will be working. In the UK, there are few hazards associated with contact with native British flora other than that resulting from an allergenic response to pollen. Those not familiar with British flora should, however, familiarize themselves with common genera such as *Urtica sp.* and the effects such 'stinging' plants may have.

iv. Rating the hazard

Having identified all the hazards that you may encounter in your work, you need to rate them. The institution you are affiliated with should give guidance on this. In general, however, a major hazard would be one that may cause death or serious injury, a serious hazard may cause injuries or illness resulting in short-term disability, and a slight hazard may cause less significant injuries or illnesses.

 In Example 9.1, the investigator was studying *Littorina* species on the mid to low areas of a rocky shore. What hazards might she encounter, and how would you rate them?

4.4.2 Activity

How might you become exposed to the hazard and what is the possible scale of exposure?

The way in which you encounter, use, and/or generate hazardous substances will determine the probable extent to which you may become exposed to the hazard. Some substances you handle may be in a powdered form and so you may risk exposure through inhalation. Some substances can penetrate the skin, whilst others damage the skin.

i. Hazards arising from the activity

Activities can be potentially hazardous in their own right and these need to be recognized as part of your evaluation of health and safety. Examples include where you or others need to move equipment, to sit or stand in awkward or cramped conditions, or to make repetitive movements.

Hazards in some activities will be the result of a particular substance being exposed to a particular treatment, and the potential hazard will differ from either the hazard related to the substance or the hazard relating to the activity alone. For example, agar in its crystalline powder form is believed to present a negligible risk to health. If you then wish to make agar plates, the protocol requires the agar to be placed in a container (usually a glass beaker or bottle) with distilled water and heated to 100 °C. The agar needs to be poured when it is still hot, at about 80 °C. Splashes from the hot liquid agar could cause serious scalds, which will be hard to treat because the agar sticks to the skin. Therefore, the hazard as a result of the activity is both different from the substance and activity alone and more serious.

ii. Scale of exposure

Having considered in what way you may become exposed to the hazard, you then need to grade this possible exposure. You should be given guidance as to how to grade the likelihood of exposure. However, in general you would assign the hazard to a 'high' category where it is certain harm will occur, 'medium' when harm will often occur, and 'low' where harm will seldom occur. For example, concentrated ethanoic acid is caustic (see Q1/A1) and would be rated as a major serious hazard. You may be provided with 1 ml of a very dilute solution of this (e.g. 5%, w/v) of which you will pipette $1 \mu l$ into a small tube. As you are using small volumes of a solution with a low concentration, the scale of exposure would receive a 'low' rating. (In fact, this dilution of ethanoic acid is commonly found in kitchens as vinegar.)

iii. Storage and disposal

With some potential sources of hazards, such as chemicals and micro-organisms, you need to consider more than just your use of them in an investigation. These substances need to be stored and safely disposed of. Your consideration of hazards and risks needs to extend to the full lifetime of such substances. This is of particular importance, as it is during these times that other people are likely to become exposed to the potential hazard. Usually, technical staff can advise you about safe storage and disposal, and regulations such as the Special Waste Regulations 1996 (and as amended) and the Dangerous Substances and Explosive Atmospheres Regulations 2002 provide statutory guidance.

4.4.3 Probability of harm

Given the nature of the hazard, the possible effect on health, and the possible scale of exposure, what is the probability of harm?

i. Probability of harm

This, the risk assessment step, is not easy and is usually subjective and qualitative. Most institutions provide guidelines to help you. For supplied substances, the HSE has developed a generic web-based guide to risk assessment called *COSHH Essentials*.

In a risk assessment, you need to make a judgement about each hazard you have identified using the following information:

- The hazard and the nature of its possible effect on health (major, serious, or slight).
- The possible scale of exposure as a result of the way in which the hazard will be encountered or used (high, medium, or low).

The probability of harm is the first of these combined with the second. Therefore, in Q1/A1, the major caustic hazard of ethanoic acid will only be present in a form where the risk of exposure is low. In this example, we would grade the overall probability of harm as low.

ii. Variation among individuals

Human beings all differ, and some of this variation can increase or decrease the risks from any one hazard. Points that often need to be considered include height, handedness, sensitivity to allergens, a suppressed immune system, the side effects from medication (such as antihistamines causing drowsiness), and pregnancy. For example, you may not be pregnant, but you may need to alert those who are to a hazard that is of more significance to them than to yourself.

iii. Workplace Exposure Limits (WELs)

Some substances have Workplace Exposure Limits. As an undergraduate, you are unlikely to be using such substances, although as a graduate this is possible, e.g. radioactive substances. Further information about Workplace Exposure Limits is available from the HSE.

4.4.4 Minimizing the risk

Decide what precautions are needed to control or reduce this risk.

It is impossible to cover in this section all possible types of hazard with suggestions as to how you may reduce the risk from each hazard. We have concentrated on the most common, but for each of the hazards you have listed, you must consider how to prevent the hazard or reduce the risk (i, ii) and then re-evaluate the probability of harm (iii).

i. Prevention

No risk to health is really acceptable; therefore, ideally you should prevent exposure to the hazard. Prevention can be achieved by changing the process or activity (e.g. using a different field site) and/or by using a safer alternative (e.g. using a pelleted form of a substance rather than a powder).

ii. **Reduction**

If prevention is not reasonably practicable, you must adequately control exposure and therefore you need to put in place control measures suitable for the activity and consistent with your risk assessment.

a. *Wear personal protection*

You may wear and/or provide personal protection appropriate to the nature of the hazard. In the biosciences, this protective clothing most often consists of laboratory coats, general safety glasses or face masks, UV light shields, and gloves known to exclude the particular hazard you may be exposed to (thermal or chemical). In the field, additional items providing personal protection include strong boots, waders, waterproof clothing, sun cream, and insect repellent. It should be noted, however, that the sun cream and insect repellent are substances and as such their use should also be considered under COSHH.

Additional medical protection may be appropriate, for example vaccinations (e.g. tetanus or hepatitis), medication to control allergies, and medical/biological monitoring. If you are ever in doubt or feel unwell, you should always seek medical assistance and draw to the attention of the medical staff the nature of your research and the hazards you have identified.

b. *Reduce quantities*

You may reduce the hazard by ensuring, for example, that the quantity of substances you are storing, using, or will need to dispose of are limited. For example, you may keep most of your stock solution in safe storage and only take out the volume required at the time.

c. *Control ventilation*

You may use controlled ventilation as in a fume hood or laminar flow hood to limit exposure to a hazard. This is common practice where volatile substances are used or produced, and in microbiology.

d. *Organize space*

The most common method used in risk reduction is to organize your space. Eating and drinking should be prohibited in all laboratories to prevent accidental ingestion of a hazardous substance. In the field, the equivalent practice is to ensure that you are able to wash your hands with clean water before you eat. Other examples of good practice are to keep flammables away from a Bunsen burner and to keep electrically powered equipment away from liquids. For some research, a dedicated area to which access is restricted is appropriate (e.g. microbiology areas and areas where radioactive sources are used and stored). For some substances such as flammables and poisons, special storage should be provided. If you are in doubt your supervisor or technical staff will be able to advise you.

e. Take a break

This is especially important if you are doing a repetitive task or if you are using light sources such as a microscope.

f. Inform

It is essential that you alert other people to potential hazards. With stocks of substances this is usually already done for you by the manufacturers. Solutions or mixtures that you pre-pare yourself must be similarly labelled with details of the contents including concentrations, the hazard(s), and the date. There are internationally recognizable symbols that are used to indicate certain hazards. Laboratories often carry stocks of sticky labels with the symbols for chemical hazards such as 'explosive' (Fig. 4.1) or 'oxidizing' (Fig. 4.2). These symbols can be stuck onto the containers holding your solutions or substances.

If you will be working in the field, then providing information is important in a number of ways. For example, you may wish to leave markers protruding from the ground to indicate the location of permanent quadrats. If so, you should consider the need to alert other people to their presence so that they will not fall over them. When working in the field, you should

Fig. 4.1 Standard symbol for explosive hazard

Fig. 4.2 Standard symbol for oxidizing hazard

ensure that at least one responsible person knows exactly where you are working and when you are expected back. It is also advisable to take a phone so that you can keep in touch.

g. Company

It is general practice in most institutions where students may be working independently to require or at least to recommend that the researcher always has someone else with them or nearby. In laboratories this can most easily be achieved by agreeing a schedule of work that is compatible with another student also working in that laboratory. For those working in the field (both urban and rural areas), this may not be so easily arranged but should be achieved if at all possible.

h. Be informed

Part of the process in risk assessment is planning for emergencies. This level of planning is a form of risk reduction. We consider this in 4.4.6.

iii. Re-evaluate the probability of harm

Having put into place control measures to reduce the risk, you can now re-evaluate both the hazard and the nature of its possible effects on health (major, serious, or slight) and the possible scale of exposure as a result of the way in which the hazard will be encountered/used (high, medium, or low). Most control measures work by addressing the second of these. You will now have two measures of the probability of harm, one where there are no control measures (4.4.3) and one where the control measures are in place and effective (4.4.4).

4.4.5 **Risk evaluation**

Is this an acceptable level of risk?

Using these two sets of risk evaluations, you are now in a position to determine whether the level of risk is acceptable. We cannot provide detailed guidance on this as there are so many differing potential hazards. You must therefore seek specific guidance from your supervisor. However, in general, undergraduate work will normally be allowed to continue if the risk, with the controls in place, is low and occasionally moderate and where the controls are not likely to fail, and where if they did the risk would be no more than moderate. Graduates may be allowed to accept a higher risk, but in our experience a severe risk of injury to health is not usually deemed acceptable for any student.

4.4.6 **Emergencies**

Be aware of, or prepare procedures to deal with, accidents and emergencies.

There are two types of emergency that you need to consider before you may finally complete this part of your research preparation. The first point you need to consider is what to do if one or more of your measures introduced to control the hazards fail. The second is what to do if there is a fire or other external emergency that means you must leave your research immediately.

i. **Failure of control measures**

You must consider here both in what ways these protective measures may fail and what needs to be done in response. For example, if a laminar flow hood fails, you are more likely to become contaminated with the micro-organisms you are handling. You may therefore need to carry out biological monitoring or be aware of symptoms of any illnesses that may develop as a result. You may also wish to satisfy yourself that the laminar flow hood is appropriately maintained. A second example of such an emergency would be if you spilt a hazardous substance. You will need to know: who needs to be informed and how; how quickly anyone else needs to be informed; who can tidy the spill and how; whether the area needs to be evacuated; and if so who is responsible for this. In the field, this preparation can include ensuring that you have emergency phone numbers, additional food, water, and clothing, and basic first aid, and to have worked out an emergency escape route.

ii. **External emergencies**

In preparation for such events as a fire alarm sounding, you should discuss with your supervisor and technical staff what needs to be done to ensure that the hazard(s) in your work continue to be contained and other hazards do not arise. There may be a need, for example, to turn off a Bunsen burner or safely store hazardous substances. As part of their emergency procedures, your institution will have guidelines as to who should carry out such actions: it may not be you.

iii. **Information**

If your control measures fail or you experience an external emergency, then the risk assessment that you carried out will inform you and those around you as to what action is required.

Therefore, it is essential that you keep it with you. For example, if you splash yourself with something, you may urgently need to be able to communicate details about that substance and its hazard to others. Therefore, you should keep a copy of your risk assessment including your emergency plans in your field or laboratory notebook and keep the notebook with you.

4.4.7 Action

Ensure these precautions are in place and are followed.

Having carried out your risk assessment and determined that the risk is acceptable, you can then carry out the research. However, you will need to:

- Ensure that the precautions to control the hazard(s) are in place, are followed, and that they are effective.
- Ensure that your emergency planning information is with you when you are working and that others in the vicinity are aware of this.
- Contact medical assistance immediately if you start to feel unwell, and take your risk assessment details with you.

4.5 Access and sampling

In this section, we consider the law that relates particularly to the studying or sampling of wildlife. The legislation relating to access and to handling plants and animals is extensive. We cover some of the key areas most often encountered by undergraduates, but have restricted ourselves largely to land-based work.

4.5.1 Access

The laws relating to access in England and Wales are fundamentally different from those in Scotland and Northern Ireland. In particular, the concept of trespass is virtually absent from Scottish law and Scottish legislation has codified and simplified access rights.

i. Access in England and Wales

The laws relating to access are derived primarily from national law. The statutory regulations are included in a number of Acts, such as the Environmental Protection Act 1990, the Countryside and Rights of Way Act 2000, and the Marine and Coastal Access Act 2009.

In essence, in England and Wales you may become a trespasser:

- If you enter land without the permission of the landowner.
- If you go outside the area that you are allowed on (because you have a statutory right or you have permission from the landowner).
- If you use land in a way in which you have not been authorized to do.

Trespass is not usually a criminal offence in that you cannot be prosecuted, but you can be sued. This may result in being required to pay compensation. In addition, the landowner can use reasonable force to eject a trespasser and obtain an injunction if necessary to prevent you from returning. Trespass can be a criminal offence on Ministry of Defence land, land adjacent to railways, and some nature reserves. In addition to trespass, you may also commit offences of criminal damage if you put up signs, dig holes, etc., without the permission of the landowner.

a. Public right of way

There are some areas such as registered footpaths where you have a public right of way. However, this only means that you may use the path to travel, usually on foot, from point A to point B. You may therefore be considered to be trespassing if you carry out other activities such as taking samples without the permission of the landowner.

b. Open country

There are some parts of England and Wales that are considered to be 'open country'. There are two main pieces of legislation relating to access in open country: the National Parks and Access to the Countryside Act 1949 and the more recent Countryside and Rights of Way Act 2000 (CRoW). Both of these provide for the possibility of access to 'open country' such as mountains, moors, heath, and down. The way in which these rights are managed varies slightly in that within the first Act details about access are largely set out in an agreement or order made in conjunction with the owner. In the more recent Act, access is determined more by the terms of the Act than by local agreements. Whichever Act covers a particular piece of open ground, normally there will be a right of access on foot for open-air recreation, such as walking, bird-watching, picnicking, running, and climbing, but not usually for driving a vehicle on the land, using boats, hunting, fishing, collecting anything from the area including rocks or plants, camping, or lighting fires. There are variations and exceptions to both Acts; for example, some areas within the designated 'open country' regions in the Countryside and Rights of Way Act 2000 are not subject to these access rights. These include buildings and livestock pens, land used for growing crops or planted with trees, quarries and other active mineral workings, land used as a golf course or race course, and land where military bye-laws apply.

c. Waterways

There are a number of different types of waterways: tidal and non-tidal rivers, streams, canals, natural and man-made lakes, and reservoirs. Ownership and responsibility for the management of these areas varies. However, none of these areas has a general right of way unless they are included within the regions of 'open country' identified in the Countryside and Rights of Way Act 2000. However, many bodies responsible for waterways allow some rights of navigation and access. This permissible access and associated activities will be laid out under agreements, codes of conduct, or bye-laws. Key bodies that may be consulted in relation to waterways are the Environment Agency and The Canal and River Trust (England and Wales) or Scottish Canals.

d. Foreshore

In England and Wales the foreshore between the high- and low-water marks is the property of the Crown except where it has been sold or leased, usually to a local authority. Again access is customarily permitted, usually subject to bye-laws. The Marine and Coastal Access Act 2009 extends the right of access to some coastal land. Most beaches above the high-water mark are owned by the local authority and access is subject to bye-laws.

e. Commons and village and town greens

Commons and village and town greens are not necessarily places with a right of public access. For most of this type of land, consent is usually given by the landowner to allow access and some activities. This permission is outlined either in an agreement or in bye-laws under the terms of statutory law such as the Commons Act 1876 and 1899 and the Village and Town Greens Act 1965. Local authorities have in the last 50 years or so clarified the ownership of and public rights of way on commons, etc. Therefore, if you need to contact an owner, the council may be able to help you. Village and town greens may have increased public rights of access if local people have openly used the land for recreation for at least 20 years without the permission of the owner. These increased rights still only relate to access and recreational activities.

f. Other areas

There are some areas where you may feel you can roam at will, e.g. National Trust properties and country parks. Most of the latter are owned by the local authority. Acceptable activities in these areas will also be outlined in, for example, a code of conduct or in bye-laws. Similarly, in woodlands and forests you do not have a general right of access. There may be public rights of way such as footpaths passing through the woodland, or local agreements giving some access. Again, these local agreements usually relate to walking and a few prescribed activities such as bird-watching.

ii. Access in Northern Ireland

The basic premise in relation to trespass, as set out in 4.5.1i, applies equally to Northern Ireland. However, the law in relation to rights of way is different. Basically, Northern Ireland does not have a legally recognized concept of open country, as the Countryside and Rights of Way Act 2000 does not apply to Northern Ireland. Instead, the Access to the Countryside (Northern Ireland) Order 1983 places Northern Ireland's 26 District Councils under a statutory duty to assert, protect, maintain, and record public rights of way. For all access to the countryside, you must therefore identify the landowner and consult with them and obtain permission to carry out any study.

iii. Access in Scotland

The law in Scotland relating to access has always been fundamentally different from that elsewhere in the UK. In Scotland, the simple crossing of a piece of land to get from one point to another has never been considered trespass; damages for mere trespass are not recoverable, although an interdict could be granted by the courts. The Land Reform (Scotland) Act 2003 enshrined this 'no law of trespass' concept and laid down the formal basis for the Scottish Outdoor Access Code. The Act allows for access to land for the purposes of *'carrying on a relevant educational activity'*. This includes *'(a) furthering the person's understanding of natural or cultural heritage; or (b) enabling or assisting other persons to further their understanding of natural or cultural heritage'*.

Whilst the Scottish Outdoor Access Code is not an authoritative statement of the law, in any dispute the Sheriff will consider whether the guidance in the Code has been disregarded by any of the parties. The Code states that access rights extend to individuals undertaking surveys of the natural or cultural heritage where these surveys have a recreational or educational purpose within the meaning of the legislation. A small survey done by a few individuals is unlikely to cause any problems or concerns, provided that people living or working nearby are not alarmed by your presence. If you are organizing a survey which is extensive over a small area or requires frequent repeat visits, or a survey that will require observation over a few days in the same place, consult the relevant land manager(s) about any concerns they might have and tell them about what you are surveying, the purpose, and for how long. If the survey requires any equipment or instruments to be installed, seek the permission of the relevant land managers. In Scotland the main places where access rights do not apply are: houses and gardens, and non-residential buildings and associated land; land in which crops are growing; land next to a school and used by the school; sports or playing fields when these are in use and where the exercise of access rights would interfere with such use; land developed and in use for recreation and where the exercise of access rights would interfere with such use; golf courses (but you can cross a golf course provided you do not interfere with any games of golf); places such as airfields, railways, telecommunication sites, military bases and installations, working quarries and construction sites; and visitor attractions or other places that charge for entry.

4.5.2 Theft

Taking things from the environment, such as stones, wood, and earth, is a form of theft unless you have been authorized to do so by the landowner. Therefore, if you wish to remove leaf litter or soil, for example, you should ensure you have discussed this first with the landowner. Many species are protected in this regard by additional legislation (4.5.3), but wild plants and fish not covered by the additional legislation are the property of someone (e.g. Theft Act 1968). Therefore, it is illegal, for example, to uproot any wild plant for commercial purposes without authorization from the landowner.

4.5.3 **Plants, animals, and other organisms**

Currently, the key legislation in England and Wales that provides statutory protection for organisms is the Wildlife and Countryside Act 1981 (with amendments in each country) ('the 1981 Act'), the Conservation (Natural Habitats, &c.) Regulations 1994 (with amendments in each country), and the Countryside and Rights of Way Act 2000 (with amendments in each country). Some species are subject to additional legislation, for example the Protection of Badgers Act 1992. In Scotland, the current key legislation is the Nature Conservation (Scotland) Act 2004 (as amended), which integrated and updated previous wildlife legislation. In Northern Ireland, the legislation relating to the protection of species is primarily contained within the Nature Conservation and Amenity Lands (Northern Ireland) Order 1985, the Wildlife (Northern Ireland) Order 1985 (as amended), Conservation (Natural Habitats, etc.) Regulations (NI) 1995, the Environment (Northern Ireland) Order 2002, and the Wildlife and Natural Environment Act (NI) 2011.

Wildlife protection falls into two categories: protection of specific species and the protection of habitats. In this section, we consider the protection of specific species under three subheadings: birds, animals, and plants. The Schedules in the Wildlife and Countryside Act 1981 (as amended), Wildlife (Northern Ireland) Order 1985 (as amended), and the Nature Conservation (Scotland) Act 2004 (in Scotland), which list the birds, animals, plants, and other species covered by the legislation, are reviewed every 5 years. For current lists of Scheduled species, we refer you to the websites for the Joint Nature Conservation Committee, Natural Resources Wales, Scottish Natural Heritage, and the Northern Ireland Environment Agency.

i. Wild birds

a. *General protection*

All (approximately 600) species of wild birds recorded in the UK are covered by current legislation across the UK. Unless you have a licence or are otherwise authorized under the relevant legislation, these acts make it an offence to kill, injure, take, or sell any wild bird or their nests or eggs. There are exceptions to this protection in that it is not an offence to kill a bird if it has been mortally injured or to take a wild bird to treat and release it if it has been hurt, but only where the bird has been injured in a way that does not contravene the Act (or Order). You may also be exempt if the outcome is *'the incidental result of a lawful operation and could not have been avoided'*. For example, the action is carried out by an authorized person for the purpose of preserving public health, public safety, air safety, preventing the spread of disease, or preventing serious damage to livestock, crops, fruit, growing trees, or timber, or otherwise in accordance with a licence granted by the Minister.

b. Enhanced protection

Some species have enhanced protection and for the species listed in these Acts/Orders, the penalties for the offences described above are higher. It is an offence to disturb these wild birds whilst they are building a nest, or if they are on or near a nest containing eggs or young, or to disturb dependent young of such a bird. Some of these birds are given enhanced protection for the whole year, but for some it covers only the spring and summer (February–August). The exceptions that relate to the general protection of wild birds also apply to this enhanced protection. Enhanced levels of protection are also in place for bird sanctuaries and in Scotland for the lek sites of some species. The Joint Nature Conservancy Council (JNCC) has produced a list of conservation designations for UK wildlife which you may download.

There are other regulations relating to game birds, 'pests', and swans. The regulations outline the allowed methods, times of year, and species that may be killed as 'game birds'. Some birds such as pigeons may be 'pests' and the legal control of these is also outlined in the legislation. Wild swans in England belong to the Crown and, whilst they are subject to the 1981 Act, they are also Crown property unless they have been tamed and are on private waters. In Northern Ireland, swans are included in the Wildlife (Northern Ireland) Order 1985 (as amended).

ii. Wild animals

a. General protection

Protection for wild animals comes primarily from the legislation outlined in the first paragraphs of 4.5.3. The species in these Acts at present include certain mammals, reptiles, amphibians, fish, butterflies, moths, beetles, hemipteran bugs, crickets, dragonflies, spiders, crustaceans, sea mats, molluscs, annelid worms, sea anemones, and sea horses. The Joint Nature Conservancy Council (JNCC) has produced a list of conservation designations for UK taxa which you may download. For these Scheduled species it is an offence to damage, destroy, or obstruct access to any place which such an animal uses for shelter, protection, or breeding, except within your home; and to kill, injure, take, or sell these animals. You may be exempted from these laws for the same reasons given for wild birds: for example, if you take a Scheduled animal to tend it if it has been injured due to a reason that does not contravene these Acts.

b. Enhanced protection

Some species are covered by enhanced protection either from European regulations and/or national legislation. These include badgers, deer, bats, fish, seals, and whales.

Badgers are not included in the Wildlife and Countryside Act 1981. Instead, in England and Wales, they are protected under different legislation including the Protection of Badgers Act 1992. In Northern Ireland, badgers are protected by the Wildlife (Northern Ireland) Order 1985 (as amended). There are Scottish amendments to the Badgers Act 1992 in the Nature Conservation (Scotland) Act 2004. This combined legislation provides the same type of protection given to the animals on Schedule 5 of the 1981 Act, but with additional offences relating to the use of dogs, cruelty, and enhanced protection relating to reckless disturbance and causing damage to a badger sett.

Deer are also not included in Schedule 5 of the Wildlife and Countryside Act 1981, but are covered by the Deer Act 1991, Deer (Scotland) Act 1996, and Schedule 10 of the Wildlife (Northern Ireland) Order 1985. These laws specify which species may be killed, when, and how. Deer are also treated as property and you should therefore obtain any landowner's consent for any actions involving deer on their land.

Bats are protected by both national and international legislation including that outlined in the first few paragraphs of 4.5.3, the Wild Mammals (Protection) Act 1996 (as amended), and the Habitats Directive. Under this legislation, all bats are listed as 'European protected species' and it is an offence for any person: to deliberately capture, kill, injure, or possess a bat; or to damage, destroy, or obstruct access to any place that a bat uses for shelter or protection; or to deliberately disturb a bat. Again, there are some exceptions to this legislation, which are similar to those described in relation to wild birds.

Some fish species are included in Schedule 5 of the Wildlife and Countryside Act 1981 and by the Foyle Fisheries Act (Northern Ireland) 1952 (as amended) and the Fisheries Act (Northern Ireland) 1966 (as amended). In addition to this, common law treats fish in non-tidal waters as property. Therefore, you will need the landowner's consent before carrying out any work relating to freshwater fish. In addition, the Salmon and Freshwater Fisheries Act 1975 (as amended) and the Salmon and Freshwater Fisheries (Consolidation) (Scotland) Act 2003 prohibit killing freshwater fish by certain methods, including firearms and spears, or using lights. In UK tidal waters and seas, anyone can fish, but you are subject to some statutes and bye-laws.

Seals are protected under the 1981 Act/Order. Although it is legal under certain circumstance to kill seals this generally relates to commercial fishing and is restricted to specific species in specific locations and at particular times of the year. In Great Britain, it is also an offence to intentionally or recklessly disturb a dolphin, whale (cetacean), or basking shark.

iii. Wild plants

In most UK legislation, the working definition of the term 'plants' is wide and includes algae, lichens, and fungi, as well as true plants such as mosses, liverworts, and vascular plants. Protection is provided at three levels: general, enhanced national, and international protection.

a. General protection

All wild plants are covered by legislation relating to property (4.5.2). In addition, all wild plants are given general protection in the Wildlife and Countryside Act 1981 (as amended) and the Wildlife (Northern Ireland) Order 1985, which makes it illegal to uproot or destroy any plant for any reason without the permission of the landowner, but you may legally pick plant material for the plants not covered under the enhanced protection of, for example, Schedule 8 of the Wildlife and Countryside Act 1981 (as amended).

b. Enhanced protection

Species listed under the Act/Order have enhanced protection. For these plants it is an offence to intentionally pick, uproot, or in any way destroy such a plant, and it is an offence to sell such wild plants. This enhanced protection therefore includes collecting seeds or spores of any of the species listed in Schedule 8 and there is no exception for the owner of the land on which these plants are found. Again, there is an exemption made under the terms of the 1981 Act, which is where plants are damaged or destroyed as an incidental result of a lawful operation and the damage could not reasonably have been avoided. The Joint Nature Conservancy Council (JNCC) has produced a list of conservation designations for UK plants which you may download.

In addition to the enhanced protection given to some plant species in the Wildlife and Countryside Act 1981, the local authority can place tree preservation orders, for example to protect ancient trees and/or to preserve a certain ambience in urban areas. If a tree is included in a tree preservation order, then it is an offence to cut down, uproot, or wilfully damage or destroy the tree covered by the order without the consent of the planning authority.

4.5.4 Protection in special areas

Within the UK, species are also protected by being within certain designated areas. This protection comes primarily from international conventions, European Directives, and national legislation such as the Wildlife and Countryside 1981 Act (as amended), the Nature Conservation (Scotland) Act 2004, the Nature Conservation and Amenity Lands (Northern Ireland) Order 1985 (as amended by the Environment (Northern Ireland) Order 2002), and the Council Directives on the Conservation of Natural Habitats and of Wild Fauna and Flora (EC Habitats Directive), and the Wild Birds Directive, which primarily became UK legislation through the

Conservation (Natural Habitat, &c.) Regulations 1994 (as amended in each country). The areas we consider in this chapter are those protected because of their biological importance; however, some areas are protected because they are of archaeological or geological value, e.g. limestone pavements.

Areas within England and Wales may be given one of several different types of designation, and the level of protection varies depending on the status assigned to the area. We briefly review the current types of designation and the protection offered to the species within these areas that is in addition to that outlined in 4.5.3.

i. Habitats and Natura 2000

Special Areas of Conservation (SACs)
Areas designated within the EU with the intention of improving the conservation of specific habitats.

Under the terms of the Habitats Directive and EC Birds Directive, Member States are required to put forward a number of national sites for consideration as Special Areas of Conservation (SACs). These form a network of protected areas known as Natura 2000, which aims to protect certain rare or endangered species and habitats. At present, these candidate sites are being identified within the UK by organizations such as the Joint Nature Conservancy Council (JNCC). Once a site is designated, the regulations that might reduce the impact of future use and disturbance to the habitat will be reviewed and if necessary improved.

The amount of legislative protection for habitats is increasing, as seen in the establishment of Natura 2000 sites. In particular, wetlands and woodlands are subject to additional specific legislation. Wetlands of International Importance, especially waterfowl habitat, are covered by the Ramsar Convention (1971), which was ratified in the UK in 1976. Many of the UK's Natura 2000 sites are also Ramsar sites.

In the UK, the Forestry Act 1967 originally focused on forests as purely economic enterprises. This has changed with the Wildlife and Countryside (Amendment) Act 1985 where the terms outlined the need to find a balance between commercial forestry, amenity use, and conservation.

ii. Nature reserves

These are areas that have a special conservation interest for a species or habitat or an unusual geological feature.

a. Statutory local and national nature reserves

The National Parks and Access to the Countryside Act 1949 and the Wildlife and Countryside Act 1981 (as amended) allow Natural England, Natural Resources Wales, and Scottish Natural Heritage to establish a nature reserve. In Northern Ireland, nature reserves are designated under the Nature Conservation and Amenity Lands (Northern Ireland) Order 1985. These can be in private ownership or owned by the council. If a body such as the JNCC or the Department of the Environment thinks that a nature reserve is very important, then it may be designated as a national nature reserve. Bye-laws are then established for each nature reserve, which may, for example, control access and in other ways protect the species on the nature reserve. A few marine reserves have been established in this way.

b. Non-statutory nature reserves

Under the National Parks and Access to the Countryside Act 1949, local nature reserves may be declared by local authorities after consultation with the relevant statutory nature conservation agency. Other local nature reserves may be run by non-statutory bodies such as the Woodland Trust, the Wildlife Trusts, and the Royal Society for the Protection of Birds. Some of these non-statutory nature reserves are run by County Trusts, which come together under the title of the 'Royal Society of Wildlife Trusts'. None of these management bodies can make bye-laws and they are therefore dependent on voluntary codes of conduct and national legislation to provide protection to the species within the nature reserve.

iii. National Parks, Areas of Outstanding Natural Beauty, and Sites of Special Scientific Interest

In England and Wales, the purpose of National Parks is to conserve and enhance landscapes within the countryside while promoting public enjoyment of them and having regard for the social and economic well-being of those living within them. The National Parks and Access to the Countryside Act 1949 established the National Park designation in England and Wales. In addition, the Environment Act 1995 requires relevant authorities to have regard for nature conservation. Special Acts of Parliament may be used to establish statutory authorities for their management (e.g. the Broads Authority was set up through the Norfolk and Suffolk Broads Act 1988).

The National Parks (Scotland) Act 2000 enabled the establishment of National Parks in Scotland. In addition to the two purposes described above, National Parks in Scotland are designated to promote the sustainable use of the natural resources of the area and the sustainable social and economic development of its communities. These purposes have equal weight and are to be pursued collectively unless conservation interests are threatened.

In Northern Ireland, the Amenity Lands (Northern Ireland) Act 1965 and Nature Conservation and Amenity Lands (Northern Ireland) Order 1985 made provision for the designation of National Parks in Northern Ireland. The designation of such an area is currently under discussion.

The primary purpose of the Areas of Outstanding Natural Beauty (AONB) designation is to facilitate the conservation of natural beauty, which by statute includes wildlife, physiographical features, and cultural heritage, as well as the more conventional concepts of landscape and scenery. Account is taken of the need to safeguard agriculture, forestry, and other rural industries, and the economic and social needs of local communities. AONBs have equivalent status to National Parks as far as conservation is concerned. AONBs are designated under the National Parks and Access to the Countryside Act 1949, amended in the Environment Act 1995. The Countryside and Rights of Way Act 2000 clarifies the procedure and purpose of designating AONBs in England and Wales. Originally designated in Northern Ireland under the Amenity Lands Act (Northern Ireland) 1965, AONBs are now designated under the Nature Conservation and Amenity Lands (Northern Ireland) Order 1985. In Scotland, National Scenic Areas are broadly equivalent to AONBs.

The network of Sites of Special Scientific Interest (SSSI) (Areas of Special Scientific Interest (ASSI) in Northern Ireland) has developed since 1949 as the national suite of sites providing statutory protection for the best examples of the UK's flora, fauna, or geological or physio-graphical features. These sites are also used to underpin other national and international nature conservation designations. Most SSSIs are privately owned or managed; others are owned or managed by public bodies or non-government organizations. The SSSI/ASSI des-ignation may extend into intertidal areas out to the jurisdictional limit of local authorities, generally Mean Low Water in England and Northern Ireland and Mean Low Water of Spring tides in Scotland. In Wales, the limit is Mean Low Water for SSSIs notified before 2002, and, for more recent notifications, the limit of Lowest Astronomical Tides (LAT), where the features of interest extend down to LAT. There is no provision for marine SSSIs/ASSIs beyond the low water mark, although boundaries sometimes extend more widely within estuaries and other enclosed waters.

Originally notified under the National Parks and Access to the Countryside Act 1949, SSSIs have been renotified under the Wildlife and Countryside Act 1981 (as amended). Improved provisions for the protection and management of SSSIs were introduced by the Countryside and Rights of Way Act 2000 (in England and Wales) and the Nature Conservation (Scotland) Act 2004. ASSIs are notified under the Nature Conservation and Amenity Lands (Northern Ireland) Order 1985. Measures to improve ASSI protection and management are contained in the Environment (Northern Ireland) Order 2002.

Land within these areas remains in the same, usually private, ownership as they were prior to designation. However, owners of the land and authorities such as the planning authority then have a statutory responsibility for these areas and may be limited as to what they are able to do.

iv. Ministry of Defence land and National Trust land

Land used by or owned by the Ministry of Defence is subject to access restrictions (sometimes amounting to prohibitions) and often carries the additional risk of unexploded ordnance. Other land, for example National Trust properties, land owned by the Duchy of Cornwall, the Windsor Estate, and the Malvern Hills in Worcestershire, whilst not statutory nature reserves, may also be covered by bye-laws, etc., which may control access and provide additional pro-tection for species within the area.

4.5.5 Movement, import, export, and control

There are several national statutory controls and EC regulations that control the movement of animals, plants, and other organisms. Controlling potential pests and diseases requires restrictions on the movement of materials around the globe and suitable containment if work is carried out in the UK. For example, the Plant Health (Great Britain) Order 1993 prohibits the import, movement, and keeping of certain plants, plant pests, and other material, including soil. Some species are listed in the Wildlife and Countryside Act 1981 Part II of Schedule 9, which prohibits any person from releasing and allowing to escape into the wild certain plants such as *Fallopia japonica* (Japanese knotweed). The Act also pro-hibits anyone from releasing or allowing to escape any wild animal that is not ordinarily resident or a common visitor to the UK. Some 'alien' animals have become established in the wild, but it is not considered to be desirable to add to their number by allowing further escapes/introductions. Other legislation that relates to the control and release of species are the Ragwort Control Act 2003, and European Directive 2001/18/EC on the release of genetically modified organisms (GMOs) and the EC GMO Regulations (Deliberate Release) 2002 (4.7).

At an international level, the most substantive regulation in relation to rare and endangered species is the Convention on International Trade in Endangered Species (CITES) 1975 where 160 Member States have ratified controls regulating the movement of more than 2500 animals and 25,000 plants. This has been extended by the Control of Trade in Endangered Species (Enforcement) Regulations 1997 (COTES). There are three levels of protection. Those species listed in Appendix I of the convention are those that may be threatened with extinction, where international trade is only allowed in exceptional circumstances and where the prohibition extends to dead or live individuals or derivatives such as ivory and furs. The trade in species in Appendix II is monitored through a licensing system, whilst in the final category (Appendix III) are those species not threatened on a global level but protected under national legislation within Member States.

In addition to international and national legislation, carriers within the UK, such as the Royal Mail, have regulations concerning what may be carried and the method and appropriate packaging required. These regulations extend to micro-organisms and other biomaterials such as blood or tissue samples.

4.5.6 Permits and licences

If you wish to work on a species or habitat that is protected by legislation or wish to transport a protected species, etc., it may be possible to obtain a licence or permit that allows you to carry out otherwise proscribed activities on protected species or in protected areas. The bodies responsible for granting these permits or licences are usually listed within the Act of Parliament. Most commonly, these are the Department for Environment, Food & Rural Affairs (Defra), Natural England, Natural Resources Wales, Scottish Natural Heritage (SNH), and the Northern Ireland Environment Agency. For example, taking photographs of bats is considered to be unlawful disturbance and you therefore need a licence from the relevant body to do so. More unusual arrangements exist for some species; for example, sturgeon and whales in British waters may only be taken under licence from the Crown.

4.6 Animal welfare

An ethical approach to research involving animals seeks to minimize harm to any participating individual and to do good to that animal and its or other species. The starting point therefore is to consider whether it is necessary to use animals at all in your research. There can be benefits from using an alternative model system such as having greater control over confounding variables. However the 'one step removed' approach when using such a model can lead to conclusions that are then not borne out in the actual species.

Where animals need to be used the second step is to consider how to minimize the number used. This approach is only infrequently used in human research where the ability by the participants to give informed consent usually mitigates these particular ethical concerns.

However minimizing the number of animals is a critical step in non-human animal research. The final steps are to then use guidelines, legislation, and an independent review to ensure that your work complies with best practice.

Much of the relevant legislation for working with animals in the wild has been considered in 4.5. However, for animal work that involves working with an animal in an institution such as a university, or working with a domestic or captive animal, there is additional legislation. At present, there are three main Acts that are most pertinent to this: the Protection of Animals Act 1911; the Animals (Scientific Procedures) Act 1986 (as amended); and the Animal Welfare Act 2006 (or Animal Health and Welfare (Scotland) Act 2006). If you are working with agricultural animals there is considerably more legislation that is likely to be relevant, including the Welfare of Animals (Slaughter or Killing) Regulations 1995, and the Welfare of Farmed Animals Regulations (England 2000/Northern Ireland 2000/Wales 2001).

4.6.1 Protection of Animals Act 1911

This Act has been the core of much animal protection legislation. The essence of the law is that it is an offence to be cruel to any captive animal, which in this context includes any 'wild' bird, fish, and reptile confined or in captivity, as well as domestic animals. The offence of 'cruelty' can be applied to a wide variety of circumstances and events, and this has been the strength of this law. The notion of what is cruel has changed over time in response to greater scientific understanding and changes in social attitudes. Therefore, what may have been acceptable practice when the bill was first passed may now be considered to contravene the Act.

The definition of cruelty varies but relates primarily to the idea of causing unnecessary suffering. Cruelty in the Act is defined in terms of types of conduct including:

- To cruelly beat, kick, ill-treat, override, overload, torture, infuriate, or terrify any animal; or cause, procure, or, being the owner, permit any animal to be so used.

- To wantonly or unreasonably do or omit to do any act, causing unnecessary suffering to any animal; or cause, procure, or, being the owner, permit any such act.

- To convey or carry any animal in such a manner or position as to cause it any unnecessary suffering; or cause, procure, or, being the owner, permit any animal to be so conveyed or carried.

- To wilfully, without a reasonable cause or excuse, administer any poisonous or injurious drug or substance to any animal; or cause, procure, or, being the owner, permit such administration or wilfully, without any reasonable cause or excuse, cause any such substance to be taken by any animal.

- To subject any animal to any operation that is performed without due care or humanity; or cause, procure, or, being the owner, permit any animal to be subjected to such an operation.

- Without reasonable cause or excuse, to abandon an animal in circumstances likely to cause it any unnecessary suffering; or cause, procure, or, being the owner, permit it to be abandoned.

Further legislation has introduced amendments and extensions to this Act; for example, the Welfare of Animals (Transport) Order 1997 relates to carrying animals in the course of trade and the Protection of Animals (Anaesthetics) Act 1954 adds legislation referring to anaesthetization of mammals during an operation.

4.6.2 Animals (Scientific Procedures) Act 1986 (as amended)

The Animals (Scientific Procedures) Act 1986, and its 2013 revision based on a European Directive, regulates scientific procedures that may cause pain, suffering, distress, or lasting harm to protected animals which are under the responsibility of humans. Unlike the definition of animals in 4.5 and 4.6.1, in this instance a protected animal includes all living vertebrates (excluding humans), and any living cephalopod (octopus, squid, cuttlefish, and nautilus). These species are protected from specific points in their life cycles. For example cephalopods are protected at the point when they hatch, whilst embryonic and foetal forms of mammals, birds, and reptiles are protected from the last third of their gestation or incubation period. Additional controls are in place for some groups such as primates.

A procedure becomes regulated *'if it is carried out on a protected animal and may cause that animal a level of pain, suffering, distress or lasting harm equivalent to, or higher than, that caused by inserting a hypodermic needle according to good veterinary practice'.* Some procedures are not included in the Act, such as those carried out as part of normal veterinary, agricultural, or animal husbandry practices; ringing or other methods of tagging animals to allow their identification, provided that this causes no more than momentary pain and distress; the humane killing of a protected animal by a method listed in Schedule 1 of the Act; and the administration of materials to animals as part of a medicinal test in accordance with the Medicines Act 1968. However, in these circumstances, other legislation such as the Protection of Animals Act 1911 still applies.

Under the terms of this Act, a system of licensing enables such work to be carried out when the benefits that the work is likely to bring (to humans, other animals, or the environment) outweigh the pain or distress that the animals may experience. Other criteria that have to be met before a licence is granted are that there are no alternatives, the procedure uses the minimum number of animals, involves animals with the lowest degree of neurophysiological sensitivity, and causes the least pain, suffering, distress, or lasting harm. In addition, within the terms of the Act, certain types of animal must also be obtained from designated breeding or supplying establishments.

The Act includes the arrangements for a three-level licensing system where those carrying out procedures must hold a personal licence, which will not be granted unless they are qualified and suitable. The programme of work must also be authorized in a project licence, and the work must also normally take place at a designated-user establishment. However, in specific circumstances (such as field trials), work can be carried out elsewhere with the Home Secretary's authority. Clearly, there will be considerable variation between institutions in terms of which type of licence, if any, is already held. Therefore, if you are intending to work with such animals, you will need to obtain advice about the existing licences for your institution from your supervisor.

4.6.3 Animal Welfare Act (2006) and Animal Health and Welfare (Scotland) Act 2006

These Acts review and extend legislation relating to animal welfare previously included as a small element in a number of specific Acts, for example the Animal Health Act 1981, the Zoo Licensing Act 1981, the Welfare of Animals at Market Order 1990, and the Welfare of Farmed Animals Regulations England 2000/Wales 2001/Wales 2007, and in the Animals (Scientific Procedures) Act 1986 (as amended) and Protection of Animals Act 1911.

The Welfare Acts define protected animals as those not living in the wild, which are commonly domesticated, or in some way under the control of humans. The main responsibilities outlined in the Act are to provide a suitable environment, to provide a suitable diet, to enable the animals to exhibit normal behavioural patterns, to provide suitable housing (with or apart from other animals), and to protect the animals from suffering, injury, and disease.

4.7 Genetically modified organisms (GMO)

Genetically modified organism
An organism where any of the genes or other genetic material have been artificially modified by a process which does not occur naturally in mating or natural recombination or if the organism has inherited or otherwise derived, through any number of replications, genes or other genetic material (from any source) which were so modified.

The production of a genetically modified organism and its release is controlled primarily by the Environmental Protection Act 1990, the Genetically Modified (Deliberate Release) Regulations 2002 (for each country), and the Genetically Modified Organisms (Contained Use) Regulations 2014. In the 1990 Act an organism is described as being genetically modified if any of the genes or other genetic material in the organism have been artificially modified by a process which does not occur naturally in mating or natural recombination or if the organism has inherited or otherwise derived, through any number of replications, genes or other genetic material (from any source) which were so modified.

Within the UK release of GMOs is heavily prescribed and there are significant steps that need to be taken to obtain permission for any environmental release of a GMO. Most research in the UK therefore involves studies that do not require environmental release. Institutions carrying out research that involves cloning or uses organisms that have in some way been genetically modified will have their own local policy that outlines permissions needed and actions that may and may not be carried out in relation to GMOs. You will therefore need to become familiar with these requirements before designing any experiments which involve these techniques.

4.8 Working with humans

After the atrocities of World War II, the Nuremberg Code (1947) was developed and was the first internationally recognized code of human ethics. This code has informed policies relating to work on humans since this date, and has been restated and extended in other international agreements such as the World Medical Association's Helsinki agreements of 1964 and 2000. The Nuremberg Code and subsequent agreements recognize the need during research to minimize harm, to find a balance between the risks to the individual and the benefits to the individual and to society, to recognize the importance of obtaining fully informed and voluntary consent from humans participating in research, and to ensure that the research has real validity.

Within the UK, there is a considerable body of legislation that also impinges on studies involving humans, including the Obscene Publications Act 1964, the Sex Discrimination Act 1975, the Race Relations Act 1976 and the Race Relations (Amendment) Act 2000, the Computer Misuse Act 1990, the Data Protection Act 1998, the Human Rights Act 1998 (as amended), the Freedom of Information Act 2000, the Special Educational Needs and Disability Act 2001, the Disability Discrimination Act 2005 (as in all countries), the Mental Capacity Act 2005, the Safeguarding Vulnerable Groups Act 2006, the Special Educational Needs (Information) Act 2008, and health and safety legislation (4.4). Relevant Scottish legislation includes the Freedom of Information (Scotland) Act 2002, the Scottish Commission for Human Rights Act 2006, the Protection of Vulnerable Groups (Scotland) Act 2007, Adult Support and Protection (Scotland) Act 2007, and the Equality Act 2010. Apart from this legislation, there are also extensive guidelines that come from professional bodies and codes of practice within your own and other institutions such as the National Health Service. This includes the invaluable Research Governance Framework that considers all issues relating to good practice in research that involves patients and should be read by anyone considering recruiting human participants to their study.

It is impossible for us to cover in detail all this legislation and the guidelines, and therefore in this section we look instead at the general approach taken in research involving humans that is both good science and reflects appropriate practice. Most institutions have some form of ethics scrutiny, which considers research involving human volunteers, so you are unlikely to be left making decisions about what is acceptable practice in isolation.

Additional legislation covers the use of human tissues (Human Tissues Act 2004 and Human Tissue (Scotland) Act 2006). This Act emphasizes the need for informed consent and for tissue to be used only for the research it was originally donated for. It also makes it unlawful to obtain tissues with the intention of analysing the DNA, without the consent of the person from whom the tissue came. This Act is delivered through a licensing system. Limits of working without a licence include the requirement for cells from participants to be destroyed or made incapable of dividing within a few days of collection. If you are working in a laboratory using human tissues, there should already be procedures in place that you will therefore need to follow.

4.8.1 Informed consent

One key approach taken in ethical research on humans is to ensure each participant gives informed consent. This two-step process allows a participant to judge the probability that they may be harmed and weigh this against the good that their contribution may make and to whom. This notion of the participant making this assessment is called 'autonomy'. Autonomy is being able to make choices and decisions for oneself and by oneself. In practice this means the researcher must provide sufficient detailed information about the research in an accessible format to allow volunteers to make an informed decision when giving or withholding their consent and exerting no pressure on a volunteer to take part in the research programme.

Autonomy
Being able to make choices and decisions for oneself and by oneself.

i. Participant information sheet

The information you provide to a potential participant should make it clear what you are asking the volunteer to do, why you wish to carry out this research, and what you hope to gain from it. There may be some risks, for example if you are examining human physiological responses. These need to be made explicit in the information you provide. You may need to exclude certain people from your research either on the grounds of health and safety or due to your experimental design. If you have exclusion criteria these also need to be outlined and the reason for them explained. The name of the person who will carry out the work or is responsible for the project, and the location, may need to be included in the information you provide. However, it is advisable to provide your institutional contact details rather than personal contact details. Where external funding has been obtained, it is also ethical to advise any potential volunteers as to who is funding the work. Some volunteers may not wish to take part, for example, if your work is funded by a tobacco company. Information about what you are going to do with the data must also be included; for example, how will you ensure confidentiality or anonymity, to whom will the results be made available, and in what format? Finally, you need to make it clear that the volunteer does not need to take part and may withdraw at any time and you should outline what will happen to their data or samples as a result.

Information should be provided in a written format that can be taken away and referred to by the volunteer at a later date if they wish. The information should be explained in language that the informant can easily understand. We include an example format for a participant information sheet on the Online Resource Centre.

Participant information sheet Information given to potential volunteers which outlines what the participant will be asked to do and why.

ii. Consent

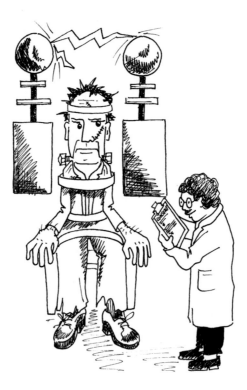

For most work involving humans, you should obtain written consent. If you do have written consent forms, you must consider where you keep this information as it is subject to the

Data Protection Act 1998 and is central to your ability to preserve confidentiality. By asking your volunteers to sign a consent form or in some other way indicate their consent, you are not passing your responsibilities for the well-being of your volunteers to them. You remain responsible for their health and safety and for the ethical conduct of your research.

To obtain honest, informed consent, you cannot directly or indirectly put pressure on (coerce) someone to take part in your research. Most coercion is negative; for example, there may be some sort of penalty for not taking part or not completing the research. But coercion can be quite subtle. For example, a department may expect all students to take part in each other's projects, or you may ask your family or friends to take part, or you may only provide the information about the project just before you intend to carry it out and so allow no time for the potential volunteer to reflect on the information. Coercion can also be positive, for example if you provide an excessive reward. This is not only unethical but can lead to bias in sampling and you may not obtain the representative sample you need for your research (1.4).

There are some circumstances where informed consent may not be obtained. These fall into two areas: either you are working with young people (under 16 years of age) or vulnerable people (e.g. mentally incapacitated adults), or the nature of your investigation is such that you do not want to provide too much information in advance as it may influence the outcome. We consider the first of these in 4.8.2. For the latter, you should first make certain that there is no alternative approach, that the benefits justify not asking for informed consent, that you will not mislead participants, and then that if you cannot provide information before the study, this should be done afterwards and the volunteers given the opportunity to withdraw retrospectively. We include an example format for a consent form on the Online Resource Centre.

Consent form
A signed statement by the participant confirming their understanding of the research requirements and indicating their willingness to take part.

4.8.2 Special cases: children and vulnerable people

For young or vulnerable people, such as children, and adults with learning difficulties, there are additional codes of practice and legal requirements in place to protect them, even if the work is indirect, for example observing behaviour at playtime from outside the school premises. If a child is under 16 years old, then you will normally be expected to obtain informed consent from the parents, carers, or guardians. If you are working with children or vulnerable adults in a school or other similar setting, then you should also obtain the consent of the school from an appropriate person such as the head teacher or chair of governors. In addition, when working with young or vulnerable people, you will be required to obtain clearance from the Disclosure and Barring Service (DBS). This disclosure scheme is designed to enhance public safety by providing criminal history information on individuals who wish to work with or in the course of their work may come into contact with vulnerable adults or children. This requirement for obtaining a disclosure includes undergraduates and graduates carrying out research projects. The institution you are affiliated with may have an established procedure for obtaining a DBS disclosure.

When working with vulnerable people you should consider whether it is acceptable to ask someone else to provide informed consent. It is therefore even more important to be sure that all elements of your project are ethical. In some areas of research, such as within the Health Service, this difficulty is recognized and it is therefore considered to be unethical to conduct research on children where there is no direct benefit to the individual child. If it is feasible, young and vulnerable people should also be asked for their consent, having been given information in an appropriate format.

4.8.3 Equality

In Chapter 1 (1.4), we discussed the notion of a representative sample. This is particularly pertinent when studying humans, where it can involve additional planning and preparation to ensure that all potential volunteers can take part in a research project. In addition, some legislation (e.g. the Equality Act 2010) makes it an offence to exclude certain groups of people through either practices or attitudes. The Human Rights Act 1998 prohibits discrimination on any grounds, such as gender, race, skin colour, language, religion, political or other opinion, national or social origin, association with a national minority, property, birth, or other status. The Disability Discrimination Act 1995 requires you to make reasonable adjustments to enable a disabled person to take part in your research. Making a 'reasonable adjustment' can involve the provision of materials that are more accessible by individuals with dyslexia or with restricted sight, or providing an environment more suitable to someone with a physical disability, for example. We touch on some of these points in 3.4. Most higher education establishments have student support service staff who will be able to provide more specific advice on this matter.

   Other rights enshrined within the Human Rights Act 1998 that need to be considered when designing and carrying out research, to ensure equality in relation to both inclusion and communication of results, are that individuals have: a right to freedom of expression; a right to freedom of thought, conscience, and religion; a right to respect for private and family life; and a right to be treated with justice and fairness.

4.8.4 Anonymity, confidentiality, information storage, and dissemination

When working with humans, you may wish to ensure that each volunteer's contribution is either anonymous or confidential. The rationale for doing so is that you may reduce potential harm to the volunteers, and participants may be more willing to give full cooperation if they will not be identified personally. To ensure anonymity no one, including the researcher, must know which data derive from which individual. This can be achieved, but it is difficult to ensure total anonymity where written informed consent is obtained or where personal information or visual records are required. Although anonymity in human research is usually the ideal, more often the research is confidential. Here, only one researcher can identify the individuals within the research, and when reporting, the researcher endeavours to ensure anonymity. To protect confidentiality or anonymity, data collection, storage, and dissemination have to be handled with care and consideration. Legislation that impinges on these activities includes the Human Rights Act 1998 and the Data Protection Act 1998. Under the terms of these Acts, you must ensure that the information will be used fairly and lawfully for a limited, specifically stated purpose; used in a way that is adequate, relevant, and not excessive; is reported accurately; is kept for no longer than is absolutely necessary; is handled according to people's data protection rights; and is kept safe and secure and not transferred outside the European Economic Area without adequate protection. Where you may collect sensitive information on ethnic background, political opinions, religious beliefs, health, sexual health, or criminal records, you are required to meet higher standards and should therefore discuss this aspect of your work with staff.

When planning human-based research you should therefore consider explicitly and with considerable care what data you will collect and how, how the data will be stored and for how long, and who will be able to access the data and through what media. Confidentiality can be broken inadvertently, either by storing information where names are linked to data and this stored data is accessed by a third person, or by communicating information in reports, etc., with enough detail for certain individuals to be identified. For example, if you have a relatively small cohort, known to a third person, and you report the data from the only individual over 50 years of age, it will be easy to identify this individual. In undergraduate and graduate work, you may therefore need to remove certain information from reports, such as location, age, or gender, to ensure that individuals cannot be identified. For example, when a study of the effectiveness of tea-tree oil as a bactericide handwash in a GPs surgery (Example 2.2) was reported, information about the name and geographical location of the GP's surgery was not included.

There may be some occasions when you may need to reveal a person's identity. This should not be done unless you have received the volunteer's written permission. You may also need to extend this to members of the public if they are included in photographs, video, or film. There is considerable sensitivity about recording and storing images of children. If your work requires this, you must discuss it with your supervisor and possibly the police first.

In undergraduate work it would normally be considered inappropriate to return personal results to your participants or work in such a way that your participants obtain any personal data that may cause them harm through that knowledge. This is particularly pertinent for undergraduate studies as you are learning techniques and so are more likely to make errors generating inaccurate or incorrect data. You will also not be qualified to comment on the results or provide counselling if it is needed. However, an exception might be made where test results indicate that more harm would be caused by withholding the results such as a result that indicates a possible medical complication previously not known by the participant. In this instance your institution's ethics committee should provide guidance on what to do.

4.9 Discussion topics

Consider the following scenarios. What issues arise from these research proposals in relation to the law?

 An undergraduate keeps sticklebacks at home. She wishes to use them in her honours year project and investigate their behaviour when presented with different food sources in a number of different environments.

 An undergraduate wishes to compare the effectiveness of two teaching media (book versus computer) in a primary school. The class will be divided and half will cover a topic from the national science curriculum using a book resource, while the other half will use a computer-based resource. The children will be tested on the topic before and after.

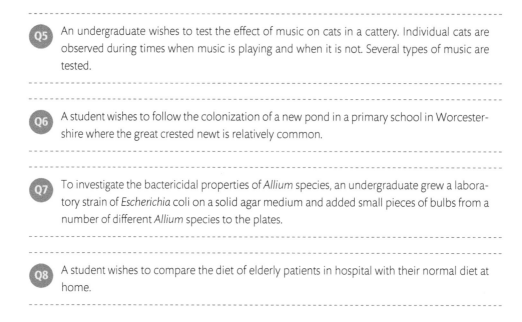

Q5 An undergraduate wishes to test the effect of music on cats in a cattery. Individual cats are observed during times when music is playing and when it is not. Several types of music are tested.

Q6 A student wishes to follow the colonization of a new pond in a primary school in Worcestershire where the great crested newt is relatively common.

Q7 To investigate the bactericidal properties of *Allium* species, an undergraduate grew a laboratory strain of *Escherichia* coli on a solid agar medium and added small pieces of bulbs from a number of different *Allium* species to the plates.

Q8 A student wishes to compare the diet of elderly patients in hospital with their normal diet at home.

The answers to these discussion topics are given at the end of this chapter.

Summary of Chapter 4

- The aim of this chapter is to provide a necessarily brief overview of the nature of law and regulations in the UK (4.1).
- The chapter provides an introduction to the law in relation to seven key areas: ethics (4.2); intellectual property rights (4.3); health and safety (4.4); access and working with wildlife (4.5); working with animals in captivity (4.6); genetically modified organisms (4.7); and working with humans (4.8).
- The nature of an ethical approach to research which underpins all aspects of research is explained (4.2).
- As a member of an institution that undertakes research and development your rights and your responsibilities in relation to intellectual property are explained (4.3).
- The information in this chapter should help you to carry out a risk assessment and to carry out your research safely (4.4).
- You are encouraged to be aware of your responsibilities in relation to research involving wild species and to be able to extend and update your knowledge of this body of law (4.5).
- You are encouraged to be aware of your responsibilities in relation to research involving captive animals and to be able to extend and update your knowledge of this body of law (4.6).
- Current issues that arise when developing, working with, or releasing genetically modified organisms are explained (4.7).

- You are introduced to some of the legislation relating to working with humans. We consider the ideas of non-maleficence and beneficence and how these principles inform the use of informed consent, approaches taken when working with children and vulnerable people, equality, and data storage and handling. Issues that need to be considered when working with human tissues are outlined (4.8).

- A number of examples from undergraduate honours project proposals are included in the chapter for you to discuss and to challenge your understanding of the topics covered in this chapter (Q3–Q8).

- The Online Resource Centre includes interactive exercises that test your understanding of this chapter with other topics, particularly those considered in Chapters 2–11.

Although every effort has been made to ensure that the information in this chapter is accurate, it should not be taken as a definitive statement of the law, nor can responsibility be accepted for any errors or omissions.

Answers to chapter questions

A1 Corrosive, very harmful if swallowed.

R10 Flammable

R20 Harmful by inhalation

R21 Harmful in contact with skin

R22 Harmful if swallowed

R35 Causes severe burns

The R (risk) codes are used as a shorthand when providing some safety information. They are recognized standard descriptions of different hazards. For example, R10 always means flammable. (There are also S (safety) codes that indicate the safety precautions that should be followed.)

A2 The hazards and ratings include: being swept out to sea and drowning (major), falling over and breaking a bone (serious), and other injury such as a cut (serious slight) or sprain (slight), effects of temperature (hypothermia, sunburn, etc.) (major–slight), waterborne pathogens causing an infection (major–slight) or toxic reaction (major–slight), and being attacked by other people or their dogs (major–slight).

A3 The two main areas relevant to this proposal are: firstly, that the student will need to complete a health and safety assessment; and secondly, the work will be subject to the laws outlined in the Animal Welfare Act 2006 and the Animals (Scientific Procedures) Act 1986 (and amendments).

A4 All the topics covered in 4.8 are pertinent. The student would have to obtain a DBS review. Decisions about consent would need to be made. To reduce harm to the children, the student would have to consider how to protect confidentiality or anonymity, how to ensure that no child's learning was affected, etc. To protect the school from harm, the student would have to consider how the school's anonymity is protected. The student would have to ensure that the 'harm' to the children is outweighed by the benefit. The student should also carry out a risk assessment. Risks might include picking up infections, head lice, etc.

A5 A health and safety assessment would need to be carried out by the student. The student would need to comply with the requirements under the licence given to the cattery. Permission would need to be sought from the cats' owners. The student would be subject to the Protection of Animals Act 1911 and the Animal Welfare Act 2006.

A6 This project has three elements. Firstly, the student will need to carry out a health and safety assessment, particularly in relation to waterborne pathogens. Secondly, the student will need permission from the school to ensure that they are not trespassing. He or she will need a DBS disclosure as they are working in an area where children may be present. The great crested newt is a European Protected Species. The student may continue to sample from the pond until such time as they have confirmed that great crested newts have colonized the pond. Should this happen, he or she will need to apply for a licence under the terms of the Wildlife and Countryside Act 1981 from English Nature or the Natural Resources Wales.

A7 One *Allium* is currently listed on Schedule 8 of the Wildlife and Countryside Act and the student should not therefore be using this species in the research without confirming explicitly that they may do so. In addition, the student must carry out a health and safety assessment.

A8 The student must consider the topics and review the legislation introduced in 4.8 such as obtaining consent, ethics, and data handling. The student would probably require DBS clearance. In addition, the student must complete a risk assessment for themselves both in relation to working in a hospital and in relation to visiting other people's homes (4.4).

Section 2

Handling your data

5 What to do with raw data

 In a nutshell

Knowing the type of measurements you will record is critical when planning your experiment and when summarizing it for presentation. We start in this chapter by explaining the terms that are used to describe your data, the scales they are measured on, and the type of distributions seen in your data. This allows you then to decide whether your data are normally distributed or not and to calculate the correct summary statistics.

So far, we have looked at the general terms that you may encounter when reading about or carrying out research relating to experimental design (Chapter 1) and the steps you need to follow when planning an experiment or evaluating other people's research (Chapters 2 and 3). In this chapter, we consider what to do with the data you have collected from your investigations; these ideas are then developed in Chapters 6–12.

As we explained in 2.2.8, the first time to think about the data that your research may generate is while you are planning your experiment. This is critical. Most research uses statistics as a tool to help identify the trends in the data. But each statistical test has certain requirements that need to be met. For example, to use the Mann–Whitney U test (11.1), you need to have between five and 20 observations in each sample. If you have not decided on which statistical test to use before you start your investigation, your sample size may be too small and you will not be able to analyse your data. This is a waste of your time and reflects badly on you as a scientist, as it is clear you have not planned your work properly in the first place. To choose the correct test, you first need to understand terms such as 'qualitative' and 'parametric'. These are explained in this chapter (5.1, 5.8, and 5.9). You also need to understand about 'distributions' (5.2) and 'transforming data' (5.10). We provide an overview of how to choose the correct statistical test based on this information in Chapter 7.

The second time you need to think about your raw data is after you have completed your investigation. You will need to identify the trends in your data and to communicate these to other people. When you come to communicate your findings, you may wish to use a figure

(12.1.10), a table (12.1.9), or summary statistics (5.3–5.7). In learning how to summarize your data, you will also be introduced to some of the central steps in statistics, which are the calculation of a sum of squares, variance, and standard deviation (see Box 5.1).

If you work through the exercises in this chapter it will take about two hours. The answers for these exercises are at the end of the chapter. Nearly all of our examples are based on real undergraduate research projects. If these examples are not in your subject area, you will find more in the Online Resource Centre.

5.1 Types of data

Qualitative

Categories of a treatment variable or observations that are descriptive and non-numerical. If categories are qualitative they are mutually exclusive and therefore discrete.

Quantitative

Categories of a variable or observations that are numerical. Data may be either discrete or continuous.

Discrete

Observations measured on a discontinuous scale or a variable that falls into a series of distinct categories and where the number of categories is limited.

Continuous

Observations that do not fall into a series of distinct categories and may take any value in the scale of measurement, e.g. height (cm).

Rankable

Observations that consist of named categories or values that have a consistent order to them.

When carrying out an investigation, you will generate data as a series of observations measured on a particular scale. For example, if you are investigating the change in human body temperature in relation to exercise, the dependent variable is measured on the scale of degrees Celsius (°C). There are several terms that are used frequently to describe the scales of measurements used when collecting data. You need to become familiar with these terms: they are essential in helping you decide how to design your investigation and how to communicate your findings. These terms are: qualitative, quantitative, discrete, continuous, rankable, nominal, ordinal, interval, and derived variables.

Qualitative refers to information that is not numerical but descriptive. In Chapter 3, we refer to methods such as focus groups and interviews, which may gather qualitative responses, opinions, and thoughts expressed in words. The term 'qualitative' can also be used when a variable that is being studied has named, descriptive categories that are mutually exclusive and non-numerical, for example the number of *Lotus corniculatus* (bird's-foot trefoil) with a yellow or a red keel. The categories yellow and red are qualitative. These scales of measurement are always discrete.

Quantitative refers to the use of numbers. A quantitative scale of measurement is one where the observations are assigned to ordered numerical categories; for example, the height of adult males (cm) is measured on a continuous quantitative scale.

Discrete observations would be those where you have a discontinuous scale. For example, if you recorded the number of eggs in a clutch you cannot have half an egg in this context. These are therefore quantitative discrete values. You may also have treatment variables that are organized into a series of distinct, mutually exclusive categories where the number of categories is limited. For example, the Royal Horticultural Society's (RHS) scale for recording petal colours (e.g. red, pink, white). This is a qualitative discrete scale of measurement.

Continuous scales are ones where observations do not fall into a series of distinct categories and may take any value within the scale of measurement. For example, height (cm) can be measured from 0cm upwards with no limit. Some scales, such as percentages and pH, are also continuous, but only within prescribed boundaries. For example, the percentage scale is restricted to 0–100%.

Rankable applies to scales of measurement where the categories can be ranked (put in a consistent order). These can include qualitative, quantitative, discrete, and continuous, and therefore ordinal and interval scales. For example, if a questionnaire included the question, 'How warm are you?', the answers (very hot, hot, warm, cool, cold), although qualitative, would be ranked consistently. Similarly, a numerical but discrete scale, such as the number

Table 5.1 Frequency classes for height (cm) of 87 male students. (Observations of height are recorded on a continuous scale but these have been organized into nine discrete categories.)

	Height of male students								
	150.0–154.9	155.0–159.9	160.0–164.9	165.0–169.9	170.0–174.9	175.0–179.9	180.0–184.9	185.0–189.9	190.0–194.9
Frequency	3	4	12	15	19	15	12	4	3

of eggs in a clutch, can be ranked (e.g. 0, 1, 2, 3, etc.). Continuous data are ranked in two ways: either by numerical order or by category order. For example, if the heights of five men were recorded, the observations could be ranked 166.0cm, 166.5cm, 168.0cm, 168.2cm, and 170.0cm. Where you have many observations recorded on a continuous scale, these can be arranged into categories or classes (e.g. Table 5.1). These categories can also be ranked. For some statistical tests observations are ranked and then 'assigned a rank order'. We explain this process in 5.9.1.

Nominal scales of measurement fall into discrete categories with no particular order to these categories. The number and nature of the categories can be either an inherent property of what is being measured or imposed by the investigator. The categories should ensure that every observation in the data set can be classified. For example, the RHS scale used for recording petal colours is an artificial device that provides a method for categorizing flower colours; these colours have no particular order and cannot therefore be ranked. If a closed question is included in a questionnaire (3.2) where the prescribed answers are 'yes', 'no', or 'don't know', these mutually exclusive categories are an inherent property of what is being measured but there is no rationale for ranking them: they are therefore nominal.

Ordinal scales of measurement are similar to nominal scales in that they are also based on discrete, mutually exclusive categories. However, in this case, the categories can be ranked but the points between the categories are not meaningful and may not be consistent. The categories can be qualitative or quantitative. In ecology, a common scale that is used when examining species abundance is the ACFOR scale, which comprises the categories 'abundant', 'common', 'frequent', 'occasional', and 'rare'. These are a qualitative measure but have an inherent order, can be ranked, and are therefore ordinal. However it would not be possible to comment on any point between 'abundant' and 'common', as it has no inherent meaning. The number of eggs in a clutch also falls into discrete categories. These quantitative categories can also be ranked and are therefore ordinal. Again a mid-point between each measure is not sensible—in this context you cannot have one and a half eggs.

Interval data are measured on a continuous and rankable scale, and unlike data measured on an ordinal scale, it is possible to measure the difference between each observation. Examples of interval scales include temperature (°C), distance (m), mass (g), and time (min).

Derived variables are scales of measurements that are the result of a calculation; they are therefore quantitative and usually continuous. Four types of derived variables that you are likely to use are ratios, proportions, percentages, and rates. Some of these scales, such as percentages, are constrained within certain limits (e.g. 0–100%); others, such as a rate (e.g. km/h), are not.

Nominal
Observation or categories of a variable that are discrete but where the values or categories cannot be ordered.

Ordinal
Observations or categories of a variable that are discrete and where the values or categories can be consistently ordered. The interval between each observation or category has no meaning.

Interval data
Measured on a continuous scale; the data are rankable and it is possible to measure the difference between each observation.

Derived variables
The unit of measure for the observations is the result of a calculation, e.g. ratio, proportion, percentage, or rate.

You will usually find that more than one term can be used to describe any one measurement. For example, if an investigation examines the height of students within a higher education institution, the scale of measurement will be centimetres. This is a quantitative, continuous, rankable, interval scale.

 Which terms best describe these data:

    a. The number of prickles on holly leaves

    b. Percentage (%)

    c. pH

    d. Grams

The terms we have considered so far all relate to the scale on which the measurements are made. Understanding which type of scale is being used is important in helping you choose the type of statistical test you should use when you design your experiment and analyse your data, and which figure or table you could use when communicating your findings. There are two more terms (parametric and non-parametric) that are also critical. These terms relate to your observations and the characteristics of the distribution of your data. To be able to tell whether your data are parametric or non-parametric, you have to first understand distributions (5.2) and how to calculate a mean and variance (5.3 and 5.4). We therefore consider the terms parametric and non-parametric in 5.8 and 5.9 and explain how to tell whether your data are parametric in Box 5.2.

5.2 Distributions of data

Distribution

The shape seen on a graph when data are plotted.

In investigations examining the effect of one variable, the data may be plotted, for example as a bar chart or histogram (e.g. Fig. 5.1). These figures can be seen to have a particular shape or distribution. Some distributions are well known, such as the normal (5.2.1), binomial (5.2.2), Poisson (5.2.3), and exponential (5.2.4) distributions. The shapes of these distributions have been described mathematically and these equations have been used to demonstrate other relationships, including the tendency of a distribution to have a central point (e.g. a mean) and the spread of the data around the central point, such as the variance. We explain how you may confirm that your data have a particular distribution in 8.4 and 5.8.

Normal distribution

A symmetrical, 'bell-shaped' distribution with a single central peak (unimodal) which can be described by the Gaussian equation. The data are measured on an interval scale.

5.2.1 Normal distribution

This is a very important distribution, which can best be explained using an example. In the study of the height of 87 male students, the raw data are first organized into a frequency table (Table 5.1). As the data are measured on an interval scale, they can be plotted as a histogram (Fig. 5.1). If you collected more observations, then this increased sample size would allow you to reduce the sizes of the classes you are using and the distribution looks much smoother when plotted (Fig. 5.2). If you could extend this study still further, the distribution becomes even smoother, and more symmetrical and 'bell-shaped' (Fig. 5.3), with a single central peak (unimodal). This is typical of a normal distribution.

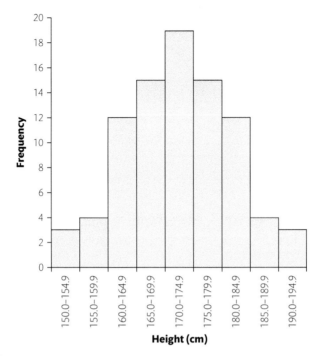

Fig. 5.1   Height of 87 male students

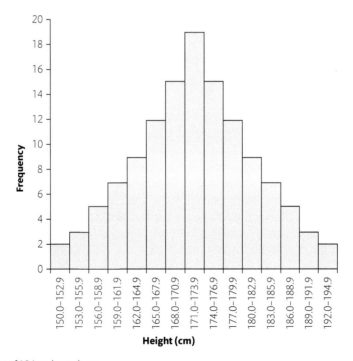

Fig. 5.2   Height of 124 male students

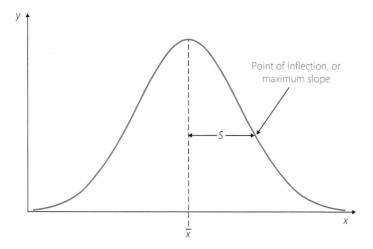

Fig. 5.3 Normal distribution with mean $(\bar{x})$ and standard deviation (s)

If your data are normally distributed like this, then there is a mathematical relationship between the x values and y values. This mathematical description is called the Gaussian equation, where:

e
A number that is encountered often in mathematics so is given a letter as a short-hand way of conveying the number. e = 2.72 (rounded to two decimal places).

$$y = \frac{1}{\sqrt{(2\pi s^2)}} e^{-h}$$

and

π
(Lower case, Greek letter 'pi') A constant found often in mathematical relationships so the number is replaced with a letter for ease of communication. π = 3.14 (rounded to two decimal places).

$$h = \frac{(x - \bar{x})^2}{2s^2}$$

The terms in this equation may not be familiar to you. The symbols e and π are particular numbers (constants) that occur so often in maths that they have been given letters to refer to them. Their values are 2.72 and 3.14 respectively (rounded to two decimal places).

The $\bar{x}$ (mean) and s^2 (variance) are two of the summary statistics that we consider in 5.3 and 5.4. x is any one observation in your sample and y is the y value calculated for any given x value. This equation is considered again in 8.4 where we show you how to test whether your data can be described by the Gaussian equation, and Appendix B where we explain the symbols in this equation in more detail. A normal distribution has several features that are widely exploited in statistics, and we refer back to this distribution throughout the rest of the book.

Binomial distribution
This may be obtained if each item examined can have only one of two possible states.

5.2.2 Binomial distribution

The second distribution we consider, the binomial distribution, is one you would expect to obtain if each item examined can have only one or another state. For example, a seed can either germinate or not germinate, or in a T maze, a rat can only turn either left or right. For

Table 5.2 The probability of having a given number of female (or male) children in ten births

Number of female (or male) children (x) in ten births	Probability (y)
0	0.001
1	0.010
2	0.044
3	0.117
4	0.205
5	0.246
6	0.205
7	0.117
8	0.044
9	0.010
10	0.001

humans, under normal circumstances, the outcome at birth is to be either male or female. The chance of being born male is 1 in 2 and the chance of being born female is 1 in 2.

The chance of a family having only three girls is $1/2 \times 1/2 \times 1/2 = (1/2)^3 = 0.125$. This calculation of probability, the probability of having three girls in a three-child family, can be extended and it is possible to write a mathematical equation that allows you to work out the probability for any particular number of males and females in any family of a particular size using the binomial equation:

$$y = \frac{n!}{x!(n-x)!} \times p^x \times q^{(n-x)}$$

This equation describes the mathematical relationship between y (the probability of obtaining a particular number of children of one gender in a given number of births) and x (the selected number of children of one gender). If you are not familiar with the terms we have used in this equation (e.g. !), we give more details and some worked examples in Appendix B.

We know that when a child is born, the probability that they will be female (p) is 1/2 and the probability of the child being male (q) is also 1/2. If we examined ten births (n), we can use this information to work out the probability of obtaining any number of girls within those ten births (Table 5.2). For example the probability of obtaining four girls and six boys is 0.205; and the probability of all the children being female is very small, only 0.001.

When the data are plotted (Fig. 5.4), you can see that this distribution also has a symmetrical shape with a single (unimodal) high point. If you collect data where you know there can only be two alternative outcomes, where the distribution has this shape, and where the data can be described by this mathematical equation, then your data are said to be binomial and they have a binomial distribution.

!
The symbol used to indicate that a factorial calculation is required. A factorial is the value that results from multiplying a whole number (integer) by all the integers less than itself. A factorial of 3 is $3! = 3 \times 2 \times 1 = 6$.

p^x
The 'power' symbol used to indicate that in this example p needs to be multiplied by itself x number of times.

5.2.3 Poisson distribution

A Poisson distribution is often found where events are randomly distributed in time or space, such as the distribution of cells within a liquid culture or the dispersal of pollen or seed by wind. This distribution is unimodal, but is often very asymmetrical with a protracted tail either

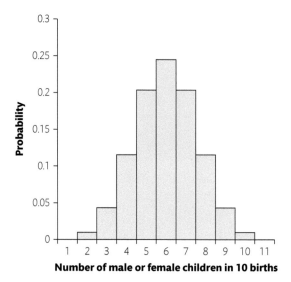

Fig. 5.4 Binomial distribution

to the right (a positive skew) or to the left (a negative skew). Figure 5.5 illustrates four Poisson distributions that vary in their degree of skewness. The Poisson distribution can be described by the equation:

$$y = e^{-\bar{x}} \times \frac{\bar{x}^x}{x!}$$

As we explained when we looked at the Gaussian equation, e is a constant with the approximate value of 2.72 and $\bar{x}$ is the mean of the sample. x is a particular observation in the sample and y is the number of events for a given x. The symbols ! and e^{-x} are defined in the margin

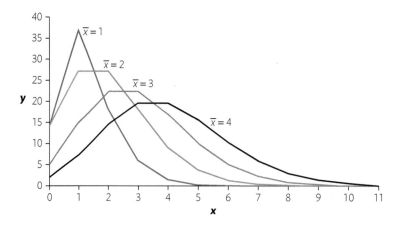

Fig. 5.5 An example of a Poisson distribution

Table 5.3 The distance seeds are dispersed from the canopy edge of one *Taxus baccata* (yew) tree

Distance seeds are dispersed from the canopy edge of one *Taxus baccata* tree	
Distance (m)	Number of seeds
0	13
1	27
2	27
3	18
4	9
5	4
6	1
7	0
8	0
9	0
10	0

in relation to the binomial equation. For example, y might be the number of 1cm^3 samples containing x cells in a liquid culture, or y might be the number of seeds that have travelled a distance x metres from the parent plant during the dispersal process. One Poisson distribution is shown in Table 5.3.

Poisson distribution
A unimodal distribution, which is often very asymmetrical with a protracted tail either to the right (a positive skew) or to the left (a negative skew) and is often found where events are randomly distributed in time or space.

5.2.4 Exponential distribution

The exponential distribution, like the Poisson distribution, considers events that are randomly distributed in time or space. But where the Poisson distribution describes the density of events (number of events per unit time or distance or volume), the exponential distribution describes the intervals between adjacent events. For example, the Poisson distribution describes the number of mutations per million bases in DNA, but the exponential distribution describes the distances between adjacent mutations. The Poisson distribution describes the number of oak trees (*Quercus robur*) per kilometre of roadside hedge, but the exponential distribution describes the distance between adjacent oak trees in that roadside hedge. The Poisson distribution describes the number of daisy plants (*Bellis perennis*) found within quadrats thrown randomly onto a large lawn, but the exponential distribution describes the distance from each plant to its nearest neighbour (Fig. 5.6). The exponential distribution can be described by the equation:

$$y = \lambda\, e^{-\lambda x}$$

where λ (lower case, Greek letter 'lambda'), is called the rate parameter of the distribution. The mean and standard deviation of an exponential distribution are both equal to $1/\lambda$.

This distribution is commonly used to model the growth of bacteria in liquid cultures during their peak growth where the numbers of individuals keep doubling, or the increase in the number of copies of DNA during a polymerase chain reaction (PCR).

Exponential
Where the rate of increase in y (e.g. numbers of an item) becomes greater with an increase in x so that when plotted the change in the curve becomes steeper or shallower as x increases.

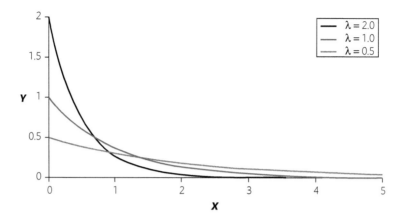

Fig. 5.6 An example of an exponential distribution

5.3 Summary statistics

Summary statistics have three roles in research. In the first, summary statistics are used when you first examine your data to enable you to get an idea as to what your data indicate about the factors you have been investigating. The second role of summary statistics is that many of the numbers, particularly the mean and variance, are terms that appear in statistical calculations. Finally, when you are at the point where you wish to communicate your results you use summary statistics to avoid the inclusion of extensive tables of raw data and to allow you to present the main trends in your data more succinctly. It is therefore important to appreciate types of summary statistics, their relative strengths, and how they are calculated.

Summaries of data focus on two things: the central point of the distribution and the variation around this central point. The calculations of the central point can be derived from the mathematical relationship, such as the Gaussian equation or Poisson equation, and in these instances this central point is called the mean. Where the distribution is not known, then other measures, such as the median and mode, can be used. A similar distinction occurs in relation to estimates of the spread of the data about the central point. Some distributions are not symmetrical about the central point and this deviation from symmetry can be indicated by using measures of skewness (5.4.4). The shape of the peak can also vary and a useful measure of this kurtosis is given in 5.4.4. The central point, skew, kurtosis, and estimates of variation together are used to provide a summary of your raw data. The appropriate summary statistics for different types of data is a hotly debated topic, especially in relation to ordinal data and where the underlying distribution is not known. The approach we take is summarized in Table 5.4.

In Example 5.1 (Table 5.5), the data for both years are numbers of seeds per umbel, which is a discrete ordinal scale. From Table 5.4, we can see that these data may be summarized using a median, mode, and range. In Table 5.6, we have replicate samples and here we can summarize within each category of 'number of seeds' and/or within one sample.

Table 5.4 Appropriate summary statistics for different types of data

Summary statistic	Type of data			
	Nominal	Ordinal	Interval–unknown distribution	Interval–known distribution
Central tendency	Mode	Mode Median	Mode Median	Mean
Spread of data (Variation)		Range Interquartile range Percentiles	Range Interquartile range Percentiles	Standard deviation Coefficient of variation Confidence limits

Example 5.1 The number of seeds per umbel in *Allium schoenoprasum*

A single *Allium schoenoprasum* (chive) may produce one inflorescence called an umbel containing many flowers. Each umbel usually therefore produces many seeds. We recorded the number of seeds produced per umbel over two consecutive years (Table 5.5), examining nine umbels in 2004 and eight in 2005.

Table 5.5 The number of seeds per umbel in *Allium schoenoprasum* in 2004 and 2005

Year	Number of seeds per umbel in *Allium schoenoprasum*								
2004	13	14	15	15	15	17	19	19	20
2005	13	14	15	15	17	19	19	20	–

In 2004, a number of separate samples were examined and the number of seeds recorded (Table 5.6).

Table 5.6 The number of seeds per umbel in four random samples of *Allium schoenoprasum* in 2004

Sample	Number of seeds per umbel in *Allium schoenoprasum*								
	13	14	15	16	17	18	19	20	21
a	1	1	3	0	1	0	2	1	0
b	2	0	3	1	2	0	3	2	4
c	0	3	1	3	1	3	1	0	2
d	2	2	1	2	3	0	0	2	0

The data from Table 5.5 have been reorganized into a frequency table (Table 5.7) so that it is easier to identify the mode and median.

(continued)

Table 5.7 The number of umbels of *Allium schoenoprasum* producing a certain number of seeds (data recorded in 2004 and 2005)

	Number of umbels producing a certain number of seeds							
Year	Number of seeds							
	13	14	15	16	17	18	19	20
2004	1	1	3	0	1	0	2	1
2005	1	1	2	0	1	0	2	1

5.4 Estimates of the central tendency

There are three common methods used to measure central tendency: the mode, median, and mean. The mean should not be confused with the average. A mean is a measure of central tendency that can be calculated using a specific formula when the distribution is known. The average is an estimate of the mean using the same method of calculation as the mean for a normal distribution but applied to data whose distribution is not known. Although this is mathematically incorrect, calculations of averages are in common usage.

Mode

One measure of central tendency. When data are organized into categories, the mode is the category that contains the greatest number of observations or the highest frequency.

Unimodal

A distribution that has a single high point or peak.

Bimodal

A distribution that has two distinct high points or peaks.

5.4.1 Mode

The mode is the category that contains the greatest number of observations. For example, the scale of measurement used in Example 5.1, the 'number of seeds', is ordinal. If the data are summarized as shown in Table 5.7, it is clear that the mode is 15 seeds in 2004 (i.e. unimodal). In 2005, there are two modes: 15 seeds and 19 seeds (i.e. bimodal).

If the data are measured on an interval scale but the underlying distribution is not known, then you may use the mode as a measure of central tendency but you will first need to organize your data into a frequency table. The modal class is the class with the highest frequency and the mode is the mid-point of this category. Clearly, as you have imposed these categories on the data, there is a degree of artificiality in relation to the mode; grouping the data into other size classes may generate a different mode. You should be aware of this both when choosing the classes for the frequency table and when interpreting your results.

5.4.2 Median

When data can be ranked (i.e. ordinal or interval data), a simple measure of the central tendency is to take the 'middle' value. This is the median.

The median is dependent on the number of observations in the data set (n). When n is an odd number, then the median is the middle value. When n is an even number, then the median is calculated as half the sum of the two middle values. In Table 5.5 for the year 2004 ($n = 9$), the data are already in numerical order and the middle value is 15 seeds per umbel. In 2005 ($n = 8$), here the median is $(15 + 17)/2 = 16$ seeds per umbel.

Median
One measure of central tendency, calculated as the middle value of an ordered data set.

5.4.3 Mean

The mean is used for interval data where the distribution is known. Here we give the methods for calculating a mean for a normal distribution (described by the Gaussian equation), a binomial distribution, and a Poisson distribution. For other distributions, you will need to refer to other texts and/or computer software.

Mean
The measure of central tendency for interval data where the distribution is known. The sample mean for normally distributed data is calculated as the sum of all observations divided by the number of observations in the data set.

For any one distribution, a sample mean ($\bar{x}$) or a population mean (μ) may be calculated. The sample mean is used as an estimate of the population mean when not all items in the population have been measured. A value called the confidence interval can be calculated to demonstrate the area around a sample mean in which the population mean will probably fall (5.7).

i. Normal distribution

When calculating a mean for a normal distribution, most often you will have data that are not organized in categories. However, if you have data that are organized into categories a mean may still be calculated using a different method that provides a reasonable estimate of the value.

Σ
(Capital Greek letter 'sigma'). Mathematical symbol for 'sum'. The terms that follow this symbol indicate what needs to be summed.

$\bar{x}$
Mathematical symbol used to indicate the sample mean. The sample mean is often used as an estimate of the population mean.

μ
(Lower case Greek letter 'mu') Mathematical symbol used to denote the population mean which can be calculated if every item in the statistical population is measured.

a. Your data are not grouped in categories

In normally distributed data, the mean is calculated as the sum (Σ) of all observations (x) divided by the number of observations in the data set (n). If you had collected measurements from all items in the population, then the calculation is the same, but you would refer to N rather than n and the mean would be referred to as μ. If you are not familiar with these terms, you may also wish to refer to Appendix B. The calculation of the mean for a sample is written as:

$$\bar{x} = \frac{\sum x}{n}$$

We demonstrate in Box 5.2 that the data in Example 5.2 appear to be normally distributed and therefore we can calculate a mean assuming that the data are described by a Gaussian equation. For this example, if you add all the x values together (i.e. Σx = 1 + 5 + 2 + 5 + ... + 3 + 4 + 2 + 5), this equals 254 and there are 50 observations. Therefore, the sample mean is:

$$\bar{x} = \frac{254}{50} = 5.08mm$$

b. Your data are grouped into categories

Some data may be from a normal distribution but may be grouped. In this case, the mean can be estimated in a different way. For this calculation, you need the mid-point for each category (m) and the frequency within that category (f). Then:

$$\bar{x} = \frac{\sum mf}{\sum f}$$

Example 5.2 Length (mm) of two-spot ladybirds (*Adalia bipunctata*)

An investigator was interested in the length of two-spot ladybirds (*Adalia bipunctata*) collected at random from a garden (Table 5.8; Fig. 5.7).

Table 5.8 The length (mm) of 50 *Adalia bipunctata* sampled in a garden

Length of *Adalia bipunctata* (mm)									
1	5	2	5	7	8	3	6	7	4
4	5	6	4	5	5	7	5	3	5
4	5	1	7	9	2	6	5	6	3
3	6	8	6	4	6	6	8	5	6
7	4	8	9	5	4	3	4	2	5

(continued)

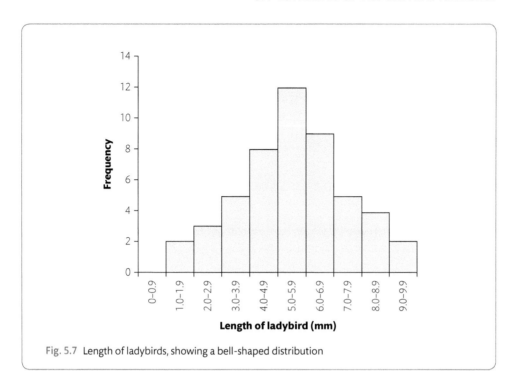

Fig. 5.7 Length of ladybirds, showing a bell-shaped distribution

Table 5.9 is a frequency table for the data from Example 5.2, where the mid-point for each class and the mid-point multiplied by the frequency (*mf*) are given.

We know that there are 50 observations, so $\Sigma f = 50$. From Table 5.9, we can add all the *mf* values together $(\Sigma mf) = 276.5$. Therefore, $\bar{x} = 276.5 / 50 = 5.53$. This method will tend to overestimate the mean and therefore should only be used when you do not have the raw data.

Table 5.9 Frequency table of the data from Example 5.2 on the length of two-spot *Adalia bipunctata* (ladybirds) showing how to calculate a mean for grouped data

	Size classes for length (mm) of *Adalia bipunctata* (ladybird)								
	1.0–1.9	2.0–2.9	3.0–3.9	4.0–4.9	5.0–5.9	6.0–6.9	7.0–7.9	8.0–8.9	9.0–9.9
Mid-point of class (*m*)	1.45	2.45	3.45	4.45	5.45	6.45	7.45	8.45	9.45
Frequency (*f*)	2	3	5	8	12	9	5	4	2
mf	2.90	7.35	17.25	35.60	65.40	58.05	37.25	33.80	18.90

Q2 Assume the data in Table 5.1 are normally distributed. Calculate the mean and mode.

ii. Binomial distribution

In 5.2.2, we illustrated the idea of a binomial distribution in relation to the gender of children. The total number of girls you would expect in your sample would be the number of births (n) multiplied by the probability that each child was a girl (p). If you had N families each with n children, then you would expect to get $n \times N \times p$ girls. For example if we collected data from six families each with five children, then we would expect the total number of girls will be $5 \times 6 \times 1/2$ girls = 15 girls in total. The mean for this binomial data will be this total number divided by the number of families (N):

$$\bar{x} = \frac{n \times N \times p}{N}$$

The two N values cancel each other out, so the mean can be worked out simply as $\mu \approx \bar{x} = np$, and so for six families with five children, $\bar{x} = 5 \times 1/2 = 2.5$ girls. (The symbol $\approx$ means 'approximately equals'.)

iii. Poisson distribution

The mean in a Poisson distribution relates both to the skew of the distribution and the variation in the data. The mean for a Poisson distribution is calculated as:

$$\mu \approx \bar{x} \approx \frac{\sum xy}{\sum y}$$

The terms Σxy and Σy appear in many statistical calculations and therefore we have shown you in detail how these terms are calculated. The steps in this calculation are shown in Table 5.10

Table 5.10 Calculating a mean and variance for data with a Poisson distribution: the distance seed dispersed from the canopy edge of one *Taxus baccata* (yew) tree

Distance seed dispersed from the canopy edge of one *Taxus baccata* (yew) tree		Mathematical steps in the calculation of a mean and variance for data with a Poisson distribution	
Distance (m) (x)	Number of seeds (y)	xy	x^2y
0	13	$0 \times 13 = 0$	$0^2 \times 13 = 0$
1	27	$1 \times 27 = 27$	$1^2 \times 27 = 27$
2	27	$2 \times 27 = 54$ etc.	$2^2 \times 27 = 108$, etc.
3	18	54	162
4	9	36	144
5	4	20	100
6	1	6	36
7	0	0	0
	$\sum y = 99$	$\sum xy = 197$	$\sum x^2y = 577$

in the first three columns. The final column (x^2y) relates to the calculation of a variance for data with a Poisson distribution, which we explain in 5.5 and which can be ignored for now. The mean for this example is:

$$\mu \approx \bar{x} \approx \frac{\sum xy}{\sum y} = \frac{197}{99} = 1.99m$$

5.4.4 Skew and kurtosis

There are two further measures that may usefully be used to describe features of the central tendency. These are skew and kurtosis. We have already seen an example of variation in the degree of skew in a distribution. In data with a Poisson distribution, as the mean decreases the distribution becomes less symmetrical (Fig. 5.5). A measure of this degree of skew is the relationship between the mean, median, and mode. If a distribution is symmetrical, such as the normal distribution, then the mean should equal the mode, which should equal the median. However, this guide to a symmetrical distribution is rarely absolute and you are often left wondering just how much of a difference between values for the mean, median, and mode can be tolerated before you accept that you are not looking at a symmetrical distribution. If there is a doubt in relation to a normal distribution, you can use the additional criteria outlined in Box 5.2 to determine whether the distribution is normal or use other statistical tests for normality (5.8).

Skew
When the distribution of the plotted observations is not symmetrical.

Kurtosis
A measurement that indicates how sharp the peak of the central point is.

- -

Q3 Does the mean = median = mode for the data from Example 5.2?

- -

There are several measures of skew of which the most common is:

$$Skew(\gamma^3) = \frac{\sum(x - \bar{x})^3}{(n-1)s^3}$$

The value for skew for a perfectly symmetrical distribution should be zero. If the distribution has a positive skew (Fig. 5.8), then the value of γ^3 will also be positive, and for a negative skew (Fig. 5.9), the value of γ^3 will be negative. An assessment of skew is one of the tests for normality (Box 5.2) and may also be used as a descriptive statistic with which to compare distributions.

Kurtosis (γ^4) indicates how sharp the peak of the central point is. A frequency distribution with a pointed narrow peak is called leptokurtic (Fig. 5.10), a moderate peak is called mesokurtic, and a very flat peak is called platykurtic (Fig. 5.11). Kurtosis can be measured as:

$$kurtosis(\gamma^4) = \frac{\sum(x - \bar{x})^4}{(n-1)s^4}$$

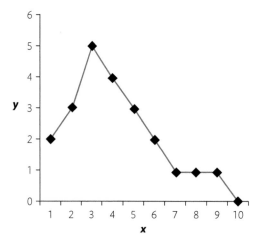

Fig. 5.8 Distribution skewed to the right (positive skew)

Normally distributed data should have a mesokurtic appearance with a kurtosis value of about 3; a leptokurtic distribution will have a kurtosis value of more than 3, and a platykurtic distribution will have a kurtosis value of less than 3.

These two methods for calculating skew and kurtosis assume that the mean and variance have been calculated using the equations outlined for a normal distribution. Therefore, for data with a distribution other than a normal distribution, you may use these methods for indicating skew and kurtosis, having first calculated a mean and standard deviation using the equations for a normal distribution (5.4.3i and 5.5.3i). The seed dispersal around *Taxus baccata* is known to have a Poisson distribution (Table 5.10). If the mean and standard deviation for this data are calculated using the normal equations, then $\bar{x} = 2.29$, $s = 1.13$, $\gamma^3 = 2.8$, and $\gamma^4 = 6.24$. This supports the notion that this distribution is skewed to the right and is very leptokurtic.

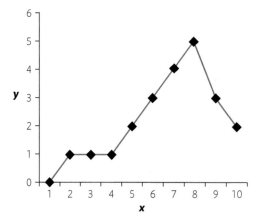

Fig. 5.9 Distribution skewed to the left (negative skew)

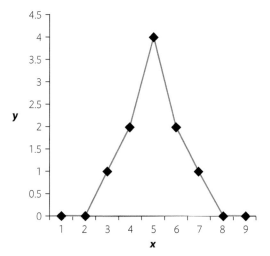

Fig. 5.10 Leptokurtic distribution

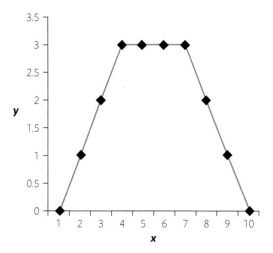

Fig. 5.11 Platykurtic distribution

5.5 Estimates of variation

For nominal data that cannot be ordered, there is no appropriate method for indicating the variation in the data (Table 5.4). For ordinal and interval data the calculation of skew and kurtosis can give some information about the variation about the central point of a distribution, particularly the general shape of the distribution. However, there are a number of additional measures that are used extensively to obtain an idea of how data are distributed around the central point.

5.5.1 Range

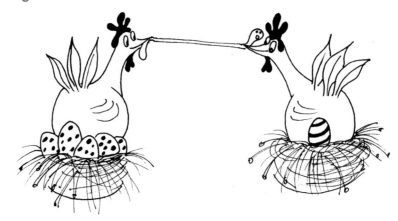

For ordinal and interval data where the underlying distribution is not known, the range is the simplest indication of the spread of data around a central point. The range is the distance between the highest and lowest observations and is therefore calculated by subtracting the value of the lowest observation from the highest observation. For example, the range for the data illustrated in Table 5.5 for both 2004 and 2005 is 20 seeds − 13 seeds = 7 seeds per umbel. The disadvantage of this measure of variation is that it can be substantially altered by the presence of a single 'outlier' and the range does not take into account the shape of the distribution between the two ends. The range as calculated does not make any allowance for sample size, yet commonly an increase in sample size will tend to increase the range.

Outlier
An observation that is noticeably different from all other observations, one that does not follow the apparent trend.

5.5.2 Interquartile range and percentiles

Interquartile range
A measure of range which spans from the observation that falls at the 25% point in a data set to the observation at the 75% point in a data set. This range indicates the variation in the data without being unduly influenced by outliers.

The effect of outliers can be reduced by taking a range not from the smallest to the largest observation but at some specified point within the range. Most commonly, the points used are the 25th centile (where 25% of the observations are smaller) and the 75th centile (where 25% of the observations are larger). This is known as the interquartile range. Alternatively, you may select another set of points, such as the 90th centile and 10th centile. These are known as percentiles. The interquartile range is the difference between the observation at the 25% point and the observation at the 75% point. In the data from Table 5.7 for the year 2004, the interquartile range is from 15 seeds to 19 seeds = 4 seeds per umbel. Only 50% of the observations fall within the interquartile range. Within a percentile range of 10–90%, 80% of observations are included.

5.5.3 Variance and standard deviation

One of the most useful and frequently used measures of variation is the variance. This term may be calculated for interval data where the distribution is known. We illustrate how to calculate the variance for data with a normal, binomial, or Poisson distribution. For other distributions, you will need to find other sources of information. If you have collected observations from every item in your population, then you should use the population equation to calculate the population variance (σ^2) and its square root, the population standard deviation (σ). For samples you calculate the sample variance (s^2) and/or sample standard deviation (s).

Unlike the mean, the calculations for the variance and standard deviation differ depending on whether you are working out the population parameter or the sample statistic. As most biological research works with samples, here we only explain in detail the calculation for samples. One equation for calculating the population variance is given in 1.5.

i. Normal distribution

Normally distributed data can be evaluated using a particular set of statistics called parametric statistics. A common component of these statistical tests is the calculation of a sum of squares ($SS(x)$), the variance (s^2) and the standard deviation (s). Therefore, we have included these calculations in a box for ease of cross-referencing from future chapters. The box is a format we use throughout the chapters on statistics and is based on two columns: one with the general information about the calculation and a second column with a specific worked example. In this case, we use the data from Example 5.2, the length (mm) of the two-spot ladybird (*Adalia bipunctata*).

a. Your data are not grouped into classes

The variance for a sample is calculated as the squared average deviation of an observation from the mean:

$$s^2 = \frac{\sum (x - \bar{x})^2}{n - 1}$$

The numerator in this equation is the sum of squared deviations from the mean $\left(\sum (x - \bar{x})^2 \right)$, a term often abbreviated to 'sum of squares' or $SS(x)$. This is a term you will often encounter when using parametric statistics and is integral to calculations for regressions and ANOVAs.

The denominator reflects that this is a calculation for the sample variance in the use of $n - 1$ rather than N. It has been found that if you take every possible sample from any given population and calculate the variance as described here, the average of these sample variance estimates is the same as the population variance. If, however, this exercise is repeated and N is used instead of $n - 1$ then the average of the sample variances does not equal the population variance. The equation that uses $n - 1$ as the denominator in the estimate of sample variance is therefore known as 'unbiased' and is the one in common use.

It is possible to use this equation as it is. However, there is an alternative form of this equation that provides a very close approximation to the answer. This equation simplifies the calculation as you do not need to first calculate the mean. Here the sample variance for normally distributed data (s^2) is:

$$s^2 = \frac{\sum x^2 - \dfrac{\left(\sum x \right)^2}{n}}{n - 1}$$

Variance
A measure of the variation in the data. The calculation differs depending on the underlying distribution and whether it is for a population variance (σ^2) or a sample variance (s^2).

Sum of squares
$SS(x)$ The sum of squared deviations from the mean which is the numerator in a calculation of the variance for normally distributed data.

And the sum of squares (SS(x)) is:

$$SS(x) = \sum x^2 - \frac{\left(\sum x\right)^2}{n}$$

The variance is, as the symbol (s^2) indicates, a squared value and the units of variance are also squared. Therefore, it is common to report the square root of this measure of variation, the standard deviation (σ or s), which has the same units as the original observations. The standard deviation for normally distributed data (s) is:

Standard deviation
A measure of the variation in data where the distribution is known. For data with a normal distribution, it is the square root of the variance and is a value with the same units as the original observations.

$$s = \sqrt{\frac{\sum x^2 - \frac{\left(\sum x\right)^2}{n}}{n-1}}$$

We take you through the method for calculating the sums of squares, variance, and standard deviation for sample data from a normally distributed population in Box 5.1. Working out the variance by this method may give you a very slightly different answer to one calculated by a computer or calculator that is programmed with the other equation. However, using the 'estimation' method is accepted as standard practice as the difference between answers is so small.

BOX 5.1 How to calculate a standard deviation and variance for normally distributed (parametric) data

GENERAL DETAILS	EXAMPLE 5.2
To calculate the sums of squares (SS(x)). Where $$SS(x) = \sum x^2 - \frac{\left(\sum x\right)^2}{n}$$	This calculation is given in full in the Online Resource Centre. For presentation purposes only all values have been rounded to five decimal places.
a. First add all the observations in a sample together ($\sum x$) and square the total ($\sum x$)^2. Divide this value by n: $$\frac{\left(\sum x\right)^2}{n}$$	a. Table 5.8 $\sum x = 1 + 5 + 2 + \dots + 4 + 2 + 5 = 254$ $\left(\sum x\right)^2 = (254)^2 = 64516$ $n = 50$ $\frac{\left(\sum x\right)^2}{n} = \frac{64516}{50} = 1290.32$
b. The next step is to square each of the observations (x^2) and add these squared values together ($\sum x^2$).	b. $\left(\sum x^2\right) = 1^2 + 5^2 + 2^2 + \dots + 4^2 + 2^2 + 5^2 = 1474$

(continued)

c. Subtract the value you worked out at step (a) from the value you worked out in step (b). **To calculate the sample variance** where $$s^2 = \frac{\sum x^2 - \frac{(\sum x)^2}{n}}{n-1} = SS(x)/n-1$$	c. Sum of squares $SS(x) = 1474 - 1290.32 = 183.68$
d. Divide SS(x) by $n-1$. This is the sample variance (s^2). **Calculate the standard deviation** (s) where $$s = \sqrt{\frac{\sum x^2 - \frac{(\sum x)^2}{n}}{n-1}}$$	d. Variance$(s^2) = \frac{183.68}{50-1} = 3.74857\text{mm}^2$
e. The sample standard deviation (s) is the square root of the variance $s = \sqrt{s^2}$	e. Sample standard deviation (s) $s = \sqrt{3.74857} = 1.93612\text{mm}$

Further examples relating to the topic including how to use statistical software are included in the Online Resource Centre.

In many distributions, including the normal distribution and the Poisson distribution, there is a relationship between the mean and the standard deviation. For a normal distribution, this is shown in Table 6.2 and Fig. 6.2. To briefly introduce this relationship, we can consider the height recorded for 87 men (Table 5.1). Clearly, most of the men have an average or near average height. A few are much smaller or much taller. It is possible to use the mathematical description of the distribution to work out how many people you would expect to fall within a particular size class. To do this, you use the standard deviation. The relationships we use most often in normally distributed data are approximately that:

- 68% of observations should fall in the range $\bar{x} \pm 1s$
- 95% of observations should fall in the range $\bar{x} \pm 1.96s$
- 99% of observations should fall in the range $\bar{x} \pm 2.58s$

This relationship can be used as one check to decide whether your data are normally distributed. A worked example showing this is included in Box 5.2.

 In Example 2.3, the leaf length (mm) of *Secale cereale* was recorded for 15 seedlings. In her notebook, the researcher noted down the following: 12.0, 12.0, 11.5, 18.0, 14.0, 11.0, 14.5, 11.5, 10.0, 10.0, 19.5, 19.0, 21.0, 15.5, 14.5.
Determine whether these data are interval, ordinal, or nominal and then summarize the data appropriately.

----

b. Your data are grouped into classes

In most research, you will have access to the full data set, but there may be some occasions when you are only provided with data already summarized in a frequency table. In this case, you may use the following to obtain an estimate of the variance and standard deviation, assuming that the data are normally distributed.

The sample variance is derived by first calculating the sums of squares of x ($SS(x)$) from the mid-point of each class (m) and the frequency in each class (f) and where n is the total number of observations:

$$SS(x) = \sum m^2 f - \frac{\left(\sum mf\right)^2}{n}$$

The variance is then:

$$s^2 = \frac{SS(x)}{n-1}$$

When this approach is applied to the length of ladybirds (Table 5.9), $\Sigma mf = 276.5$ (as calculated in 5.4.3i(b)). Squaring each mid-point (m^2) and multiplying this by its frequency gives a series of values for $m^2 f$. Adding all these values together for this example, $\Sigma m^2 f = 1712.725$. Therefore, $SS(x) = 183.68$, $s^2 = 3.7486$, and $s = 1.93612$. These values for the variance and standard deviation are a very close match to those we calculated in Box 5.1 using the same data but not grouped into classes.

As the mean, variance, and standard deviation for normally distributed data are common components in statistics, calculators and computers invariably include software to allow these terms to be calculated quickly; it is worth becoming familiar with these tools. However, you must make sure that if you have a sample you use sample statistics not population parameters (1.5).

ii. Binomial distribution

In a binomial distribution, the variance is determined by the number of observations (n) and the probability of obtaining each outcome (p, q). Unlike data with a normal distribution, however, the population variance is approximately the same as the sample variance so that $\sigma^2 \approx s^2 = npq$. In our example in 5.2.2, $s^2 = 10 \times 1/2 \times 1/2 = 2.50$ girls^2; the standard deviation (s) is the square root of the variance (s^2), which for our example is $s = 1.58$ girls.

iii. Poisson distribution

A measure of both the population and sample variance in a Poisson distribution can be estimated by the following equation:

$$\sigma^2 \approx s^2 \approx \frac{\sum x^2 y}{\sum y} - \left[\frac{\sum xy}{\sum y}\right]^2$$

Using the terms $\Sigma x^2 y$, Σy, and Σxy from Table 5.10 and 5.4.3iii, the variance and standard deviation are:

$$s^2 \approx \frac{577}{99} - (197/99)^2 = 1.86858 \text{ and } s = 1.3669$$

In data with a Poisson distribution, the mean should equal the variance. For the data in Table 5.3, the mean distance travelled by the *Taxus baccata* seeds was 1.99m. Using the equation to calculate the variance gives us a value of 1.87m. These estimates are similar to each other although not identical. This means that if you believe you have data with a Poisson distribution, you can initially confirm this by comparing the mean and variance. The definitive test is to 'fit' the Poisson equation to your data. We show you how to do this in principle in 8.4 and in practice in the Online Resource Centre.

5.6 Coefficient of variation

The standard deviation is the most commonly used measure of dispersion. If, however, you wish to compare variation in one set of data with that of another set of data, then we use a relative (scaleless) measure of variation called the coefficient of variation (V). The coefficient of variation is especially useful when data you wish to compare are of different orders of magnitude or if they are measured on different scales. V is determined as a ratio between the mean and the standard deviation expressed as a percentage:

$$V = \frac{s}{\bar{x}} \times 100$$

Coefficient of variation (V) The ratio between the mean and the standard deviation expressed as a percentage. Used if you wish to compare variation in one set of data with that of another set of data.

The coefficient of variation can only be used for interval data with a known distribution and where a standard deviation can be calculated. In Chapter 10, we consider a survey carried out at Aberystwyth, Wales, UK, in 2002. Periwinkles from the mid- and lower shore were sampled and the height of their shells (mm) was recorded (Table 10.3). The coefficient of variation for the periwinkles from the lower shore was $(1.89477/5.54467) \times 100 = 34.17282\%$ and the coefficient of variation for the periwinkles on the mid-shore was $(2.02419/7.47) \times 100 = 27.09759\%$. This means that there is relatively more variation in the shell height (mm) of the periwinkles sampled from the lower shore than in those sampled from the mid-shore.

5.7 Confidence limits

Confidence limits for a mean The values derived from a sample between which the population mean probably falls and is calculated using the standard error of the sample mean.

In most research, you do not collect observations from every item in your statistical population; instead, you collect observations from a sample. A useful measure, therefore, is to know how good a predictor of the population parameter your sample statistic is. We illustrate the principles behind these 'confidence limits' in relation to the means of samples, but confidence limits can be calculated for many population parameters including regression coefficients.

Standard error of the mean (SEM)
For a given statistical population, you can take many samples, which can all be described in terms of a mean. Using these means as your data, you can calculate a standard deviation of these mean values, more commonly known as the standard error of the mean.

The confidence limits for a mean are derived from the standard error of the sample mean. The idea behind this is that, for a given statistical population, you may take many samples and these can all be described in terms of a mean. If you then used these means as your data, you could calculate a standard deviation of these mean values, thus indicating how much the means from the many samples varied. This standard deviation of the mean is more commonly known as the standard error of the mean. The term standard deviation is usually restricted to the standard deviation you calculate from the items in one sample. The standard error of the mean (SEM) is:

$$SEM = \frac{s}{\sqrt{n}}$$

The distribution of these sample means will tend to be normal (even if the data in each sample are not normal), and so the standard error of the mean has the same relationship that we introduced in 5.5.3i in that there is a 95% probability that the population mean will fall within the range $\bar{x} \pm 1.96\,SEM$ and there is a 99% probability that the population mean will fall within the range $\bar{x} \pm 2.58\,SEM$. Most often these confidence limits for means are included as vertical bars around sample mean values on figures. This is illustrated on Fig. 12.3.

If you have a small sample (less than 30), then the variation in the data due to confounding variables (sampling error) will become relatively large and this can have an undue effect on the relationship between probability and the standard error of the mean. Therefore, a factor that recognizes the effect of the small sample size is included in the calculation of these confidence limits for small samples, so that the confidence limits are defined by the range $\bar{x} \pm (t \times SEM)$. The value for t can be found in Appendix D, Table D6. We explain these tables in Chapter 6. To find a particular t value, you need to choose the probability you wish to use in your confidence limits. A probability of 95% will be indicated by $p = 0.05$; a probability of 99% is indicated as $p = 0.01$. You also need to know the degrees of freedom. For these values of t, the degrees of freedom are $n - 1$. Using your chosen p value and the degrees of freedom, you can identify the value for t, which you then use when calculating the confidence limits. Calculating confidence limits in this way is acceptable for interval data even from non-normal or unknown distributions.

 What is the 95% confidence interval for the mean from Example 5.2 (Table 5.8) for the length (mm) of two-spot ladybirds (*Adalia bipunctata*)?

5.8 Parametric data

The statistical tests that are used in hypothesis testing fall into two groups: 'parametric' and 'non-parametric'. As a scientist you will need to be able to identify parametric data to enable you to select a statistical test you may use. In this section we explain why this is the case and how to confirm if your data are parametric.

5.8.1 What are parametric data and why do we need them?

As we discussed in 5.2, a set of observations can be found to have a particular distribution and the relationship between the x and y values can be described by a particular equation. In the normal distribution the equation is called the Gaussian equation. From this equation it is possible to work out other relationships such as the mean and variance. There is a group of statistical tests called 'parametric' tests that are derived from these relationships (parameters). The underlying assumption of these parametric statistical tests is that the statistical population is normally distributed, i.e. if you measured a characteristic of interest such as height in every item of that population the data would have all the characteristics of a normal distribution. However, it is very rare that in practice any researcher measures every item in a population—we all tend to sample. It is standard practice to make the general assumption that if the sample has the characteristics of a normal distribution then the population from which the sample is derived will also have the characteristics of a normal population and so these parametric statistics may be used. Parametric statistics are more powerful and should always be used where possible. Powerful tests are those that are more effective at detecting a real biological difference when one exists. It is therefore important to be able to check whether your data are normally distributed so that you can then use the parametric statistics.

Parametric data
Data that can be confirmed to have the characteristics of a normal distribution.

5.8.2 How to confirm that data are parametric

Confirming whether data are parametric can be done by looking at the characteristics of the distribution, usually by graphing the data or by carrying out a statistical test to confirm that the data you have collected are not significantly different from a normal distribution. Looking at the distribution of your observations is the more subjective approach to take but nonetheless this tends to be more effective when your sample sizes are small, as statistical tests then tend to be less discriminating. Using statistical tests to confirm the normality of data has the advantage in that it is objective but also tends to be more conservative. We therefore suggest whenever possible that you take both approaches.

5.8.3 Using the shape of the distribution to confirm that your data are parametric

As we saw in 5.2 and 5.3, data with a normal distribution have a number of features. For example data with a normal distribution are measured on an interval scale and so are quantitative, continuous, and rankable. The distribution should be symmetrical and therefore the mean = median = mode. We have also introduced you to the relationship in normally distributed data between the number of observations in a particular region of the distribution, the probability, and the standard deviation, so that, for example, 95% of all observations should be found in the range $\bar{x} \pm 1.96s$ and 68% of your observations should be found in the range $\bar{x} \pm s$. As you will often need to check whether your data are parametric, we have summarized the key features of data with a normal distribution in Box 5.2 and illustrated this using the data from Table 5.8.

BOX 5.2 The features of a normal distribution

GENERAL DETAILS	EXAMPLE 5.2
If your data meet the following characteristics this indicates that the population from which the data are derived has a normal distribution and you may consider using parametric statistics. If you are designing an experiment you can only confirm the first characteristic at this stage. You then need to check your data when you have collected it.	This calculation is given in full in the Online Resource Centre. For presentation purposes only all values have been rounded to five decimal places.
a. Are the data measured on an interval scale, such as mm or grams, therefore quantitative and continuous? Data that are measured on an ordinal or nominal scale will not be parametric. Derived variables recorded on restricted scales such as percentages, where the scale of measurement only ranges between 0% and 100%, are not usually normally distributed. Some derived variables measured on a continuous scale such as rates (e.g. m/s) may be normally distributed.	a. Yes. The scale is mm, which is an interval scale.
b. Does the distribution appear to be a bell-shaped curve (e.g. Fig. 5.3)? This criterion is often not very convincing as you may have few observations in your sample.	b. Using the frequency table (Table 5.9), the data can be plotted as a histogram (Fig. 5.7) There does appear to be a bell shape to the distribution.
c. Do about 68% of your observations fall within the range $\bar{x} \pm 1s$?	c. For this example, the mean and standard deviation are: $\bar{x} = \dfrac{254}{50} = 5.08\text{mm}$ $s = 1.9361228\text{mm (Box 5.1)}$ Therefore: $\bar{x} + s = 5.08 + 1.9361228 = 7.01612\text{mm}$ $\bar{x} - s = 5.08 - 1.9361228 = 3.14388\text{mm}$ If you examine Table 5.8, it is clear that 34/50 observations in this sample (68%) fall within the range 3.1–7.0mm.

(continued)

d. Does the mean = median = mode? This can be a difficult criterion to use, as it is not clear how much difference there can be between these values before they indicate that the data are not parametric.	d. The mean has already been calculated ($\bar{x} = 5.08mm$). Median: Arrange all the values in numerical order. $n = 50$ so the median lies halfway between two values of 5.0mm. Therefore, the median is 5.0mm. Mode: If you examine Table 5.8, it is clear that the most frequent value is 5mm. Therefore, the mean, median, and mode are all very similar.
e. You can now draw a conclusion as to whether the distribution of the data appears to be normal based on features a–d.	e. We would conclude that these data appear to be normally distributed.

Further examples relating to the topic including how to use statistical software are included in the Online Resource Centre.

You should check your own data set to see if they have some or all of these features. However, if you are at the stage of designing your experiment and so do not have any data then the number of features you can check is limited to the first characteristic only.

 Q6   In Q4, we calculated the summary statistics for data from Example 2.3. Use the four criteria detailed in Box 5.2 to decide whether the data are parametric.

5.8.4 Using statistical tests to confirm that your data are parametric

Statistical tests for normality are also widely used to tell if your data are suitable for the use of parametric statistical tests (if all other criteria are also met). For example a normal distribution may be mathematically described by the Gaussian equation (5.2.1). Using this equation and your own x values, you can calculate the corresponding y values. These are an 'expected' data set and can be compared with your 'observed' data set by a χ^2 goodness-of-fit test. This has been done in 8.4 for the ladybird data from Example 5.2 and it was found that there is no significant difference ($\chi^2_{calculated} = 4.40$, $p = 0.05$) between the observed lengths of ladybirds (mm) compared with that expected if the data are normally distributed. We conclude therefore that the data are normally distributed and can be described as being parametric.

Some commercial statistics software includes tests for normality as a standard. One of the most common of these tests is the Shapiro–Wilk test. However, this test is adversely affected by tied values where more than one of your observations has the same value. Modifications of the Shapiro–Wilk test are available that are more robust in the presence of tied data.

5.9 Non-parametric data

Non-parametric data
Data where either the distribution is not normal or, more usually, the distribution is unknown.

Non-parametric data are data where either the distribution is not normal or more usually the distribution is unknown. When testing hypotheses and you have non-parametric data, you may either attempt to normalize the data by transforming them (5.10) or use non-parametric statistics. Parametric data may be analysed using non-parametric statistics, but not the other way round.

5.9.1 Ranking data

When using parametric statistics (Chapters 7–10), you will need to know how to calculate a sum of squares, variance, and standard deviation, and we have explained these steps in Box 5.1. If you are using non-parametric statistics, you will usually need to know how to rank data. This is a central skill for most non-parametric statistical tests and we have provided a step-by-step guide in Box 5.3. To rank data, the values are placed in their logical, usually numerical, order and each observation is given a 'ranking' depending on where in this order they fall. It is then the ranks that are used in the calculation rather than the observations. Examine the data from Example 2.3 on the growth rate (mm) of rye seedlings. These interval data can be organized into a numerical order. (You did this to answer Q6 when finding the median.) Ordinal data may also be ranked. For example, the qualitative ACFOR scale used in ecology can be ordered as shown: A–C–F–O–R. However, nominal data cannot be ranked. For example, if you graded petal colours as red, light red, and spotted red, there is no reason for these colours to be organized in any particular order.

One of the most important aspects of ranking data is what to do when you have more than one observation with the same value. The ordered data from Example 2.3 on the growth rate (mm) of *Secale cereale* seedlings is: 10.0, 10.0, 11.0, 11.5, 11.5, 12.0, 12.0, 14.0, 14.5, 14.5, 15.5, 18.0, 19.0, 19.5, 21.0. There are two observations with the value 10mm, two with the value 11.5mm, two with the value 12.0mm, and two with the value 14.5mm. We show you in Box 5.3 how to rank tied values for statistical analysis but you do need to be very wary as there is a commonly available spreadsheet software that will rank values for you but does so incorrectly. We review this in the Online Resource Centre.

BOX 5.3 How to rank data for non-parametric statistics

1. If you have one sample.

Sample   1 4 5 5 6 3 4 6 5 5 7 9

i. Arrange these values in numerical order.

Sample 1 3 4 4 5 5 5 5 6 6 7 9

ii. Assign *possible* ranks.

The first number in order is rank 1. The second number in rank order is 2, etc.

Sample 1 3 4 4 5 5 5 5 6 6 7 9
Rank   1 2 3 4 5 6 7 8 9 10 11 12

2. If you have more than one sample.

Many statistical tests require you to combine samples when assigning ranks.

e.g. Sample 1   1   3   5   3   2   2   4
      Sample 2   1   1   4   6   7   8   7

i. Arrange all the observations in numerical order.

e.g. Sample 1   1   2   2   3   3
      Sample 2   1   1   4
e.g. Sample 1   4   5
      Sample 2   6   7   7   8

(continued)

iii. Where the data has the same value, these observations all take the average rank.

e.g. In this set of data, there are two 4s. In theory, they should be ranked 3 and 4.

Sample 1 3 4 4 5 5 5 5 6 6 7 9
Rank   1 2 3 4

But since the observations have the same value, they should have the same rank.

The average rank for the 4s is therefore $\frac{3+4}{2}=3.5$

The ranks are now

Sample 1 3 4 4 5 5 5 5 6 6 7 9
Rank   1 2 3.5 3.5

This process is repeated for the four 5s and the two 6s.

Rank for 5s $=\dfrac{5+6+7+8}{4}=6.5$

Rank for 6s $=\dfrac{9+10}{2}=9.5$

The ranks are now

Sample 1 3 4 4 5 5 5 5 6 6 7 9
Rank   1 2 3.5 3.5 6.5 6.5 6.5 6.5 9.5 9.5 11 12

ii. Assign ranks. Where values are 'tied', assign the average rank.

e.g. Sample 1   1   2 2 3 3
     Sample 2   1 1   4
     Rank   2 2 2 4.5 4.5 6.5 6.5 8.5
e.g. Sample 1   4 5
     Sample 2   6 7 7 8
     Rank   8.5 10 11 12.5 12.5 14

iii. When the samples are then analysed, they will have the following ranks.

Sample 1 2   2   3   3   4   5
Rank   2 4.5 4.5 6.5 6.5 8.5 10
Sum of ranks $(\Sigma r_1)=2+4.5+4.5+6.5+6.5+8.5+10=42.5$

Sample 2 1 1   4 6   7   7   8
Rank   2 2 8.5 11 12.5 12.5 14
Sum of ranks $(\Sigma r_2)=2+2+8.5+11+12.5+12.5+14=62.5$

These sums of ranks are often used in non-parametric statistical tests, e.g. Mann–Whitney U test (11.1).

5.10 Transforming data

If your data are normally distributed (Box 5.2), then you are able to use the more powerful and generally more flexible parametric statistics to test your hypotheses. If you have data that are not normally distributed, you should always consider transforming them so that they become normalized. To transform your data you select a particular method for the recalculation of your x values. In carrying out this process this transformation of the x values squeezes and/or stretches the scale used so that the distribution takes on the appearance of a bell-shaped, 'normal' curve. When applied successfully to samples you wish to compare you are then able to consider using parametric statistics.

In some investigations where you have collected data relating to a number of variables you may be able to use the general linear or general linearized models to test hypotheses. In this case the statistical software or program that you use will determine which model to use to test your data. This avoids the need for you to make this decision and removes the need to transform your data. Explanations relating to how and when to use these tests are outside the scope of this book and you will therefore need to refer to other texts for details.

Transforming data
A mathematical calculation that has the effect of squeezing and/or stretching the scale you used when making your measurements so that the distribution takes on an alternative shape.

5.10.1 How to choose a suitable transformation

It is not easy to decide on which transformation may successfully normalize your data. See Table 5.12 for a list of a number of common transformations and when to try them. A further approach to help in making this selection is to look at published research papers in the area you are working in to see which transformations have been found to be successful. The choice of transformation method depends on a number of factors:

a. Transformations are only appropriate for ordinal or interval data.

b. Transforming two variables is not necessarily complex; however, interpreting the results from the analysis of transformed data following, for example, a regression analysis, can be complex and is outside the scope of this book.

c. When you are considering transforming data the first step is to plot your data and examine the original distribution. If the data are not normally distributed, then the distribution will have one or more of the following features: the data are very spread out and the distribution may be platykurtic (Fig. 5.11); there is little variation and the distribution may appear to be leptokurtic (Fig. 5.10); the distribution is skewed to the right (Fig. 5.8); the distribution is skewed to the left (Fig. 5.9). These features can be used to determine the most appropriate transformation to try.

d. The presence of zeros in your data can have an undue influence in some transformations. Therefore, it is common practice, if you have zeros, to add one to all observations before transforming the data.

5.10.2 How to carry out the transformation

The process of transformation requires you to first identify the calculation most likely to normalize the data. You then apply this calculation to your x values. We illustrate this in Table 5.11 with some data measured on an ordinal scale of 1–10. When plotted the distribution is mesokurtic and skewed to the right (Fig. 5.12). Table 5.12 indicates that a square root, $\log_{10}$, or natural log transformation may normalize these data. Each x value has been recalculated and replotted (Table 5.11 and Figs 5.13–5.15). Of the three processes it appears that taking the $\log_{10}$ has been the most effective transformation.

5.10.3 Did the transformation work, and what to do if it didn't

When you have transformed your data you can recheck whether the normalization has worked using features such as those described in Box 5.2. If using the suggested methods and any others you encounter does not appear to normalize your data you should use non-parametric statistical tests.

Table 5.11 Demonstration of transformation of data to normalize it

Original x	y	Transformed values of x		
	Frequency of x	Square root of x	Log of x	Natural log of x
1.0	0	1.00000	0.00000	0.00000
2.0	74	1.41421	0.30103	0.69315
3.0	98	1.73205	0.47712	1.09861
4.0	94	2.00000	0.60206	1.38629
5.0	84	2.23606	0.69897	1.60944
6.0	69	2.4495	0.77815	1.79176
7.0	52	2.6457	0.84510	1.94591
8.0	36	2.82842	0.90309	2.07944
9.0	22	3.00000	0.95424	2.19723
10.0	10	3.16228	1.00000	2.30259

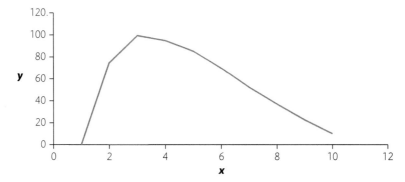

Fig. 5.12 Distribution of data measured on an ordinal scale from Table 5.11 before transformation

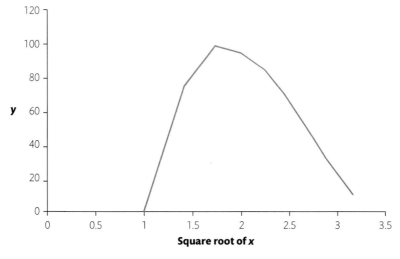

Fig. 5.13 Distribution of data from Table 5.11 after taking the square root of each x value

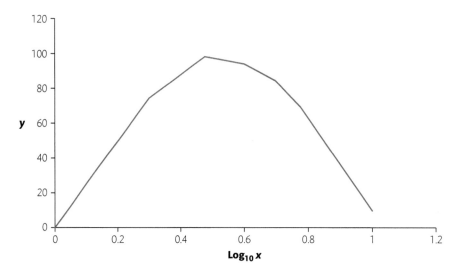

Fig. 5.14 Distribution of data from Table 5.11 after taking the log of each x value

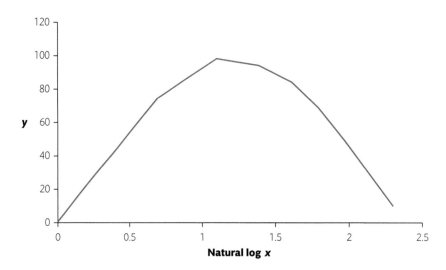

Fig. 5.15 Distribution of data from Table 5.11 after taking a natural log of each x value

5.10.4 How to report analyses that have used transformed data

As your original observations are non-parametric you should not use the mean and variance but present descriptive statistics suitable for your data such as the median, mode, and range. When reporting the results from most statistical tests you will not need to take further action as the test statistic (this term is explained in 6.1) is usually independent of the original scale of measurement and so you may report the evaluation as you would normally do.

Table 5.12 How do I transform my data: one variable?

Type of distribution	Other features	Transformation
There is a lot of variation in the data so that the distribution appears to be spread out and may appear to be platykurtic.	The original measurements are on a scale where the values are greater than 10.	Take the $\log_{10}$ of each observation.
	Some of the original values are 0.	Add 1 to each observation and then take the $\log_{10}$ of each observation.
The variation is limited and the distribution is therefore narrow and may appear to be leptokurtic. Or the scale of measurement is limited such as percentages (0–100%) or proportions (0.0–1.0).	The observations are recorded as a percentage and there is a spread of values.	Divide each value by 100. Take the square root of each observation. Then find the angle whose sine equals each value or 'inverse sine' $(\sin^{-1})$.
	None of the original measurements is 0.	Take the square root of each observation.
	Some of the original values are 0.	Add 1 to each observation and then take the square root of each observation.
	The observations are recorded as percentages calculated from counts with a common denominator (e.g. between 0% and 20%.	Take the square root of each observation.
	The observations are recorded as percentages calculated from counts with a common denominator (e.g. 45/50, 49/50) and most values lie between 80% and 100%.	Subtract each observation from 100 and then take the square root of each of these values.
	The data is recorded as a proportion.	(Arcsine) Take the square root of each observation. Then find the angle whose sine equals each value or 'inverse sine' $(\sin^{-1})$.
The distribution is skewed to the right (positive).	There are a number of suitable transformations. The one to use depends on the degree of skewness. The following is in the order of little skew (try a) to considerable skew (try d or e). If you have zeros in your original data, you should add 1 to each observation before transforming the data.	a. Take the square root of each observation. b. Take the $\log_{10}$ of each observation if none of the x values are zero. Add one to each x value and then take $\log_{10}$ if any of the values are zero. c. Take the natural logarithm (ln) of each observation. d. Divide 1 by each observation (reciprocal, $1/x$). e. Take the square root of an observation and then divide 1 by this value (reciprocal square root, $1/\sqrt{x}$).
The distribution is skewed to the left (negative).	Again, there are a number of transformations that may be used depending on the degree of skew from least (try a) to considerable (try b, etc.).	a. Square each observation. b. Cube each observation, etc.

Summary of Chapter 5

- When designing an investigation or choosing a statistical test, you will need to be familiar with the terms used to describe scales of measurement. These terms are quantitative, qualitative, discrete, continuous, rankable, nominal, ordinal, interval, and derived variable (5.1).

- Observations have a particular shape or distribution when plotted. Some of these distributions can be described mathematically including the normal distribution, Poisson distribution, and binomial distribution (5.2).

- Distributions may be described in terms of the central point (mean, median, mode, skew, and kurtosis) and spread of data around the central point (range, interquartile range, percentile, variance, and standard deviation). The methods used depend on whether the data are measured on a nominal, ordinal, or interval scale (5.3, 5.4, 5.5, 5.7).

- The coefficient of variation can be used to compare the relative amount of variation in different data (5.6).

- Data may be normally distributed and these data are also known as parametric data (5.8). If data are not normally distributed or the distribution is unknown, then the data are non-parametric (5.9).

- Non-parametric data may sometimes be normalized by transforming them. The choice of transformation depends on the original shape of the distribution (5.10).

- When using parametric statistics, you often need to know how to calculate the sums of squares, variance, and standard deviation (Box 5.1). If you are using non-parametric statistics, you will need to know how to rank data (Box 5.3).

- The Online Resource Centre includes interactive exercises that test your understanding of this chapter with other topics, particularly those considered in Chapters 7–12.

Answers to chapter questions

A1 1) Quantitative, discrete, rankable, ordinal.

2) Quantitative, continuous, rankable, derived variable.

3) Quantitative, continuous, rankable, derived variable.

4) Quantitative, continuous, rankable, interval.

A2 $\Sigma mf = 15\ 003.15$, $n = 87$, $\bar{x} = 172.45$cm. The mode = 172.45cm. See the Online Resource Centre for the full calculation.

A3 In 5.4.3i, we calculated the mean ($\bar{x}$) = 5.08mm. To calculate the median, arrange all the values in numerical order. The mid-position lies halfway between two values of 5.0mm. Therefore, the median is 5.0mm. If you examine Table 5.9, it is clear that the most frequent value is 5.0mm. This is the mode. Therefore, the mean almost equals the median and the mode.

A4 The data are measured on an interval scale (cm), but there are few observations and it will be difficult to establish the underlying distribution. The data have been organized into a frequency table in Chapter 12 (Table 12.4). This indicates that the data are skewed and are unlikely to be normally distributed. Therefore, the data can be summarized using the median, mode, and a range. The data can be written in numerical order:

10.0 (×2), 11.0, 11.5 (×2), 12.0 (×2), 14.0, 14.5 (×2), 15.5, 18.0, 19.0, 19.5, 21.0.

There are 15 observations in total; therefore, the median is the eighth observation, which is 14.0cm. If the raw data are considered, there are four values (10.0cm, 11.5cm, 12.0cm, and 14.5cm) that occur twice and are therefore modes. If the data are organized into a frequency table (Table 12.4), the modal class is 10.0–11.9cm, with mid-point 10.95cm. The range is between 10cm and 21cm = 11cm. The interquartile range is 18.0cm − 11.5cm = 6.5cm.

A5 There are 50 observations in this sample; therefore, we can calculate the confidence limits as $\bar{x} \pm 1.96SEM$ where $SEM = 1.93612 / \sqrt{50} = 0.27381$ and the 95% confidence limits are 4.54 to 5.62mm.

A6 a. Yes. The observations are measured on an interval scale.

b. No. When plotted, the data does not have a bell shape to the distribution.

c. The range determined by $\bar{x} \pm s$ is 10.6mm to 17.9mm. Only 60% of observations fall within this range. However this difference can be accounted for by one observation only. This criterion is not convincing.

d. No. The mean (14.26cm) is approximately the same as the median (14.0cm), but these differ from the mid-point of the modal class (10.95cm).

e. Using these criteria, the data (Example 2.3) do not appear to be normally distributed. (You could confirm this using a statistical test such as the chi-squared goodness-of-fit test.) If you would like to see the calculations in full, they are given in the Online Resource Centre.

6 An introduction to hypothesis testing

 In a nutshell

In this chapter, we consider the general principles behind hypothesis testing. There are six steps that you follow in all hypothesis testing:

1) Confirm which distribution is the best one to use.

2) Select and correctly phrase the hypotheses to be tested and determine which is/ are the correct statistical test(s) to use.

3) Work out the calculated value of the test statistic.

4) Find the critical value of the test statistic.

5) Reject or do not reject the null hypothesis depending on the rule relating to the critical and calculated test statistics.

6) Decide what the outcome of the statistical test means in terms of the biology of the system.

We explain these steps here and introduce you to a number of key terms such as p values, Type I and II errors, the power of a test, and degrees of freedom. In the chapters on statistical testing (Chapters 8–11), we follow these six steps for each statistical test.

In this book so far, we have thought about general principles relating to the planning of an experiment (Chapters 1–3) and introduced you to the ways in which you might start to handle raw data (Chapter 5). What we have not yet done is shown you how to answer the questions your experiment may have set out to examine.

Some investigations, often those producing baseline data, or a preliminary investigation, will not set out to test a hypothesis. For example, a survey of the flora and fauna of a wood is not 'asking' a specific question: you will be finding out what is there. The data from this type of investigation are usually evaluated using analysis based on information-theoretic models (1.8). They may generate hypotheses at the end of the evaluation, but they do not start out with one. In this chapter, we are thinking about experiments that are designed to answer a specific question; for example, is there a difference in mineral content between organic and non-organic potatoes? This is what hypothesis testing is all about.

We first explain how hypothesis testing works by using one approach as an example. We then cover a number of other important principles that relate to hypothesis testing.

If you work through this chapter it should only take one hour to read and complete all the exercises. The answers for these exercises are at the end of the chapter. All our examples are based on real undergraduate research projects. If these examples are not in your subject area, you will find more in the Online Resource Centre.

6.1 What is hypothesis testing?

Hypothesis testing focuses on the extreme values of distributions and uses mathematics to determine where to divide the distribution into what is extreme and what is not. The tests for a hypothesis then determine if samples or single values fall into these extreme regions.

6.1.1 Extreme rare values in a population

To start to understand what is happening when you test hypotheses you first need to think about extreme or rare values in a population.

Example 6.1 Shell length (mm) of a population of *Cepaea nemoralis*

In preparation for a study on *C.nemoralis* all individuals from a population of this snail were measured. The observations were organized into a frequency table with size classes (Table 6.1 and Fig. 6.1) and it is clear that some shell sizes are more common than others.

Table 6.1 The length of *Cepaea nemoralis* shells (mm) in a population

Length of *Cepaea nemoralis* shells			
Size classes (mm)	Number of observations	Size classes (mm)	Number of observations
0.00–0.45	0	5.50–5.95	11
0.50–0.95	0	6.00–6.45	9
1.00–1.45	0	6.50–6.95	8
1.50–1.95	1	7.00–7.45	5
2.00–2.45	1	7.50–7.95	3
2.50–2.95	2	8.00–8.45	2
3.00–3.45	4	8.50–8.95	1
3.50–3.95	5	9.00–9.45	1
4.0–4.45	6	9.50–9.95	1
4.50–4.95	8	10.00–10.45	0
5.00–5.45	10	N = 78	

(continued)

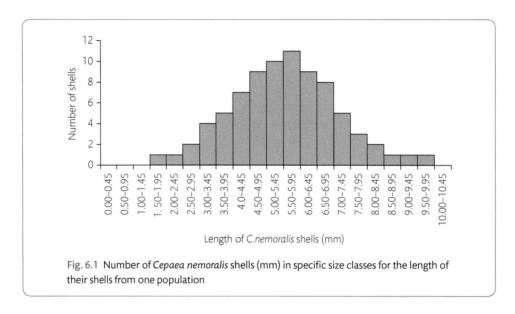

Fig. 6.1 Number of *Cepaea nemoralis* shells (mm) in specific size classes for the length of their shells from one population

i. Describing extreme values in terms of their frequency (%)

It is possible to describe how extreme a value is by calculating how often any one value appears in a population. For example in size class 1.5–1.95mm (Table 6.1) there is only one shell at this length, accounting for 1.28% (1/78) of the population, which makes this an extreme or rare observation. By comparison 10 shells fall into the size class 5.00–5.45mm and so these are relatively common in the population and account for 12.8% (10/78) of the population. Using this information we can subdivide our data into groups described in terms of how many observations fall into these 'common' and 'rare' groups (Table 6.2). (We have looked at this idea before when we considered the interquartile range (5.5.2).)

Regions of rejection and acceptance
A distribution can be divided into regions using mathematical relationships such as between the mean and standard deviation. These regions separate rare or extreme values from more common values. In hypothesis testing the regions with the rare values are called the regions of rejection and the region with the common values is called the region of acceptance.

ii. Describing extreme values in terms of the population mean (μ) and standard deviation (σ)

An alternative approach to take when describing extreme values is to use the range. We illustrate this using a normal distribution and the range measured in relation to the mean and standard deviation. If you calculate the mean and standard deviation then you would expect 68% of your observations from your population to fall within the lower boundary value calculated as the mean minus the standard deviation ($\mu - \sigma$) and the upper boundary of the range calculated as the mean plus the standard deviation ($\mu + \sigma$). Alternative boundaries that become important shortly include the mean ±1.96 standard deviation ($\mu \pm 1.96\sigma$) which will enclose 95% of your observations in the central section and leave 2.5% of observations in each of the two tails (Fig. 6.2) and the mean ±2.58 × standard deviation ($\mu \pm 2.58\sigma$) which will enclose 99% of your observations in the central section and leave 0.5% of observations in each of the two tails. These values such as 1.96 and 2.58 are called z values and can be found in Appendix D, Table D5. As you will learn shortly, it is possible to use other distributions and so other sets of values to place these boundaries on a distribution.

Table 6.2 Length of *Cepaea nemoralis* shells (mm) in 21 size classes and groupings of classes indicating common and rare observations based on their relative frequency

Length (mm) of *Cepaea nemoralis* shells in a statistical population			
Size classes (mm)	Number of observations	Frequency (%)	Groupings of common and rare shell lengths in the population based on relative frequency
0.00–0.45	0	0.00	
0.50–0.95	0	0.00	
1.00–1.45	0	0.00	
1. 50–1.95	1	1.28	Only 1.28% of all observations are in this size class / 2.56% of all observations are in these two combined classes (1.5–2.45)
2.00–2.45	1	1.28	
2.50–2.95	2	2.56	
3.00–3.45	4	5.13	
3.50–3.95	5	6.41	
4.0–4.45	6	7.61	97.44% of all observations are in these combined classes (2.0–9.45) / 94.88% of all observations are in these combined classes (2.50–9.45) / 66.59% of all observation are in these combined classes (4.0–6.9)
4.50–4.95	8	10.26	
5.00–5.45	10	12.82	
5.50–5.95	11	14.10	
6.00–6.45	9	11.54	
6.50–6.95	8	10.26	
7.00–7.45	5	6.41	
7.50–7.95	3	3.85	
8.00–8.45	2	2.56	
8.50–8.95	1	1.28	
9.00–9.45	1	1.28	Only 1.28% of all observations are in this size class / 2.56% of all observations are in these combined classes (9.00–9.95)
9.50–9.95	1	1.28	
10.00–10.45	0	0.00	
	N = 78		

Critical value The point on the x axis where the extreme values are divided from the common values in a distribution using mathematical relationships such as the mean ± standard deviation.

These methods create points on the *x* axis that separate a distribution into regions. In hypothesis testing these points divide a distribution into regions of acceptance and rejection. The point on the *x* axis where the distribution is divided is called the critical value.

In Example 6.1 we can do the same and divide our distribution into regions defined by the likelihood of obtaining a value. In Fig. 6.3 we will use the boundary of $\mu \pm 1.96\sigma$ and so have marked off the most extreme 2.5% of our observations in each end (tail).

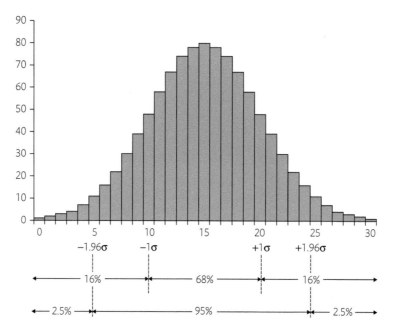

Fig. 6.2 Normal distribution, $n = 1000$, $\mu = 15$, $\sigma = 5$. The first bars indicate that when the range is defined by $\mu \pm \sigma$, then 68% of observations fall into this range and 16% of observations are outside this range in each of the two tails. The second bars indicate the number of observations that would be expected to fall into the centre of the distribution and two tails, when the range is defined by $\mu \pm 1.96\sigma$, and are 95% and 2.5% respectively. The boundary points are called the critical values and separate the region of rejection (the tails) and acceptance (the central portion).

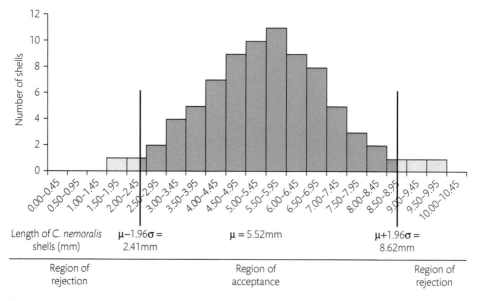

Fig. 6.3 Using a critical value of 1.96 standard deviations to partition the distribution into a 'common' region with 95% of all observations and two regions in the tails, each with 2.5% of observations. The probability of an observation being in **only** one of these tails is 0.025 or 2.5% and the probability of an observation being in **either** one of these two tails is 0.05 or 5%. These are the regions of rejection.

In Example 10.2 we present the findings from another mollusc-related study. In this case the species are periwinkles. The following questions are based on the extract from Table 10.2 reproduced here. The mean ($\bar{x}$) and the standard deviation (**s**) of the periwinkles from the lower shore are $\bar{x} = 5.89231$ and $s_1 = 1.4186$; $n = 13$.

Extract from Table 10.2 Shell height in periwinkles from the lower shore at Porthcawl, 2002

Shell height (mm) for sample of periwinkles on the lower shore ($n = 13$)			
4.0	5.0	6.0	8.4
4.8	5.0	6.2	
5.0	5.5	7.7	
5.0	5.6	8.4	

a. How many observations fall within the range $\bar{x} \pm s$, i.e. 4.5mm to 7.3 mm?
b. How many observations fall outside this range?
c. Now divide the data so that 95% of your observations are included in the central block and only 5% of the observations are included in the tails. How many observations are in this central block and how many are in the tails?
d. What is the probability that the next periwinkle you measured also falls within these tails?

6.1.2 What are null and alternate hypotheses?

If we are sent a *Cepaea nemoralis* shell in the post from another interested biologist who tells us they collected this from the same location from which the original data was collected in Example 6.1, how can we tell if this single specimen does indeed come from the same population? We can establish a hypothesis that the single shell does come from the population we measured. If this is the case the parameters should be the same, as we are arguing that the single shell and our earlier measures are from the same population. In hypothesis testing the parameter that is used is usually either the mean or median although other parameters could be used. In our example we are considering observations with a normal distribution so we will use the mean. If the single shell does come from the same population we can assume that the population means would be the same and we would formally record this in our null hypothesis as: $\mu = \mu_s$ (where μ is the population mean and μ_s is the mean of the population from which the single shell was derived). This statement is telling us that there should be no difference between the two means and any variation in the data is due to sampling error (1.7.4).

We now have to make a decision as to whether to reject the null hypothesis, but if we do so what are we 'accepting' in its place? There are three generic alternate hypotheses. In the first alternate hypothesis we say that the populations and hence the means from these populations are not the same, they are different ($\mu \neq \mu_s$). This is a straightforward statement as it makes no comment about the nature of the difference. However, it may be that we know that if the single shell does not come from our population then it can only come from

Null hypothesis (H_0)
An assumption that two or more samples or observations are derived from the same population. In a population with a normal distribution this model may be described in terms of the means, $\mu = \mu_s$.

Alternate hypotheses (H₁)

In hypothesis testing we choose between the null hypothesis and an alternate hypothesis. There are three possible generic alternate hypotheses which can be illustrated as models relating to the mean: the samples/observation do not come from the same populations ($\mu \neq \mu_s$); the samples/observation come from a population whose mean is smaller than the original population ($\mu > \mu_s$); the samples/observation come from a population whose mean is larger than the original population ($\mu < \mu_s$).

Two-tailed test

Where the hypothesis testing is against both ends of a distribution. Most general tests of hypotheses are usually two-tailed. The alternate hypothesis can be illustrated as a model relating to the mean as $\mu \neq \mu_s$.

One-tailed test

When the hypothesis testing is such that only one end of a distribution is being considered. One-tailed tests are most often encountered when testing specific hypotheses. The alternate hypothesis can be illustrated as a model relating to the mean and will be either $\mu > \mu_s$ or $\mu < \mu_s$.

a population with a larger mean. This can be written formally for our alternate hypothesis as $\mu < \mu_s$. Finally the opposite might be possible where the only other population that the single shell can come from has a smaller mean (i.e. $\mu > \mu_s$). If we select the first of these alternate hypotheses ($\mu \neq \mu_s$) this is called a two-tailed test of hypotheses as we are not making any assumptions about the alternate population. This means that the size of the single shell could be in either the upper tail or the lower tail of our population distribution. However if we use either $\mu > \mu_s$ or $\mu < \mu_s$ as our alternate hypothesis then this is called a one-tailed test as we are choosing between a null hypothesis of no difference ($\mu = \mu_s$) and only a larger mean for our sample population ($\mu < \mu_s$) or between no difference ($\mu = \mu_s$) and only a smaller mean for our sample population ($\mu > \mu_s$). In biology is it very unusual to know about the alternate population and just because a value is either larger or smaller than our original population this does not provide any evidence about the alternate population. Therefore most tests of hypotheses are two-tailed tests.

6.1.3 Testing the hypotheses

If our single shell is 9.2 mm in length we would consider this extreme in relation to our original population (Table 6.1) and very rare. It is therefore indeed possible that this single shell does come from a different population, but it just may come from our original population, though the likelihood is small. Our null hypothesis will be that the single observation comes from the same population ($\mu = \mu_s$). Even though this single value is large we know nothing about any possible alternative populations so we will choose the two-tailed hypothesis which states generally that the single observation is derived from a different population with a different mean ($\mu \neq \mu_s$). We now need a mechanism that allows us to decide which is the most probable of these two possible outcomes.

The first step is to use the measure of 'extremeness' we described in 6.1.1ii and calculate the number of standard deviations between the mean from the original population and the single value. In our example this is found as the difference between the single value and the population mean as a proportion of the population standard deviation:

$$\frac{Y - \mu}{\sigma}$$

Where Y is the single observation, μ is the population mean, and σ is the population standard deviation. In our example $\mu = 5.51875$, $\sigma = 1.584525$, $Y = 9.2$

$$\text{Therefore} \quad \frac{9.2 - 5.51875}{1.584525} = \frac{3.68125}{1.584525} = 2.32325$$

Our single observation is 2.32 standard deviations away from the population mean (Fig. 6.4). We will now use the regions we selected in Fig. 6.3 defined by the range $\mu \pm 1.96\sigma$, and therefore a critical value of 1.96 and the probability of 0.05. Our single value lies in the region of rejection with its calculated value of 2.32 clearly indicating that this is an extreme value. We will therefore reject our null hypothesis ($\mu = \mu_s$) and not reject our alternate hypothesis ($\mu \neq \mu_s$).

Had we known due to the biology of our system under investigation that if our shell did not come from our original population and could only come from a population which has a

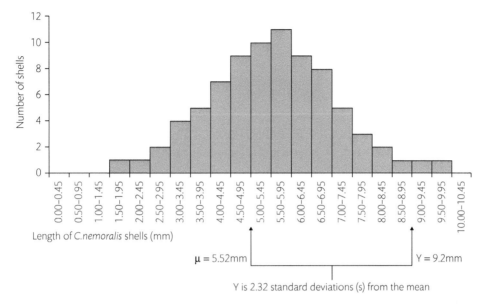

Fig. 6.4 The number of standard deviations between the length of shell from a single individual of *Cepaea nemoralis* and mean of the population (Example 6.1)

larger mean then we would test the one-tailed alternate hypothesis of $\mu < \mu_s$. We would use the critical value found in Appendix D, Table D5 for z values, where for a probability of 5% in a one-tailed test our critical value is 1.65 (Fig. 6.5). Our calculated value (2.32) is clearly in the rejection region for the probability of $p = 0.05$ (5% observations in this one tail) and we

Significant difference

If a sample or observation is found to fall into the region of rejection then the sample or observation is said to be significantly different from the original population. This is usually reported along with the test statistic and the probability.

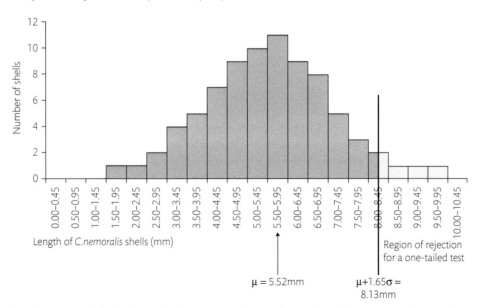

Fig. 6.5 The one-tailed rejection region for Example 6.1 determined by 5% of observations in the population being in one tail only ($p = 0.05$) so that the critical value is 1.65

would again reject the null hypothesis and not reject the alternate hypothesis. In both these instances we would decide there was a significant difference between the single observation and the original population.

6.1.4 Choosing the correct critical value

Test statistic

The value calculated during a test of hypothesis using the data in your experiment and based on a specific distribution. The test statistic is named after the distribution such as z, t, or χ^2.

We have shown you how the difference between a single observation and a population mean can be converted to 'standard deviation units' and this is used as a calculated value which is compared to a critical value. The calculated value is called the test statistic and depends on the distribution the calculation is based on.

The method for working out a calculated value from your data and finding a critical value is based on a known distribution and a level of probability. You therefore need to know how to choose each of these factors to enable you complete your test of hypotheses.

6.1.5 Choosing a distribution

The formula for estimating the calculated value and the identification of the critical value depend on the specific distribution you decide to use. In our example on *C. nemoralis* we have used the formula $\frac{Y-\mu}{\sigma}$ to work out a calculated value of z and Appendix D, Table D5 to find the critical values. These z values are derived from a normal distribution. However, when population or sample sizes are small the normal distribution changes so t values would be more appropriate. For non-normal data the approach to hypothesis testing follows the same principles and different distributions provide the critical values, such as the U values. You choose the distribution based on the characteristics of your data and experimental design. It is therefore these characteristics that help you determine which statistical test to use (Chapter 7). We list these characteristics at the start of our explanations of each statistical test in Chapters 8–11; we then explain how you work out the calculated value and find the critical values.

6.1.6 Choosing the *p* value

The important question remaining is which p value to use. Why use $p = 0.05$ which means the distribution is divided into a central region with 95% of the observations and two tails, the regions of rejection, with 2.5% observations in each? Why not use $p = 0.50$ and divide the distribution into a region with 50% of common observations and 25% of less common observations in each of the two tails? There is a simple answer to this in that the p values are determined by convention. In the biological sciences the standard level at which a null hypothesis will be rejected is at $p \leq 0.05$. In medical science the p value is often set at a lower level than this, at $p \leq 0.01$ or less. There are reasons for this difference which can be explained by considering what risks you are taking and the possible consequences of this.

There are four possible outcomes in hypothesis testing depending on the actual underlying biological system and the outcome from the hypothesis test (Table 6.3). The two possible correct outcomes would be that (a) the statistical test leads to a rejection of the null hypothesis, and the factors you are testing are really affecting the biology of the system; and (b) the test does not reject the null hypothesis when there is no biological effect. There are then two incorrect outcomes: a Type I error and a Type II error.

Table 6.3 The four possible outcomes from a test of hypothesis. Two outcomes are correct. The incorrect outcomes are called Type I and Type II errors and occur with probabilities α and β respectively.

	The hypothesis testing does:	
Biologically there really:	detect a difference. (The null hypothesis is rejected.)	not detect a difference. (The null hypothesis is not rejected.)
is a difference. **(The null hypothesis should be rejected.)**	Correct outcome Determined by $1 - \alpha$	Incorrect outcome Type II error Determined by β
is not a difference. **(The null hypothesis should not be rejected.)**	Incorrect outcome Type I error Determined by $\alpha\,(=p)$ % Significance level. Determines the region of rejection/acceptance	Correct outcome The power of a test Determined by $1 - \beta$

i. Alpha (α) and Type I errors

To understand the interplay between these four outcomes and the p value it is easiest if we first consider the Type I error. A Type I error is said to have occurred if we reject the null hypothesis even when it is really true and there really is no effect of the factors we are testing. In general scientists are cautious and so would prefer not to make this error. This can be understood if we give a medical example where, for example, a test of hypothesis may determine that a treatment is effective when in fact this is not the case.

Clearly we would prefer not to have any Type I errors and we can reduce the likelihood of Type I errors occurring by reducing the size of our rejection region. If we use a p value of 0.05 then we would expect 2.5% of our observations from our population to fall into the tails of the distribution, in our region of rejection. Despite this we would in fact conclude in our test of hypotheses that means of samples or single observations that fall into this region are from a different population. We are prepared to take the risk that these are not rare observations from our original population but are values from a different population, and we describe this risk in terms of p. If we lower the p value then we take less risk as it is even less likely that we will come across observations from our original population that do fall into these very extreme ends of our distribution. We therefore accept that any means from samples or single observations that do fall into this region of rejection are more likely to come from a different population than from the original population. To reduce the likelihood of a Type 1 error, the best p value would be as small as possible.

Statisticians refer to the probability of a Type I error occurring as α. However in practice we refer to it as p, the probability we have chosen to use to determine the size of our regions of rejection. When written as a percentage this is then called the level of significance. When $p = 0.05$ is used to determine the size of our regions of rejection this is a 5% level of significance.

Type I error

If we reject the null hypothesis even when it is really true and the factors we are testing are not affecting the biology of the system. The probability of a Type I error occurring is $\alpha\,(= p,$ level of significance (%)).

ii. Beta (β) and Type II errors

Type II error

Occurs when a test of hypothesis does not reject a null hypothesis when the factor being tested actually is having an effect on the biological system. The probability of this happening is called β.

The second error that can occur is when a statistical test may not reject a null hypothesis when the factor being tested actually is having an effect on the biological system. This is known as a Type II error. The probability of this happening is called β. To estimate β you need to know the distribution of the 'alternate' population as indicated by the alternate hypothesis. We rarely have any knowledge of this and so it is difficult to quantify this. We do know that as α decreases then β increases, so although you may wish to use a small p value to limit the probability of Type I errors occurring, as you do this you increase the probability of Type II errors occurring. β is important when completing power calculations to determine sample size (6.2.4). Here β is generally estimated as between 0.1 and 0.2.

iii. The correct outcomes and p

Therefore your choice of p is a balance between the likelihood of a Type I error occurring versus a Type II error. Depending on the needs of your investigation you can tip the balance towards reducing the risk of a Type I error occurring by using a smaller p value (e.g. use $p = 0.01$ rather than $p = 0.05$) and vice versa.

iv. Cumulative Type I error

A further factor that needs to be considered in relation to Type I errors in particular is that they are cumulative. With a p value of 0.05, we would expect that 5/100 times the decision we come to will be incorrect. We could extend our experiment on C.nemoralis (Example 6.1) and collect samples of snails from four different populations (A–D). In our comparisons we could then compare population A to population B, population A to population C, etc., making six comparisons in total. Each of these has a probability of a Type I error occurring and therefore if we made these six pairwise comparisons we will have six times this probability of making this error whilst analysing these data. The more of these pairwise comparisons we make the more likely we are to make a Type I error. Fortunately there are some statistical tests that have been designed to overcome this issue (e.g. ANOVAS and *post hoc* tests, Chapters 10 and 11). When selecting the correct statistical test you therefore also need to select one that will keep the cumulative likelihood of a Type I error to a minimum.

v. The power of a test

Power of a test

An indicator of how effective a statistical test is in not detecting a difference when one is not present as a result of the action of the factors under investigation.

It is possible to assess different statistical tests to see how effective they are at correctly not detecting a difference when one is not present. This capacity of a statistical test to achieve the correct outcome like this is known as the power of a test (Table 6.3). Each statistical test is most powerful over some of the range of conditions in which they could be used. For example a test may be effective with a sample size of 100 but not when the sample size is less than 20. Statisticians have identified these limits and these form some of the criteria that are used when selecting a statistical test (Chapters 7–11).

One major feature that determines the power of a test is whether it is based on a known distribution such as a normal distribution or is 'distribution free'. As a consequence parametric tests are more powerful as a group compared to non-parametric tests. Therefore where you have an opportunity to use parametric tests you should do so.

vi. Statistical software and *p* values

As mentioned earlier, the approach we are taking in this book and explaining in this chapter is where you select a *p* value that determines the size of your region of rejection and then see whether the mean from your sample or single value is within the region of rejection. An alternative approach is to calculate the exact probability of obtaining a calculated value which is derived from the mean of your sample or single observation from the original population. This is the approach taken by the statistical software you will encounter and the output is therefore a calculated value for a test statistic and an exact *p* value for that calculated value. We give examples of these actual *p* values in the output for each statistical test included in the book in the Online Resource Centre. Having selected the *p* value that is to determine your significance level, you then decide if the exact *p* value is more or less than the significance level. For example if you choose to use a *p* value of $p = 0.05$ to determine your region of rejection and the software tells you that the probability of getting the calculated value is $p = 0.3$ then you do not reject the null hypothesis. Clearly your calculated value is in the region of 'acceptance'. However if your exact *p* value is 0.004 then this is less than your significance level and you can reject the null hypothesis.

6.1.7 Overview of hypothesis testing

There are therefore six steps that you follow in all hypothesis testing.

1) Confirm which distribution is the best one to use. To confirm which distribution and therefore which statistical test is the correct one to use depends on the characteristics of your experiment and observations.

2) Select and correctly phrase the hypotheses to be tested and determine which is/are the correct statistical test(s) to use.

3) Work out the calculated value of the test statistic.

4) Find the critical value of the test statistic.

5) Reject or do not reject the null hypothesis depending on the rule relating to the critical and calculated test statistics.

6) Decide what the outcome of the statistical test means in terms of the biology of the system.

We explain in general how you choose the correct distribution and select the correct statistical test in Chapter 7. Further details are provided for each specific statistical test in Chapters 8–11. Steps 3–6 are those that relate to the statistical analysis itself and these points are illustrated in the 'Boxes' for each test.

Although we have explained the principles behind hypothesis testing in this section there are a number of other points that relate to this process that you also need to be aware of. These we cover in the remainder of the chapter and link these to the six steps in hypothesis testing outlined above.

6.2 How to choose a sample size

Ideally we would measure every item in a statistical population in every investigation. In practice we use samples and we therefore are left with the question as to what size our sample should be. There are many factors that can enable you to come to a reasonable conclusion about the sample size, some of which we have referred to previously. A number of these approaches involve what are called 'power' calculations. There are many power calculations, each relating to the specific statistical test you intend to use. We explain where power calculations are possible and the information you need to be able to carry one out.

6.2.1 A representative sample

The sample needs to be 'representative' and therefore we can reasonably expect it to provide a good estimate of population parameters such as the population mean and standard deviation (1.4.1). If you are expecting to collect parametric data and have numerical information about your statistical population, then it is possible to estimate how large a sample you need to make it likely that the sample will provide a good estimate of your statistical population parameters. These 'power' calculations are based on the idea of confidence limits (5.7), which indicate the range within which a population parameter such as the mean will fall, given a certain level of probability.

6.2.2 Meeting the criteria of statistical tests

i. All tests have different criteria

When choosing a statistical test you are selecting a particular distribution and a mathematical relationship that can be used to identify regions of acceptance and rejection for your test of hypotheses. Specific distributions and relationships have criteria that need to be met by your data for the test of hypothesis to then work correctly. These criteria can include minimum sample sizes. Where this is the case we have listed these in the specific criteria for a test. We also discuss the limitations of sample sizes in relation to chi-squared tests in particular in 9.4.

ii. Parametric and non-parametric tests

Parametric tests are more 'powerful' than non-parametric tests (6.1.6) Therefore if you are planning to collect and analyse non-parametric data you should increase your sample size to improve the likelihood of detecting a real difference.

6.2.3 Maximum sample sizes

In some tests, such as the Mann–Whitney U test (11.1), either there is a ceiling beyond which the test is not reliable or more observations will not add to the capacity of a test to identify a difference. Where maxima like this exist we have included this in the criteria for using the test.

6.2.4 **Power calculations**

There are numerical methods that can also be used to estimate a suitable minimum sample size. These calculations are sometimes referred to as power calculations as they are derived from the relationship between the power of a test (6.1.6) and the sample size. Generally these calculations require you to have an estimate of the population parameters such as the mean (μ) and standard deviation (σ), a rough idea of a suitable sample size that you think you may use, the minimum difference you wish to detect, the probability at which you wish to detect this (p), and the probability of avoiding a type II error ($1 - \beta$). All these factors therefore affect the sample size needed to detect a real biological difference.

Some of the numerical values you need such as a population standard deviation or mean can be obtained either by you carrying out a large pilot study and using these values in your power calculation or by using the data from published studies which have examined the same statistical population using a similar method. If you have this information then this approach is the most direct way of determining the sample size to use in an investigation. The formulae for the specific power calculations can be found in texts such as Zar (2009) or by using at least two reliable internet sources.

i. Minimum difference you wish to detect

It is possible to obtain an estimate of this difference from the value found in other studies. For example you may wish to compare two samples and the means of two similar samples may already be known. From this a difference between these means can be calculated and used as a guide for choosing a minimum difference. In general the smaller the difference you wish to detect as a real biological difference the larger the sample size required to detect this difference. However, clearly here a balance must be struck. There may indeed be a small biological difference between two samples and your sample size may be large enough so that the statistical test detects this difference, but is this difference meaningful in the context of the system you are investigating? In medicine a small difference may be important as you may detect a treatment that improves the quality of life even if just for a few individuals. However, this level of discrimination has to be balanced against the other factors we outline in this section.

A further point to consider when choosing a minimum difference for these calculations is that you also need to consider the accuracy of your measurements. There is no point in using a very large sample to detect a tiny difference if you know that you cannot be that accurate when recording your observations. For example there is no point in testing for a difference of 0.5mm if you can only measure to an accuracy of 1mm.

ii. Variation in your data

An indication of the variation in your data comes from a standard deviation and an estimate can be derived either from a pilot study or from the literature. If you are planning on comparing two or more samples, some power calculations differ depending on whether the variation in these samples is similar or dissimilar and you therefore need to ensure you are using the correct power calculation. In general if the variation within the samples is large then to detect a real difference will require a larger sample size.

Table 6.4 Approximate sample sizes required to select a critical t value at a given probability when the sample sizes are equal and the degrees of freedom are $v = (n_1 - 1) + (n_2 - 1)$

	Indicative minimum sample size (n)			
p	Critical t values from Table D6			
	12.706	4.303	3.182	2.776
0.05	3	4	5	6
0.01	4	7	12	29
0.001	5	15	>100	>100

iii. *p*

In Table 6.3 it is clear that in any design we wish to minimize the likelihood of a Type I error occurring and increase the likelihood of a biological difference being confirmed as such by the statistical test. A small p value decreases the likelihood of a Type I error occurring. However, there is a link between the sample size (N) and p. We can illustrate this inter-relationship between p and N using a Student's t-test and the t critical values from Appendix D, Table D6. t critical values are found using the probability value you have chosen to determine your region of rejection and the degrees of freedom, which for this test are calculated as $v = (n_1 - 1) + (n_2 - 1)$. The degrees of freedom therefore depend on sample size. If we assume that the two sample sizes are equal we can reorder the information from Table D6 in terms of sample size (Table 6.4). If to avoid a Type I error you use a small p value, this can only be achieved by increasing the sample size. For example if you were using the Student's t-test and you wish to reject the null hypothesis at a critical t value of 4.3, your sample size has to increase from 4 at $p = 0.05$ to 15 at $p = 0.001$ to expect to detect a difference at the same critical value.

iv. β and the power of the test

In power calculations an estimate of β is required. Although it is difficult to gain any real estimates of this, as we rarely know anything about our alternate hypothesis, we can select values to aspire to. β is the probability of a Type II error occurring and we would prefer to keep this small. However, as β decreases the likelihood of a Type I error occurring increases so most investigators use a value of around $\beta = 0.1$ to 0.2. In the same way as for the p value (α), a relationship exists between β, critical values, and sample size, so that if β is small the critical values will be large and large sample sizes are required to detect a real difference.

6.2.5 **Ethics**

We discuss the principles behind ethical working practices in Chapter 4. There are two issues to consider in relation to sample size. Firstly, if your study has too small a sample you are not likely to detect a difference and this is unethical in that the commitment of those taking part is to no purpose. You should not undertake a study unless you know that all aspects are adequate for a successful experiment. Similarly, if your sample is larger than needed this involvement of more participants than is necessary is also unethical.

6.2.6 Legal and practical constraints

A further group of factors you must consider are the legal and practical issues. You may be working under the terms of a specific license or working on a protected species. This can impose explicit or implicit limitations to your study which can affect your sample size. Similarly, practical constraints can arise from the budget you are working within, access to equipment, and the time it takes to carry out parts of your investigation. Where sample sizes are limited for these reasons you need to consider the impact this may have on the ability of your analysis to come to a correct conclusion.

6.3 Select and correctly phrase the hypotheses to be tested

In 6.1 we explained the principles behind hypothesis testing and showed that your null hypothesis is generally stated as there is no difference between the means ($\mu = \mu_s$). You then decide if you are carrying out a one-tailed test ($\mu \leq \mu_s$ or $\mu \geq \mu_s$) or a two-tailed test ($\mu \neq \mu_s$). Having made this choice you need to consider the type of hypothesis you are testing, which is determined by the nature of the comparisons you are making and the data you have.

6.3.1 Types of hypothesis

There are three types of hypothesis you may use:

- Do the data fit expected values?
- Is there an association between two or more variables?
- Do samples come from the same or different populations?

i. Do the data fit expected values?

There are some occasions when we have reasons to make a mathematical prediction about our observations and we wish to compare our observations with our expectations. This can arise when:

- You can reasonably argue that all samples or observations should have the same value; for example, if you sampled the numbers of beetles falling into different-coloured pitfall traps, you might expect that there should be the same number of beetles collected in each trap.
- You have carried out a genetic cross or sampled from a population and have reason to expect a particular segregation ratio, population genetic ratio, or sex ratio; for example, you may use the Hardy–Weinberg theorem to predict the allele frequencies in the F1

generation of a population of *Drosophila melanogaster* exposed to a known selection pressure.

● You wish to confirm that your data have a particular distribution, such as a normal or a Poisson distribution (5.2).

Expected values
Values calculated using a mathematical or biological model against which the observations can be compared.

The term expectation can be used in two ways, only one of which is correct. You may have a personal hunch ('expectation') about the likely outcome, but this is not a mathematical prediction, also called an 'expectation'. When we talk about expectations or expected values, it is the second of these that we mean. Chapters 8 and 9 introduce you to chi-squared tests, which can be used to test this type of hypothesis.

ii. Is there an association between two or more variables?

Association
Where one variable changes in a similar manner to another variable. The reasons for this are not necessarily known.

Two types of analysis may be carried out in relation to a possible association.

a. The presence or absence of an association

Typical examples might be: Is there an association between blood group and eye colour in people living in north Wales? Is there an association between temperature and the growth rate of bacterial colonies in liquid culture?

When you wish to test for the presence of an association, there are two groups of tests that may be useful: a chi-squared test and a correlation or regression (Chapter 9). There is some overlap between these two groups of tests but Chapters 7 and 8 will help you decide which test is the most appropriate.

b. The nature of the association

Some statistical analyses also allow an association to be described in a mathematical way: for example, in fitting a straight line to data and confirming the linear relationship between flower number and bulbil number in *Allium ampeloprasum* var. *porrum* (leek).

If you wish to examine the nature of an association in this way, then go to Chapters 7 and 9 and look at regression analyses.

iii. Do samples come from the same or a different population?

Usually, you will not collect data from all items in the statistical population. Instead, the population will be represented by a sample. Many experiments are designed so that you can compare two or more samples. In some cases, these will be samples from different ecological or genetic populations. In other cases, you may wish to compare different treatments. For example, you may wish to examine the effectiveness of different bactericide treatments on *Escherichia coli*, or you may wish to compare phosphate concentrations in water samples collected above and below an outfall.

There are many tests that address this type of hypothesis. Your choice depends on the numbers of observations in each category or sample, the number of categories or samples, whether the data are parametric or not, and the simplicity or complexity of the experimental design. Chapter 7 includes a key that can be used to direct you to the correct test for this type of hypothesis; these tests are covered in Chapters 10 and 11.

 If you designed an experiment to find out whether there is a difference in mineral content between organic and non-organic potatoes, which type of hypothesis would you be testing?

6.3.2 General and specific hypotheses

Hypotheses may be either general or specific. For example, a general hypothesis might be 'There is no difference between the means of samples A, B, and C'. Clearly any outcome from the hypothesis testing may tell you if there is probably a difference, but will not tell you anything more about that difference. Some hypothesis testing is more specific. For example, when using a chi-squared goodness-of-fit test, you compare your observations to a ratio determined by biological principles, such as Mendelian laws of segregation or the Gaussian equation describing a normal distribution (8.4). The test of hypotheses here is more specific, asking whether your observed values are significantly different from this *a priori* ratio. In Chapters 10 and 11, there are statistical tests that allow you to make other specific comparisons. Tukey's test (10.8) allows you to compare pairs of means, in experiments with more than two samples, enabling you to gain a more detailed insight into the trends in your data. A similar test is considered in Chapter 11 for non-parametric data.

General hypothesis
This is phrased in a general way; for example, there is no difference between the samples.

Specific hypothesis
Some statistical tests allow you to ask more discerning (specific) questions. Specific tests of hypotheses are usually one-tailed tests.

a priori
Before the event. Usually refers to a theoretical expectation of the outcome of an experiment, which can therefore be stated before any data are gathered.

6.3.3 How to write hypotheses

A hypothesis is the formal phrasing of your objective (1.1 and 1.8). You may have general or specific hypotheses. A hypothesis when written correctly should have a number of features, as the following checklist shows.

1) Hypotheses come in pairs, the null hypothesis (H_0) and an alternate hypothesis (H_1). The null hypothesis (H_0) is the negative hypothesis. The null hypothesis assumes that any variation in the data is due to sampling error. We explain what is meant by sampling error in 1.7.4. The alternate hypothesis (H_1) is worded in such a way that if H_0 is true, then H_1 must be false. The alternate hypothesis indicates that the variation in the data is probably due to the factor(s) you are investigating in your experiment (6.1.2).

2) For a general hypothesis, the phrasing indicates the type of hypothesis being tested, i.e. 'expectation', 'association', or 'difference', and/or uses a term relating to the test itself, e.g. 'heterogeneity', or 'correlation'. A specific hypothesis includes details of the particular trend you believe is present in the data.

3) The hypothesis being tested relates to the statistical population under investigation rather than the sample (1.3, 1.4).

4) Hypotheses usually indicate the statistic being examined. For example, most parametric statistical tests of hypotheses are written in terms of the means of the data. When non-parametric tests of hypotheses are used, these refer to the medians.

5) Hypotheses include details of the experiment, such as the number of samples or items, the variable(s), and the species being studied.

Example 6.2 Germination of *Allium schoenoprasum* on three soil types

A researcher was interested in the effect of soil types on seed germination in the chive (*Allium schoenoprasum*). She sowed 100 seeds on each of three soil types and after two weeks recorded the number of seeds that had germinated. She expected that the number of seeds germinating on the three soils would be the same. In this instance, the hypotheses would be written as:

H_0: There is no difference in the median numbers of seeds from *Allium schoenoprasum* germinating on the three soil types compared with the expected value.

H_1: There is a difference in the median numbers of seeds of *Allium schoenoprasum* germinating on the three soil types compared with the expected value.

We can compare these hypotheses with our checklist.

CHECKLIST	EXAMPLE 6.2
1. Hypotheses come in pairs: the null hypothesis (H_0) and an alternate hypothesis (H_1).	Yes, there is a null and alternative hypothesis. The null hypothesis (H_0) is the negative hypothesis: 'There is no difference . . . '. The alternate hypothesis (H_1) is worded in such a way that, if H_0 is true, then H_1 must be false: 'There is a difference . . . '.
2. The phrasing of a hypothesis indicates the type of hypothesis being tested, i.e. 'expectation', 'association', or 'difference', and/or uses a term relating to the test itself, e.g. heterogeneity, correlation.	H_0: ' . . . compared with the *expected value.*'
3. The hypothesis being tested refers to the statistical population under investigation rather than the sample.	In this example, the statistical population is the seed produced by *Allium schoenoprasum.*
4. Hypotheses usually indicate the population parameter being examined.	' . . . difference in the *median* numbers ... '
5. Hypotheses include details of the experiment.	' . . . seeds of *Allium schoenoprasum* germinating on the three soil types ... '

EXAMPLE 8.1 The distribution of holly leaf miners on *Ilex aquifolium* (from Chapter 8)

As part of a student project, 200 leaves were sampled at each of three heights on a holly tree and the number of holly leaf miners was recorded. The student assumed that the holly leaf miners were distributed at random and so should be found in equal numbers in each part of the tree.

Write suitable hypotheses for this investigation.

6.4 Using tables of critical values

If you are carrying out statistical tests by hand, the critical values have been calculated for you from the distribution you are using to test your hypotheses. Statistical tables with these critical values are included at the end of this book (Appendix D). There is a plethora of critical values derived from any particular test distribution. You have to choose one of these critical values and to do this you will usually need a probability value (p) and a measure of sample size, which is most often in terms of the degrees of freedom, and you will need to decide whether you wish to use a one- or two-tailed test. We explain these three criteria in this section.

6.4.1 *p* values

If calculating statistics by hand you will need to find a critical value for your test statistic corresponding to a specific p value. Usually this is $p = 0.05$ in the first instance (6.1.6). Look at the critical values in the statistical table in Appendix D for the Spearman's rank test and at the excerpt in Table 6.5.

If, when testing your hypotheses, you find that you do reject the null hypothesis at $p = 0.05$, you should see how small a p value you can adopt and still reject the null hypothesis. For example, if you can reject the null hypothesis at $p = 0.001$, then you can expect your decision

Table 6.5 Excerpt from the critical values for a Spearman's rank test

| | Critical values for a two-tailed* test for probability (p) values 0.10–0.01 | | | |
	0.10	0.05	0.02	0.01
$n = 8$	0.643	0.738	0.833	0.881
$n = 9$	0.600	0.683	0.783	0.833
$n = 10$ (etc.)	0.564	0.648	0.745	0.794

*The terms one-tailed and two-tailed are explained in 6.1.2.

probably to be incorrect only 1/1000 times. At this level of stringency, you could carry out 1000 experiments and expect probably to draw the wrong conclusion in only one of them. Therefore, if you are able to make a decision in hypothesis testing at a low p value this indicates that the likelihood that the decision is correct is high (e.g. see Boxes 8.1 and 9.1). Where you do use a p value of less than 0.05, you may wish to indicate this either by giving the exact p value, such as $p = 0.01$, or by using the 'less than' symbol, i.e. $p < 0.05$.

 Q4 Using the Spearman's rank critical values table at $p = 0.05$, what are the critical values (r_s) at $n = 10$ for a two-tailed and a one-tailed test? Which value is the greater?

6.4.2 Degrees of freedom

Turn to the F_{max} table (D7) in Appendix D. There are three F tables, so you need to select the correct one. In the F_{max} table, the critical value is found using a (the number of samples or treatments) and v (lower case Greek letter 'nu', the degrees of freedom). What are these 'degrees of freedom'?

Example 10.1 has a frequently encountered experimental design where there are two categories (periwinkles from the lower shore and periwinkles from the mid-shore), and at each location the height of the shells from a sample of the periwinkles have been measured. We illustrate this design in Table 6.6 where shell height has been recorded from three snails in each location ($n = 3$).

Table 6.6 The impact of experimental design on the degrees of freedom based on Example 10.1: The evolution of *Littorina littoralis* (periwinkles) at Aberystwyth, 2002

Observations	Height (mm) of periwinkles from lower shore	Height (mm) of periwinkles from mid-shore
1	5.3	8.8
2	8.7	9.7
3		
Total	20.0	25.0

Table 6.7 The impact of experimental design on the degrees of freedom based on Example 9.2: Frequency of *Cepaea nemoralis* and *Cepaea hortensis* in a woodland and a hedgerow

| | Species of snail | | |
	Cepaea nemoralis	*Cepaea hortensis*	Total
Hedgerow			148
Woodland			
Total	97		160

If you were to fill in Table 6.6 for category 1 (the lower shore), then you could complete the first two rows in this table for observations 1 and 2, but once these are established, the third observation can only be one value. In our example the final shell height would be 6.0mm. This final value is already fixed and there is no 'freedom' for this final value. The same would be true for the observations for category 2 (the mid-shore) where the final value must be 6.5mm. It is this constraint within the experimental design that is recognized by the degrees of freedom. For this design, the degrees of freedom for the periwinkles from the lower shore are $v_1 = n_1 - 1 = 3 - 1 = 2$, and the same is true for the periwinkles from the mid-shore.

Examine Table 6.7. This represents one experimental design where there are two variables and for each variable there are two categories. This design can be seen in Example 9.2 where the numbers of two snail species were recorded in a hedgerow and a woodland. You can see that, once the grand total is fixed and you have put in one row total, the remaining row total can only be one number. (The missing woodland row total, 12 snails.) The same is true of the columns. The table could be extended to three rows and two columns and again one row total is fixed once the grand total and two row totals are determined. For this experimental design, this constraint is recognized by the degrees of freedom which are calculated as (number of rows − 1) × (number of columns − 1).

Our final illustration of degrees of freedom is from 8.4. Here an additional constraint is recognized in that a mathematical formula has been used to determine values that were then used in the statistical analysis. In this experimental design, there is one variable with several categories (Table 6.8). Usually for this experimental design, the degrees of freedom would normally be the number of categories (a) − 1. However, because a mathematical formula in which the population parameter μ has been estimated using the sample statistic ($\bar{x}$), a further degree of freedom is lost and $v = a - 2$ (see Box 8.2).

Degrees of freedom (v)

A measure that reflects the number of observations in your calculation and the experimental design.

Table 6.8 The impact on the degrees of freedom of the use of a Gaussian equation to determine expected values in a chi-squared goodness-of-fit test

| | Variable | | | | |
	Treatment 1	Treatment 2	Treatment 3	Treatment 4	Total
Observation					106

The degrees of freedom, therefore, are a measure that reflects the number of observations in your calculation and the experimental design. You will be told for each statistical test how to calculate the degrees of freedom. It is not the same for all tests.

6.4.3 More than two criteria

There are a number of critical values that can only be found using three criteria. If you look now at the Mann–Whitney U test table in Appendix D, you will see that rows are labelled n_1 and columns are labelled n_2 (Table D12). n_1 and n_2 are the numbers of observations for sample 1 and sample 2 respectively. The third criterion for choosing the critical value for the U distribution is p. When there are more than two criteria for choosing a critical value, as in this case, only one table is usually included for one p value. We have included the set of critical values where $p = 0.05$. Thus, all the critical values in this table are at $p = 0.05$. If you wish to find the critical values for another p value for this test, you will need to refer to other sources or calculate this using statistical software.

Q5 In a Mann–Whitney U test table of critical values, if $n_1 = 7$ and $n_2 = 10$, what is the critical value of U at $p = 0.05$?

Similarly, look at the F tables, either the F table that is used before a t- or z-test, or the F table that is used in a parametric analysis of variance test. Both these require a p value and two degrees of freedom, so again each table of critical values is for a single p value.

Q6 If the degrees of freedom are $v_1 = 5$ and $v_2 = 10$, what are the critical F values if you were:

a. Carrying out an F-test for homogeneity of variances before a t-test at $p = 0.05$?

b. Carrying out an F-test in a parametric analysis of variance, $p = 0.05$?

Are they the same?

What does this tell you?

6.4.4 Interpolation

Interpolation
The estimation of a value when only nearby values are known, one greater than the required value and the other less than the required value.

The final point we wish to make about finding critical values is relevant to several tables, including the F table for a parametric analysis of variance (Table D8 in Appendix D). A small excerpt is shown in Table 6.9.

To locate a particular $F_{critical}$ value, you use the two degrees of freedom, where the v_1 column intersects the v_2 row. However, it is not possible in this book to give all possible F values for all combinations of degrees of freedom. Clearly, if you needed to find the critical value at $v_1 = 13$, $v_2 = 29$, you will need to estimate it from the values provided. This process is called interpolation. It assumes that each missing degree of freedom increases the critical value in a regular way between the two known values.

Table 6.9 Excerpt from the table of critical values for an *F*-test for a parametric analysis of variance when $p = 0.05$

v_2	v_1		
	8	10	20
28	2.29	2.19	1.94
29	2.28	2.18	1.93
30	2.27	2.16	1.88

For example, to find $F_{critical}$ at $v_1 = 13$, $v_2 = 29$, the nearest $F_{critical}$ values on the *F* table are:

v_2	v_1						
	10	11	12	13	14	15...	20
29	2.18						1.93

You therefore need to fill in the gaps between $v_1 = 10$ and $v_1 = 20$.

First, take the difference between 2.18 and 1.93 = 0.25. The number of increments between 2.18 and 1.93 is ten. Divide the difference between the known $F_{critical}$ numbers by the number of increments, 0.25/10 = 0.025. Keep subtracting this number from the largest known $F_{critical}$ value until you have filled in the gaps:

v_2	v_1						
	10	11	12	13	14	15...	20
29	2.18	2.16	2.13	2.11	2.08	2.06	1.93

Therefore, the critical value for *F* when $v_1 = 13$, $v_2 = 29$ is 2.11.

6.5 What does this mean in real terms?

The final step in hypothesis testing is the most important and yet most often forgotten. This is to refer back to your experiment. A common failing in hypothesis testing is for the first five steps to be completed and for a student to conclude that there is a significant difference. In what? What does this mean? Refer back to your hypotheses to conclude your hypothesis testing. We have included this step in all our examples of hypothesis testing to encourage you to not leave this out.

For example, in the experiment to examine the effect of soil type on seed germination (Example 6.2), a chi-squared goodness-of-fit test was used to analyse the data. The outcome may be to reject the null hypothesis ($\chi^2_{calculated} = 9.00$, $p < 0.05$). But it is not enough to stop there. You should refer back to the alternate hypothesis and add details to this statement about the outcome. For this example, we might conclude that there is a significant difference ($\chi^2_{calculated} = 9.00, p < 0.05$) in the median number of *Allium schoenoprasum* seeds germinating on the three soil types.

Summary of Chapter 6

- A test of hypothesis does or does not reject a null hypothesis. The null hypothesis may be rejected in favour of an alternate hypothesis (6.1.2).

- Hypotheses are written in a specific way reflecting the statistical population, the statistical test being used, and details of the experiment (6.3).

- The decision as to whether to reject or not reject a null hypothesis is taken based on the probability of it being correct and is defined as a p value. In biological sciences this p value is usually $p = 0.05$ (6.1).

- The p value therefore establishes a zone of rejection which can also be described in terms of the critical value for a specific distribution (6.1 and 6.4).

- The specific distributions usually therefore provide the basis for the calculations in hypothesis testing. To select the correct distribution you have to confirm that your data and experimental design are suitable by checking a number of criteria (6.1.5) and ensuring you have the correct sample size (6.2).

- Determining where to set the p value can incur errors (Type I and Type II) and may involve one end or both ends of the distribution (one-tailed or two-tailed tests).

- There are therefore six steps to hypothesis testing (6.1.7). When concluding your hypothesis testing, you must refer back to your hypotheses and your original objective (6.5 and developed in Chapters 7–11).

- The Online Resource Centre includes interactive exercises that test your understanding of this chapter with other topics, particularly those considered in Chapters 7–11.

Answers to chapter questions

A1 a. 9/13

b. 4/13 (4.0, 7.7, 8.4, 8.4)

c. 12, 1 ($13 \times 95/100 = 12.35 \approx 12, 13 \times 5/100 = 0.65 \approx 1$)

d. $p = 0.05$

A2 You do not start out with any expectation nor are you looking for an association, but you do wish to compare two treatments (organic and non-organic). Therefore, you will be testing a hypothesis of type iii.

A3 This is an example that we use later in the book. The hypotheses are given in Box 8.1. Were you correct? If not, what did you miss out?

A4 Two-tailed test $n = 10$, $p = 0.05$, $r_s = 0.648$

One-tailed test $n = 10$, $p = 0.05$, $r_s = 0.564$

The value for the two-tailed test is higher. This reflects the difference between the cut-off points. In a two-tailed test where $p = 0.05$, the cut-off points are 2.5% at each end of the distribution. In a one-tailed test where $p = 0.05$, the cut-off point will be 5% at one end only.

A5 $U = 14$ at $p = 0.05$, $n_1 = 7$, $n_2 = 10$

A6 a. $F_{critical} = 4.24$

b. $F_{critical} = 3.33$

No, they are not the same. This tells you that you don't want to get these two tables muddled up!

7 Which statistical test should I choose?

 In an nutshell

Learning to choose the correct statistical test is one of the hardest skills to master and although statistical software can take over the calculations for you these will not choose the correct test for you. For you to gain this skill requires knowledge and some experience. In this chapter we show you how to choose the correct test, illustrating this with examples and additional explanations of terms to make it easier to understand the process. We also show you how to review the analysis of data in published research so that you can evaluate whether this was done well and that the conclusions that have been drawn have an appropriate basis.

As a scientist there are three occasions when you will find yourself having to decide which is the correct statistical test to use: designing an experiment; analysing data from an experiment designed by someone else; and establishing the merits of research that has been designed and evaluated by someone else. This chapter considers each of these and is divided into three sections, one for each of these occasions.

Designing an experiment: Here the choice of a statistical test should be made as part of the design process, before you have carried out your experiment. This allows you to check that your design will generate data that can be analysed and can help you choose the right numbers of samples and replicates you need. In 7.1 we outline how to choose the correct statistical test for your design. We add further details in 7.2 where we explain the best tests to consider if you have controls or similar design features in your experiment.

Analysing data: As undergraduates the most frequently encountered occasion for needing to choose a statistical test is when you have completed a practical that was designed for you and you are asked to analyse these data. Here you are limited to someone else's design and you already have your data before you choose a statistical test. Your starting point should be to examine your data by visualizing it and carrying out summary evaluations such as calculating means (Chapter 5). We discuss how to graph your data in Chapter 12 and in the Online Resource Centre statistical walkthroughs. Your next step is to identify the key trends seen in your data in relation to your objectives. We illustrate this process in each statistics chapter (e.g. 9.6.1). Finally you can use 7.1 and 7.2 to help you determine the correct statistical test(s) to use.

The critical reader: The final occasion when you will need to consider whether the correct statistical tests have been used is when you are reading published studies and you, as a critical reader, wish to make a judgement about the quality of the work you are reading. We look at the approaches you can take in 7.3.

7.1 Designing an experiment and analysing your data

In this book we cover the most commonly used statistical tests encountered by undergraduates in the biosciences. To choose a test you could just flick through the pages of the book until you find what appears to be the right one, but statistical tests have requirements which must be met by your data to ensure they correctly determine the outcome of the test of hypotheses. If your data and experimental design fail to meet these criteria it can result in a completely incorrect outcome and Type I or Type II errors (6.1.6). Many statistics books therefore direct you to keys similar to the dichotomous keys used in the identification of plants and animals. This allows you both to identify the correct test and to check as you go whether at least some of the criteria are met. Having identified a specific statistical test using the key there may then be additional criteria that need to be met. These are given in the chapter sections indicated in the key and as your final step you should check your data and/or design against these final criteria.

This book is structured around testing three types of hypotheses. A full description of these hypotheses is given in Chapter 6 whilst an abbreviated outline is presented as part of the key and additional information. Some studies do not set out to test hypotheses and instead the data are either summarized (Chapter 5) and/or a process is used that places the data into groups (e.g. TWINSPAN). Descriptions of these hypothesis-free approaches can be found in other texts.

There are a number of fundamental terms and concepts that you need to be familiar with to make use of this chapter. These terms are: 'treatment variable' (2.2.4); 'matched and unmatched' (10.4); 'parametric and non-parametric' (5.8 and 5.9); 'item' (1.2); and 'ordinal, nominal and interval' (5.1). Explanations and examples are provided in 7.1.2. These terms are given in bold when first encountered in the key (7.1.1). If you are an experienced user you may find the 'quick look up' table we have included in Appendix C to be adequate for your needs.

7.1.1 Key to determine the correct statistical test

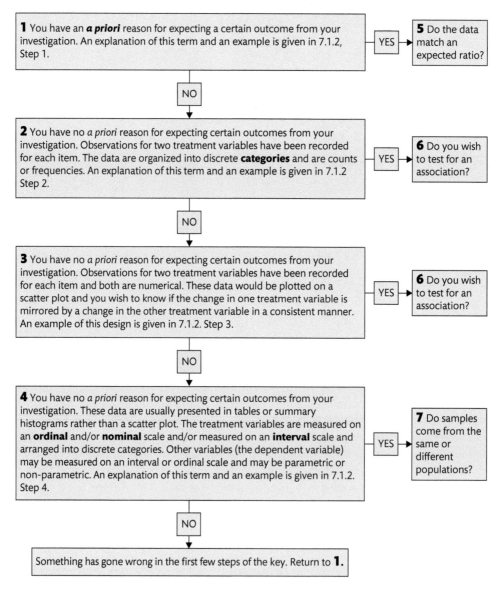

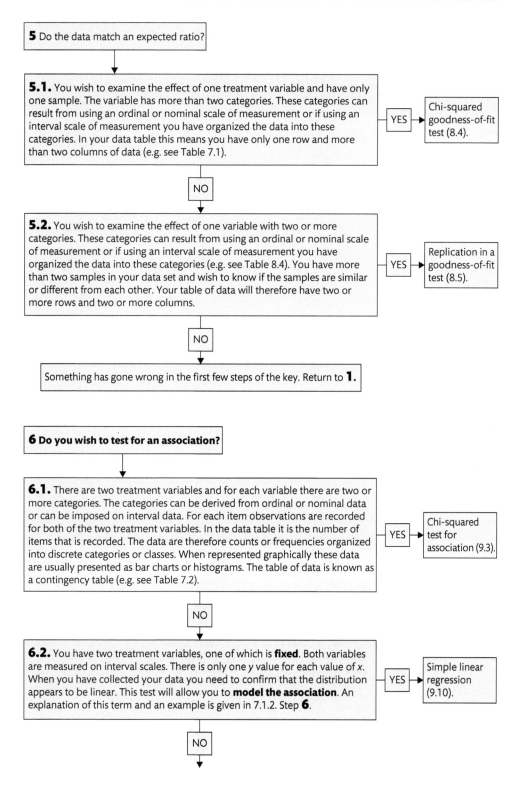

5 Do the data match an expected ratio?

5.1. You wish to examine the effect of one treatment variable and have only one sample. The variable has more than two categories. These categories can result from using an ordinal or nominal scale of measurement or if using an interval scale of measurement you have organized the data into these categories. In your data table this means you have only one row and more than two columns of data (e.g. see Table 7.1).

YES → Chi-squared goodness-of-fit test (8.4).

NO

5.2. You wish to examine the effect of one variable with two or more categories. These categories can result from using an ordinal or nominal scale of measurement or if using an interval scale of measurement you have organized the data into these categories (e.g. see Table 8.4). You have more than two samples in your data set and wish to know if the samples are similar or different from each other. Your table of data will therefore have two or more rows and two or more columns.

YES → Replication in a goodness-of-fit test (8.5).

NO

Something has gone wrong in the first few steps of the key. Return to **1.**

6 Do you wish to test for an association?

6.1. There are two treatment variables and for each variable there are two or more categories. The categories can be derived from ordinal or nominal data or can be imposed on interval data. For each item observations are recorded for both of the two treatment variables. In the data table it is the number of items that is recorded. The data are therefore counts or frequencies organized into discrete categories or classes. When represented graphically these data are usually presented as bar charts or histograms. The table of data is known as a contingency table (e.g. see Table 7.2).

YES → Chi-squared test for association (9.3).

NO

6.2. You have two treatment variables, one of which is **fixed**. Both variables are measured on interval scales. There is only one y value for each value of x. When you have collected your data you need to confirm that the distribution appears to be linear. This test will allow you to **model the association**. An explanation of this term and an example is given in 7.1.2. Step **6**.

YES → Simple linear regression (9.10).

NO

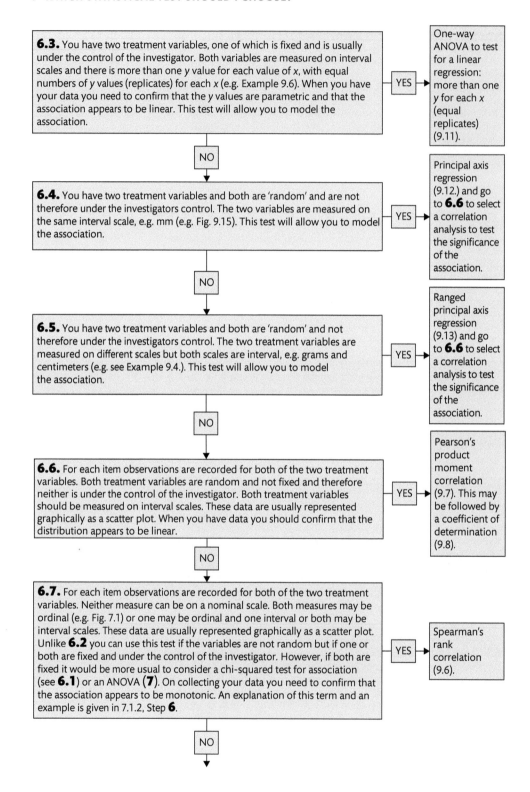

6.3. You have two treatment variables, one of which is fixed and is usually under the control of the investigator. Both variables are measured on interval scales and there is more than one y value for each value of x, with equal numbers of y values (replicates) for each x (e.g. Example 9.6). When you have your data you need to confirm that the y values are parametric and that the association appears to be linear. This test will allow you to model the association.

YES → One-way ANOVA to test for a linear regression: more than one y for each x (equal replicates) (9.11).

NO ↓

6.4. You have two treatment variables and both are 'random' and are not therefore under the investigators control. The two variables are measured on the same interval scale, e.g. mm (e.g. Fig. 9.15). This test will allow you to model the association.

YES → Principal axis regression (9.12.) and go to **6.6** to select a correlation analysis to test the significance of the association.

NO ↓

6.5. You have two treatment variables and both are 'random' and not therefore under the investigators control. The two treatment variables are measured on different scales but both scales are interval, e.g. grams and centimeters (e.g. see Example 9.4.). This test will allow you to model the association.

YES → Ranged principal axis regression (9.13) and go to **6.6** to select a correlation analysis to test the significance of the association.

NO ↓

6.6. For each item observations are recorded for both of the two treatment variables. Both treatment variables are random and not fixed and therefore neither is under the control of the investigator. Both treatment variables should be measured on interval scales. These data are usually represented graphically as a scatter plot. When you have data you should confirm that the distribution appears to be linear.

YES → Pearson's product moment correlation (9.7). This may be followed by a coefficient of determination (9.8).

NO ↓

6.7. For each item observations are recorded for both of the two treatment variables. Neither measure can be on a nominal scale. Both measures may be ordinal (e.g. Fig. 7.1) or one may be ordinal and one interval or both may be interval scales. These data are usually represented graphically as a scatter plot. Unlike **6.2** you can use this test if the variables are not random but if one or both are fixed and under the control of the investigator. However, if both are fixed it would be more usual to consider a chi-squared test for association (see **6.1**) or an ANOVA (**7**). On collecting your data you need to confirm that the association appears to be monotonic. An explanation of this term and an example is given in 7.1.2, Step **6**.

YES → Spearman's rank correlation (9.6).

NO ↓

None of these tests seems to be right for your data. For example, if you have many variables (more than three) and cannot find a test suitable for your design in this book you should consider multiple regressions, ANCOVAs, and the general or generalized linear models. If you have one or two variables then it is probable that something has gone wrong in the first few steps of the key. Return to **1**.

7 Do samples come from the same or different populations?

You have one or more treatment variables. There is a further variable, the dependent variable. The first step in choosing tests in this section is to decide if the observations for the dependent variable are 'parametric' or 'non-parametric'. In making the decision you should first read the two explanations given in the additional details in 7.1.2. Step **7** so that you understand what these terms mean and then follow this next instruction.

In using this key when planning an experiment you can assume at this point that your data are parametric if they are measured on an interval scale. Having determined which is the correct test having made this assumption you should then move to **7.2** and determine a non-parametric test as a fallback. When you have your data you must then check properly whether the criterion for 'parametric data' is met as explained in 'Are my data parametric?, (5.8, 5.9).

7.1. 'Parametric' tests   **7.2.** 'Non-parametric' tests

7.1. Do samples come from the same or different populations? Parametric tests

7.1.1. You wish to examine the effect of one treatment variable with two categories. The data are **unmatched**. The data table therefore has two columns of values. The columns may not necessarily have the same number of observations in each but both have more than one observation. For example see Table 10.1. An explanation of this term is given in 7.1.2, Step 7. — YES → Two sample t or z test for unmatched data (10.1, 10.2, and 10.3).

NO

7.1.2. You wish to examine the effect of one treatment variable with two categories. The data are **matched**. The data table therefore has two columns. There are a number of rows, each row relating to one individual or item with an observation in each column. For example see Table 10.5 — YES → t or z test for matched data (10.4).

NO

7.1.3. You wish to examine the effect of one treatment variable with two categories. You have one sample (i.e. one column of data) and wish to compare this to a single observation. — YES → One sample t test (10.5).

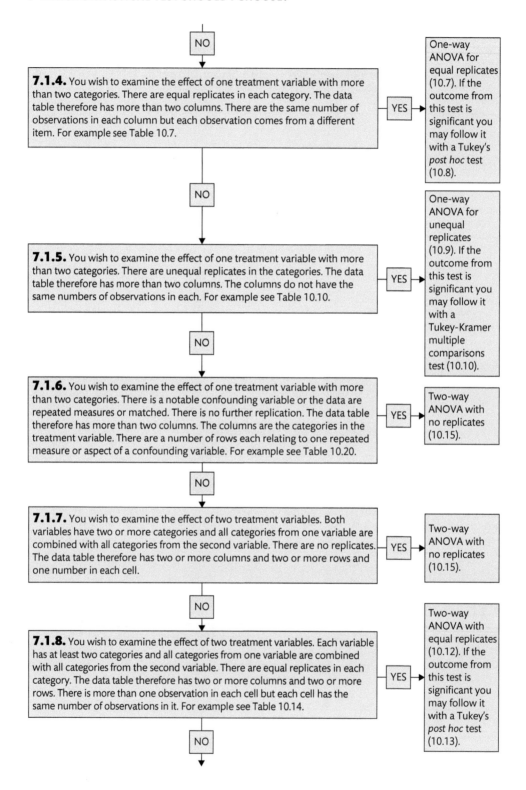

NO

7.1.4. You wish to examine the effect of one treatment variable with more than two categories. There are equal replicates in each category. The data table therefore has more than two columns. There are the same number of observations in each column but each observation comes from a different item. For example see Table 10.7.

YES → One-way ANOVA for equal replicates (10.7). If the outcome from this test is significant you may follow it with a Tukey's *post hoc* test (10.8).

NO

7.1.5. You wish to examine the effect of one treatment variable with more than two categories. There are unequal replicates in the categories. The data table therefore has more than two columns. The columns do not have the same numbers of observations in each. For example see Table 10.10.

YES → One-way ANOVA for unequal replicates (10.9). If the outcome from this test is significant you may follow it with a Tukey-Kramer multiple comparisons test (10.10).

NO

7.1.6. You wish to examine the effect of one treatment variable with more than two categories. There is a notable confounding variable or the data are repeated measures or matched. There is no further replication. The data table therefore has more than two columns. The columns are the categories in the treatment variable. There are a number of rows each relating to one repeated measure or aspect of a confounding variable. For example see Table 10.20.

YES → Two-way ANOVA with no replicates (10.15).

NO

7.1.7. You wish to examine the effect of two treatment variables. Both variables have two or more categories and all categories from one variable are combined with all categories from the second variable. There are no replicates. The data table therefore has two or more columns and two or more rows and one number in each cell.

YES → Two-way ANOVA with no replicates (10.15).

NO

7.1.8. You wish to examine the effect of two treatment variables. Each variable has at least two categories and all categories from one variable are combined with all categories from the second variable. There are equal replicates in each category. The data table therefore has two or more columns and two or more rows. There is more than one observation in each cell but each cell has the same number of observations in it. For example see Table 10.14.

YES → Two-way ANOVA with equal replicates (10.12). If the outcome from this test is significant you may follow it with a Tukey's *post hoc* test (10.13).

NO

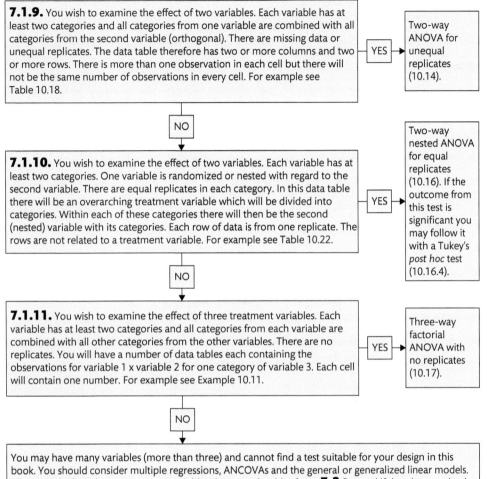

7.1.9. You wish to examine the effect of two variables. Each variable has at least two categories and all categories from one variable are combined with all categories from the second variable (orthogonal). There are missing data or unequal replicates. The data table therefore has two or more columns and two or more rows. There is more than one observation in each cell but there will not be the same number of observations in every cell. For example see Table 10.18.

YES → Two-way ANOVA for unequal replicates (10.14).

NO

7.1.10. You wish to examine the effect of two variables. Each variable has at least two categories. One variable is randomized or nested with regard to the second variable. There are equal replicates in each category. In this data table there will be an overarching treatment variable which will be divided into categories. Within each of these categories there will then be the second (nested) variable with its categories. Each row of data is from one replicate. The rows are not related to a treatment variable. For example see Table 10.22.

YES → Two-way nested ANOVA for equal replicates (10.16). If the outcome from this test is significant you may follow it with a Tukey's *post hoc* test (10.16.4).

NO

7.1.11. You wish to examine the effect of three treatment variables. Each variable has at least two categories and all categories from each variable are combined with all other categories from the other variables. There are no replicates. You will have a number of data tables each containing the observations for variable 1 x variable 2 for one category of variable 3. Each cell will contain one number. For example see Example 10.11.

YES → Three-way factorial ANOVA with no replicates (10.17).

NO

You may have many variables (more than three) and cannot find a test suitable for your design in this book. You should consider multiple regressions, ANCOVAs and the general or generalized linear models. Alternatively if you have one or two variables then you should refer to **7.2** first and if that does not lead you to the identification of a suitable test then something has gone wrong in the first few steps of the key. Return to **1**.

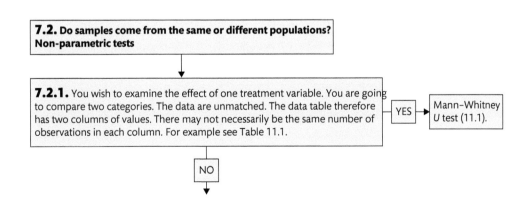

7.2. Do samples come from the same or different populations? Non-parametric tests

7.2.1. You wish to examine the effect of one treatment variable. You are going to compare two categories. The data are unmatched. The data table therefore has two columns of values. There may not necessarily be the same number of observations in each column. For example see Table 11.1.

YES → Mann–Whitney *U* test (11.1).

NO

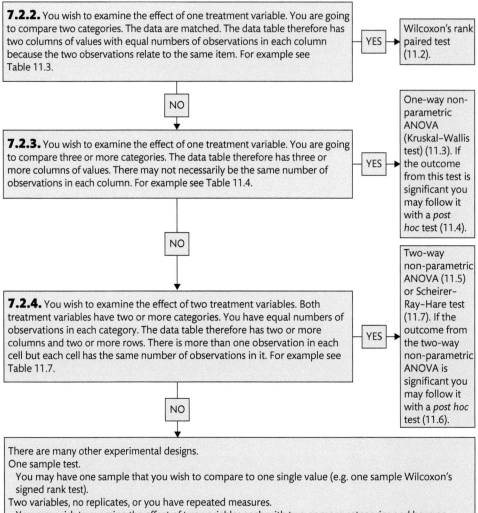

7.2.2. You wish to examine the effect of one treatment variable. You are going to compare two categories. The data are matched. The data table therefore has two columns of values with equal numbers of observations in each column because the two observations relate to the same item. For example see Table 11.3.

YES → Wilcoxon's rank paired test (11.2).

NO

7.2.3. You wish to examine the effect of one treatment variable. You are going to compare three or more categories. The data table therefore has three or more columns of values. There may not necessarily be the same number of observations in each column. For example see Table 11.4.

YES → One-way non-parametric ANOVA (Kruskal–Wallis test) (11.3). If the outcome from this test is significant you may follow it with a *post hoc* test (11.4).

NO

7.2.4. You wish to examine the effect of two treatment variables. Both treatment variables have two or more categories. You have equal numbers of observations in each category. The data table therefore has two or more columns and two or more rows. There is more than one observation in each cell but each cell has the same number of observations in it. For example see Table 11.7.

YES → Two-way non-parametric ANOVA (11.5) or Scheirer–Ray–Hare test (11.7). If the outcome from the two-way non-parametric ANOVA is significant you may follow it with a *post hoc* test (11.6).

NO

There are many other experimental designs.
One sample test.
  You may have one sample that you wish to compare to one single value (e.g. one sample Wilcoxon's signed rank test).
Two variables, no replicates, or you have repeated measures.
  You may wish to examine the effect of two variables each with two or more categories and have an orthogonal design. There is only one observation in each category and these may be repeated measures (e.g. Friedmans test).
Three or more treatment variables
  You may have many variables (more than three) and cannot find a test suitable for your design in this book. You should consider transforming your data and/or using multiple regressions, ANCOVAs, or the general or generalized linear models.
Alternatively, if you have one or two variables then something may have gone wrong in the first few steps of the key. Return to **1**.

7.1.2 **Supporting explanations and examples**

Step **1**

What does a priori *mean?*

This means you have a biological or mathematical reason for testing your data against a specific biological model and this was your intention from the outset. Common models are those derived from mathematical descriptions of distributions such as the Gaussian equation, or genetic or ecological models such as those derived from Mendelian ratios or the Hardy–Weinberg theorem. Or you may wish to test a null hypothesis that your treatment has no effect on your measurements, which would mean that your measurements should all be the same.

Example 7.1 The effect of pitfall trap size and colour on sampling large *Diptera* in a grassland at the University of Worcester, 2009

A number of pitfall traps were placed in the university grounds for a 3-hour period. The pitfall traps differed by colour. The number of large *Diptera* caught in the traps was recorded (Table 7.1). If there was no effect of the pitfall trap colour then it is to be expected that the *Diptera* would be found in equal numbers in all traps.

Table 7.1 The number of large *Diptera* caught in pitfall traps of different colours (blue, yellow, white, green, red, and orange) in grassland at the University of Worcester, 2009.

There is one treatment variable: 'Pitfall traps', with six categories. In this example the categories are based on a nominal scale and none of these are a control.

	Pitfall traps						
	Blue	Yellow	White	Green	Red	Orange	Total
Observed number of Large *Diptera*	38	260	1	4	39	29	N = 371
Expected number if distribution of *Diptera spp.* is random	N/no. categories = 371/6 = 61.83	61.83	61.83	61.83	61.83	61.83	

If the colour of the pitfall traps was not biologically important then you would expect the numbers of *Diptera* to be the same in each, i.e. a ratio of 1:1:1:1:1:1.

There are no replicates in this design.

Step 2

What does 'categories' mean?

When each observation may fall into only one of two or more mutually exclusive groups (e.g. blood groups A, B, AB, or O), these groups are known as categories.

Example 7.2 Association between polymorphic variants of the tryptophan hydroxylase 2 gene and obsessive compulsive disorder

Serotonin is an important neurotransmitter involved in regulating a number of psychological activities, such as emotion and cognition. Research indicates that polymorphism in genes involved in the serotoninergic system may be associated with a number of behavioural traits. An undergraduate therefore investigated whether small mutations (SNPs) in the tryptophan hydroxylase 2 gene (TH2) were associated with obsessive compulsive disorder. To gather evidence a self-reporting questionnaire (Foa *et al.*, 2002) was used to determine whether a participant had obsessive compulsive type behaviours. DNA was also collected from each participant and tested for the presence of the polymorphisms (Table 7.2). Alleles for this SNP are T and C. Homozygous genotypes are TT and CC, the heterozygous genotype is TC.

Table 7.2 Numbers of participants with or without obsessive compulsive behaviours (OCD) with specific genotypes of the rs4565946 SNP in the TH2 gene.

This is the second treatment variable: 'Participants with or without OCD'. There are two categories. In this example the categories are based on a nominal scale of measurement and neither of these are controls.

This is the first treatment variable: 'Genotypes of the SNP rs4565946', with three categories: TT, TC, CC. In this example the categories are based on a nominal scale of measurement and none of these are controls.

Number of participants	Genotypes of rs4565946 SNP in TH2 gene		
	TT	TC	CC
With OCD behaviours	4	15	0
Without OCD behaviours	6	18	2

You have no *a priori* reason for expecting certain outcomes from this investigation.

Each participant is the 'item'. Two variables have been recorded for each item: the genotype and the OCD assessment.

The data are counts: the number of people with a specific genotype with or without signs of OCD. There is no replication.

Step **3**

> **Example** **7.3**   Investigating whether the bioaccumulation of lead could be
> a conservation risk to the UK's European Hedgehog (*Erinaceus europaeus*)
> population
>
> The number of European Hedgehogs (*Erinaceus europaeus*) in the UK is declining. The bioaccumulation
> of heavy metals such as lead in food items such as slugs and subsequent ingestion by hedgehogs is a
> possible cause. One undergraduate study investigated the concentration of lead in slugs at eight sites
> and from this estimated the numbers of slugs that would need to be ingested for female hedgehogs to
> receive a lethal dose of lead. These location-specific estimates were plotted in relation to the numbers
> of cars passing each site per annum (Fig. 7.1).
>
>

Step **4**

What is meant by the terms nominal, ordinal, and interval?

The terms 'nominal', 'ordinal', and 'interval' describe scales of measurements (5.1). A nominal
scale is made up of descriptive terms such as 'red', 'blue', and 'green' and there is no rationale
that determines which order these should be placed in. Ordinal values are any discontinuous
values for which there is a reason for placing them in a specific order, such as the numbers
1, 2, 3, 4, and 5. However, the interval between each point on an ordinal scale may not be
consistent or mathematically meaningful. An interval scale is a continuous scale such as time
(mins) or mass (g). Any point along the scale is mathematically meaningful.

Step **6**

What does it mean by 'fixed'?

When you carry out an experiment sometimes you will set up specific treatments such as
temperatures at 0°C, 10°C, and 20°C or specific points along a transect (0m, 10m, and 20m).

Example 7.4 Space utilization within a mixed-species captive enclosure of Poison Dart Frog (*Dendrobatidae*).

A number of species of *Dendrobatidae* are housed together in an English zoo. To improve the species habitat and the visitor experience scan sampling was used to record zone utilization. As part of this study the time (mins) spent in each of six zones within the enclosure by four individual *Dendrobates leucomelas* was recorded (Table 7.3).

This is the second treatment variable: 'estimate of the number of slugs that would need to be eaten for a lethal dose of lead'. In this example y is an ordinal scale. None of these is a control. There are no replicates. These values have not been predetermined by the investigator as part of the design of the experiment; they are not 'fixed, they are "random".'

Two observations have been recorded for each item. The item is the 'location' and the two observations are 'the number of slugs' and 'the number of cars'.

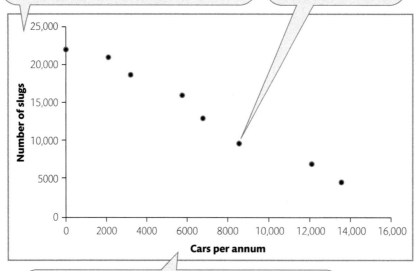

This is the first treatment variable: the numbers of cars at each location per annum. In this example x is an ordinal scale. These values have not been determined by the investigator; they are 'random'.

Fig. 7.1 An association between the number of cars per annum in an area and the number of slugs from the area that if ingested would be expected to give a lethal dose of lead to female hedgehogs (*Erinaceus europaeus*).

Table 7.3 The total time (mins) spent by individuals of *Dendrobates leucomelas* in a number of zones within one enclosure. (Zone 1: the vegetation to the left of the enclosure; Zone 2: the vegetation to the right of the enclosure; Zone 3: a water feature; Zone 4: the food dish; Zone 5: the vegetation to the back of the enclosure; Zone 6: the enclosure floor.)

> The individual frogs can be considered to be replicates but the data in all columns of one row are based on one frog so these are matched or repeated measures. The frogs are a nominal measure and there are four categories (frog A to frog D). None of these is a control.

> The treatment variable is: 'zones in the enclosure'; a nominal scale with six categories.

Individual	Total time (mins) spent in zones 1–6					
	Zones					
	1	2	3	4	5	6
A	314	113	55	42	226	755
B	349	85	72	38	192	733
C	270	107	68	36	217	807
D	384	74	77	30	160	780

> The time spent in these zones is a variable but you are not looking at the effect of this factor and it is sometimes referred to as a dependent variable. In this example this variable (time, mins) is measured on an interval scale.

These categories are going to be the same each time you use them so are not subject to sampling error and are 'fixed'. In other experiments this is not the case. For example, you may collect data on the height and weight of a number of individuals. You have not predetermined either height or weight categories, so these are 'random' and will be subject to sampling error.

What does it mean to 'model' an association?

If you select a test to model your association this means that you wish to draw a line through your data that has statistical meaning and is not just a trend line. Having drawn such a line it is then possible to predict a *y* value from any *x* value within the range covered by your investigation. These *x* values can be ones you have not actually investigated so you can predict the outcome for additional values using this statistically meaningful line.

Monotonic

'Monotonic' describes an association between two variables such that when you plot a graph of one of the variables against the other, the graph will either go up from left to right (monotonic increasing) or go down from left to right (monotonic decreasing). The association does not have to be a straight line, or even follow a mathematical equation, but it must either always increase or always decrease. An example of a monotonic distribution can be seen in Fig. 9.5d.

Step 7

Are my data parametric?

This is a frequently used way of referring to an underlying requirement for the observations making up your statistical population to have a normal distribution. However, as biologists we rarely measure every item in a population and therefore use the sample to indicate that this criterion is met. If your sample has the characteristics of a normal distribution then this is taken to indicate that the underlying population will also have these characteristics. The characteristics of a normal distribution are covered in 5.2 and 5.8. In addition the normality of data can be tested using statistical software and we cover this in the statistical walkthroughs in the Online Resource Centre. The term 'parametric' is used as the statistics tests are based on parameters derived from a specific (in this case normal) distribution.

What are non-parametric data?

Non-parametric data are therefore data which do not have a specific distribution or are data where the distribution is not known. In general for the tests described in this book non-parametric data should be transformed where possible to a normal distribution (5.10) to allow you to use the more powerful parametric statistics. Non-parametric tests may be used on parametric data but not the reverse.

What is meant by 'unmatched data'?

Unmatched data are when only one measurement for a specific treatment variable has been taken from one item. The data set is therefore made up of single values from many items. In contrast in matched data more than one observation is taken for the same treatment variable from the same item. For example if five mice were weighed once each these would be unmatched data. By contrast if one mouse is weighed five times these are matched data or repeated measures.

 Which is the correct test to analyse the data from Example 7.4 Space utilization within a mixed-species captive enclosure of Poison Dart Frog (Dendrobatidae)?

7.2 Experimental design and statistics

In 7.1 we have presented a key to help you identify the correct statistical test. In the key we have used features such as the type of data and numbers of treatment variables. The key does not take into account whether you have controls or wish to test a specific hypothesis. We therefore cover these aspects here.

7.2.1 General and specific hypothesis

You were introduced to this idea in 6.3.2 and we return to it again in 10.6. Most of the tests we have covered test a general hypothesis where the only outcome is a broad 'yes, there is a difference' or 'yes, there is an association'. Some tests, however, ask more explicit questions about the nature of the difference and so are more informative. The number of specific tests in this book is limited to two groups. In the chi-squared goodness-of-fit test the data are tested against a specific mathematical or biological ratio (8.4 and 9.3). In the *post hoc* tests the mean or median of each sample is compared to the mean or median of every other sample in the experiment. For parametric data these are first outlined in 10.8, and for non-parametric data in 11.4. By testing a specific hypothesis you are able to carry out a more in-depth analysis of your biological system so these should always be considered. The designs suitable for these tests are chi-squared goodness-of-fit test; one-way ANOVA for unequal replicates; two-way nested ANOVA for equal replicates; and two-way non-parametric ANOVA or Scheirer–Ray–Hare test.

7.2.2 Controls

The nature of a control is outlined in 2.2.7. If you have a control in your experimental design and wish to examine the difference between this and your treatment variable(s) the most appropriate statistics test will depend on the number of categories and treatment variables. If you only have the control and then one category for your treatment variable then a *t*, *z*, or Mann–Whitney *U* test is commonly used. If the experiment has a more complex design then an ANOVA followed by a *post hoc* test is effective. A common error is to use a *t* test (or equivalent) in many pairwise tests with the control versus treatment variable category 1, control versus treatment category 2, etc. This is likely to lead to a Type I error so should be avoided.

7.2.3 Replication and sample sizes

Replication always enhances the quality of an experiment. It ensures you have a more representative sample (1.4.1), it can allow you to test for interactions between two treatment variables (10.11), and can allow you to use analyses which 'allow for' the effect of confounding variables on the experiment and so are more discriminating. There are very few types of analyses that can be used when you have no replicates. The only tests included in this book are the chi-squared goodness-of-fit test and the two-way parametric ANOVA with no replicates. The output from these analyses is very limited. You should therefore always endeavour to include replication in your design.

   The criteria for some tests indicate that they are most effective with a given number of replicates or specific minimum or maximum sample size. For example if you have one treatment variable, three categories, and non-parametric data, the statistics test of choice would be a Kruskal–Wallis test. With this design there is a requirement for there to be at least five observations in each category (11.3.2). This information can be used to help you plan your investigation.

Few statistical tests cope well with unequal sample sizes. The exceptions are the tests for one treatment variable such as the *t* test, Mann–Whitney *U* test, and Kruskal–Wallis test. More complex experiments involving two treatment variables should be designed with equal numbers of replicates to allow you to use tests such as the non-parametric ANOVA and to test for interactions between the variables.

7.2.4 Matched data or repeated measures

There are a number of tests that can be used for matched data or repeated measures. If you have just two categories in the treatment variable these are the specific *t* and *z* tests and the Wilcoxon's test. If you have more than two categories in the treatment variable the two-way ANOVA with no replicates is appropriate. If you have more treatment variables with matched data then a nested ANOVA may be suitable. Designs with matched or repeated measures with two or more treatment variables usually may be analysed using an ANOVA and a statistical software package.

7.2.5 Confounding variables

Confounding variation detrimentally affects all investigations by hiding real biological effects behind background noise. The effect can be minimized using specific experimental designs such as the Latin square (10.11.3), replication (2.2.6), and by discriminating statistical tests such as the ANOVAs.

 In which of these two studies is a matched pairs test appropriate?

    a) *Allium vineale* is commonly found on road verges. Fifty bulbs were grown in normal soil and 50 bulbs were grown in soil from a road verge that was contaminated with salt. After 120 days the dry mass (g) of each plant was recorded.

    b) Ten lemurs were to be transferred into a new enclosure. To assess the effect of the move each lemur was weighed (g) before and after the move.

7.3 The critical reader of statistics and experimental design

In Chapters 1 and 2 we took you through the key points that need to be considered when designing an experiment and then applied this to a published study. Being able to appreciate the strengths, limitations, and weaknesses of a piece of published research is one of the skills of a critical reader. The second skill is being able to look at the statistical information presented in the paper and to understand what this conveys about the data analysis and its strengths, limitations, and weaknesses. To be a statistical critical reader you often have to be a detective, as important information is invariably presented in the form of one letter or a few numbers. The following is designed to help you develop these skills. We include examples of carrying out this process in the Online Resource Centre.

7.3.1 The title and introduction

First examine the Introduction in the paper. At the end of the Introduction you should find the aim of the investigation. If this is not clearly stated then the title is usually a good indication of the aim. In an extended study where a number of different aspects are investigated you should read through the whole paper and identify the objectives as well. You use the aim and objectives as the framework for your critical review of a paper. We consider aims and objectives in 1.1.

7.3.2 The method

You can see from the key (7.1.1) that to understand the analysis of the data you need to record a number of features about the experimental design. Most published papers report on more than one experiment so this process needs to be repeated for each and linked to the relevant objective. The features you need to record are:

i. How many treatment variables are being studied and what are they?

ii. What are the units of measurement of these treatment variables (e.g. grams)? Are these interval, ordinal, or nominal measures? This is an indicator of how the data should be analysed and presented. If the data are on ordinal scales then non-parametric statistical tests and the median and range should be used. If the data are measured on an interval scale then it may be parametric but this needs to be confirmed.

iii. Are there are other (dependent) variables that have been measured, what are they, and what are their units? Are these ordinal or interval measures?

iv. How many observations have been recorded for each item? Are there repeated or matched data where more than one observation has been recorded for each item for the same treatment variable? Or are observations on two or more treatment variables being recorded for each item? This information should be reflected both in the choice of statistics test and how the ideas about replication and independence are used in the paper.

v. How many replicates are there and what are they?

vi. Are there any controls? Which are they? We discuss controls in 2.2.7. As you continue to examine the paper you need to make sure the controls are treated correctly in the analysis.

vii. Have the authors recorded which statistics tests they have used? Which have been used for which data and which objective?

viii. Are there any comments about testing for normality in the data, assumptions that the data are parametric or have the data been transformed? This should then be reflected in the choice of statistics tests.

ix. What were the outcomes of the analysis reported in the text? Note down any p values or calculated values of statistics such as t, z, or F values.

Once you have identified all this information in the text you then move to the tables and figures for further details and then use all this information to identify the limitations, strengths, and weakness of the analysis.

7.3.3 **Tables and figures**

The tables and figures are very important parts of a published paper, not only as they often make it easier for you as the reader to understand the key trends in the data, but also because they invariably contain additional information about the experiment and how the data have been analysed. The following are features to record from the figures, tables, and their headings. It helps if you link each table and figure to one of the objectives you identified in 7.3.1. The features you need to record are:

i.   Look at the data in the tables and on any figures—what do you think is going on?

ii.   Have the data been presented in histograms or bar charts? This indicates that the data are categorical. If so how many categories are present for each treatment variable and what are they? What are the units of measurement? Is the scale nominal, ordinal, or an interval scale that has been divided into categories? This determines in part the correct statistical test that should be used to analyse these data.

iii. Have the data been presented on a scatter plot? If so is there a regression line with a regression equation? Is there any evidence that the significance of a regression or a correlation has been tested? If so which regression or correlation? You can use the key (7.1) to confirm that this is the correct test for the type of experimental design and data. If an incorrect statistical test has been used in what way might this affect the interpretation of the data?

iv.   Are there any confidence intervals or similar representations of a range or variance on the figures? This indicates that there is replication in the y values and/or the x values. Is there any information that explains how these are calculated? Some of these calculations require parametric data. Have these summary statistics been correctly used? If not will this potentially alter the interpretation of the results?

v.   Are there any summary statistics in the tables such as means or standard errors of the means? This can indicate whether the data are being treated as parametric data or non-parametric data. Is the choice of summary statistics correct for the data that have been collected in the experiment? If not will this potentially alter the interpretation of the results?

vi.   Look at the information presented in the headings and legends and keys in the tables and figures. Does this explain how the data were analysed and if so what is the outcome of the analysis? Look for statistical distribution terms such as chi-squared or F, and p values such as < 0.05 or NS (not significant). Is the choice of statistical test corroborated in the text? You can use the key (7.1.1) to confirm that this is the correct test for the type of experimental design and data.

vii. Are there any other symbols used to represent significance? Sometimes asterisks are used to denote levels of significance (Table 12.6) or superscripts indicate which categories are not significantly different from other categories. The use of superscripts also usually indicates that a *post hoc* test such as a Tukey's test (10.8) has been used. Is this stated in the text? Which comparisons appear to have been made? What do they indicate? You can use the key (7.1.1) to confirm that this is the correct test for the type of experimental design and data. If an incorrect statistical test has been used, in what way might this affect the interpretation of the results?

7.3.4 **Your evaluation**

Having collected all this information you now need to identify the strengths, limitations, and weaknesses of the statistical analysis. A weakness is where an action has or has not been carried out that should and could have been. A limitation is where the evaluation could not be altered due to practical difficulties or without altering the aim. A strength would be the absence of a weakness or limitation or the demonstration of good practice such as testing data for normality before using a parametric test.

i. The statistical tests

The most important place to start is to confirm that the correct statistical tests have been used and that all the criteria for using these tests are met. You should use the information you have collected in 7.3.2 and 7.3.3 to carry out a check using the key in 7.1.1. It is unlikely that you will have the raw data in the paper but by collating all the information you have gathered from the text, tables, and figures you should nonetheless be able to complete this check. If the wrong test has been used this may be a weakness in the research. Your task is to determine to what extent this might have affected the interpretation of the data. There are often helpful and interesting discussions online about the requirements that need to be met for different statistical tests which can help you with this.

ii. Common mistakes

In any evaluation like this you can also look for common mistakes. The most common are:

- Using parametric tests when the data are non-parametric and there is no evidence that the data have been checked for normality or transformed.
- In an experiment with more than two categories, carrying out tests that compare only two of the categories at a time such as using a t test over and over again rather than using an ANOVA. This unnecessarily increases the likelihood of a Type I error occurring.
- Non-parametric data should not be presented with a mean and a variance. These are parameters of a known distribution, usually a normal distribution, so should only be used if there is a reason for believing the data follow this distribution (5.3).

If you identify one or more of these mistakes then your task is to consider to what extent this matters. For example if the investigator has increased the likelihood of a Type I error occurring by overuse of t tests then they may by chance appear to detect a biologically meaningful difference when it is not really true.

iii. Sample sizes

In 2.1 we reviewed how to be a critical reader in relation to the design of an experiment. One aspect in the context of a design was sample size. Sample size is relevant again when thinking about the analysis of data. It is important as samples should be representative of the actual population; the sample is taken with the expectation that it will provide a population estimate. Consider how representative any samples are and whether the significance of the findings are not therefore overemphasized.

Sample size is often critical for the choice of statistical test. Small sample sizes and limited designs can lead to the need for specific actions when analysing the data such as those discussed in 9.4. Any investigator should therefore demonstrate (at the very least) implicitly and sometimes explicitly that they are aware of these important points.

iv. Communication

As a critical reader it is also your role to consider whether the outcomes from the analysis have been correctly and clearly reported in the main body of the paper and that there is sufficient detail to understand the analysis. The analysis should directly reflect the aim and objectives and this should be obvious in the paper. Finally you need to check that non-significant differences are considered as such in the text. There is sometimes the temptation to present trends that are not significant as having a similar importance to significant differences or to overemphasize findings based on limited designs or small sample sizes.

v. Your conclusion as a critical reader

Based on the strengths, limitations, and weaknesses of the statistical analysis you should now be able to draw a conclusion about the quality of the analysis, which, in combination with conclusions drawn about the design (Chapters 1 and 2), will enable you as a scientist to draw your own conclusions about the worth of the published research you encounter.

 Which statistical tests have been carried out if the following are reported?

    a. t

    b. U

    c. F

Summary of Chapter 7

- As a scientist there are three occasions when you will find yourself having to decide on which is the correct statistical test to use: designing an experiment; analysing data from an experiment designed by someone else; and establishing the merits of research that has been designed and evaluated by someone else.

- In 7.1 we outline how to choose the correct statistical test for your design. The key (7.1.1) takes you through how to choose the correct hypothesis and then the correct test to use. The key requires you to be familiar with a number of terms covered in Chapters 1–3. We provide some explanations and examples here to help you (7.1.2).

- The key is written as a tool to help you master the skills of choosing the correct test. Once you have gained confidence you may prefer to use the quick choice key in Appendix C instead.

- In 7.2 we take a different approach to selecting statistical tests and consider what to do if your design includes replicates, controls, and a number of other specific design features.

- In 7.3 we show you how to apply your understanding to enable you to judge the appropriateness of data evaluation in published research. This complements the points made in Chapter 2 on considering experimental designs in published research. These two skills together enable you to be a critical reader.

Answers to chapter questions

A1   Example 7.4 illustrates the design for step 4 in the key. The design is described in the annotation on Table 7.3. There is one nominal treatment variable with six categories. There are four nominal categories (frogs) which are repeated measures and could be considered to be replicates. The dependent variable is an interval measure which we will assume is parametric and there are no replicates. An extract of the key shows how the correct test can be selected.

4.   You have no *a priori* reason for expecting certain outcomes from your investigation **OK**. These data are usually presented in tables or summary histograms rather than a scatter plot (**YES, see Table 7.3**). The treatment variables are measured on an ordinal and/or nominal scale and/or measured on an interval scale and arranged into discrete categories. (Other variables may be measured on an interval scale.) **YES. Both the rows and columns are categorical.** → **7**. Do samples come from the same or different populations?

7.1   You have one or more treatment variables. **YES**. There is a further variable, the dependent variable **YES (time)**. The sample of observations for the dependent variable is parametric **MAYBE. There are very few data so it is difficult to check this formally. We have to assume that as time is measured on an interval scale that these data are parametric.** → **7.1.1** 'Parametric' data.

7.1.1 and **7.1.2** You wish to examine the effect of one treatment variable with two categories **NO. There are six categories** → **7.1.3**.

7.1.3 You wish to examine the effect of one treatment variable with more than two categories **YES**. There are equal replicates in each category **YES, the frogs can be considered to be replicates**. The data table therefore has more than two columns **YES**. There are the same number of observations in each column **YES** but each observation comes from a different item **NO** → **7.1.4**.

7.1.4 You wish to examine the effect of one treatment variable with more than two categories **YES**. There are unequal replicates in the categories **NO** → **7.1.5**.

7.1.5 You wish to examine the effect of one treatment variable with more than two categories **YES**. There is a notable confounding variable or the data are repeated measures or matched **YES—the frogs have been measured in each zone and so these measures are not independent they are 'repeated measures'**. There is no further replication **CORRECT**. The data table therefore has more than two columns **YES**. The columns are the categories in the treatment variable **YES**. There are a number of rows each relating to one repeated measure or aspect of a confounding variable **YES** → Two-way ANOVA with no replicates.

These data can be analysed using the two-way ANOVA with no replicates.

A2

  a) Each item (the bulb) is only exposed to one aspect of the treatment and measured once. Therefore the dry mass results are not matched data.

  b) Each lemur is measured before the move and then after the move so this is matched data.

A3

  a) A t test has been used but there are a number of t tests (Chapter 10) so it would not be possible to tell which has been used. The Student's t which is a commonly used t test requires homogenous variances so if this was used you might expect to see that this criterion had been checked.

  b) A Mann–Whitney U test has been used. This tells you that the researcher believes the data to be non-parametric.

  c) This is tricky as F tests are fundamental as a step in tests for homogeneity of variance prior to carrying out t tests and ANOVAs, but are also part of the parametric ANOVA test itself. You would have to look carefully at the content of the paper to find more details about the test before deciding. 'F', however, does tell you that the researcher believes their data may be parametric.

8 Hypothesis testing:
Do my data fit an expected ratio?

 In a nutshell

In this chapter, we consider statistical tests and experimental designs that will be suitable for testing the hypothesis 'Do my data fit an expected ratio?' The statistical test most often used to test this hypothesis is the chi-squared goodness-of-fit test. We explain more about this type of hypothesis, how to carry out the necessary calculations, and how to resolve some of the problems you may encounter.

The statistics tests covered in this chapter are:

8.3 Chi-squared goodness-of-fit test: one sample

8.4 How to check whether your data have a normal distribution using the chi-squared goodness-of-fit test

8.5 Replication in a goodness-of-fit test

In some investigations you may wish to compare your observations to expected proportions that are based on biological or mathematical principles and not derived from the sample. You therefore wish to test the hypothesis 'Do my data fit an expected ratio?' In these instances your observations will be assigned to two or more categories. The categories can be derived directly from the categories inherent in a nominal or ordinal variable, for example flower colour categorized as purple or white (Example 8.2), or imposed on an interval scale of measurement as classes, for example the distribution of holly leaf miners on a holly tree (Example 8.1). The data in these categories (the dependent variable) are counts, i.e. the numbers of items in a category. There are a number of tests that can be used to analyse these data. There is some debate over the relative merits of these tests. However, the consensus appears to be that they all have strengths and weaknesses and that none of their limitations seriously undermines their value. The test in most common usage is the chi-squared goodness-of-fit test and is therefore the one covered in this chapter. This test can be used when you have one sample (8.3) or replicates (8.5). To identify the correct test for your experimental design and/or data you can follow the key in Chapter 7. Each test also has additional criteria that need to be met. The full criteria including these additional requirements are included in this chapter so that you can check these against your design and/or data.

Worked examples are given for each test. If this is the first time you have used a test, you should work through these examples and the questions and then check your answers before using the test on your own data. If your answer differs considerably from that given, you should check your calculation by going to the Online Resource Centre, where we have included both the full calculation and more examples. If you work through the questions in this chapter it should take you about an hour to complete. The answers for these exercises are at the end of the chapter. Nearly all our examples are based on real undergraduate research projects. If these examples are not in your subject area, you will find more in the Online Resource Centre.

8.1 Which ratios can we fit?

There are three types of investigations where you may use a goodness-of-fit test:

1) You can reasonably argue that all samples should have the same value. This can arise when your sampling is very restricted, e.g. you may record the remains of shells from *Cepaea nemoralis* and *Cepaea hortensis* at the site where a thrush has been breaking them open (an anvil). You might hypothesize that there is no statistically significant difference between the numbers of each species, i.e. you would expect to see one *Cepaea nemoralis* shell for every one *Cepaea hortensis* shell at this thrush anvil. But beware. This sample may not be representative of thrush anvils as a whole. To examine a broader hypothesis, you would need to collect more samples and possibly use a different statistical test.

2) You wish to confirm that your data have a particular distribution, such as a normal or a Poisson distribution. In this instance, your expected values are calculated using a particular formula that describes the distribution, such as the Gaussian equation (5.2.1).

3) You have carried out a genetic cross or sampled in a population and have reason to expect a particular segregation ratio, sex ratio, allele frequency, etc.

In each of these examples, you will be able to use known underlying principles, such as Mendelian genetics or the Hardy–Weinberg theorem, to determine the 'expected' values.

A goodness-of-fit test allows you to compare the data you have collected with that predicted by an *a priori* expectation and to calculate how probable the match is. A chi-squared goodness-of-fit test is introduced in 8.3 in relation to an experimental design where the expected values are determined by a ratio resulting from 'random' events (Investigation type 1). A second example is considered in 8.4 where this test is also used to confirm that data have a normal distribution (Investigation type 2). Whilst Example 8.2 illustrates the testing of an expectation arising from a genetic cross (Investigation type 3).

8.2 Expected values

In these tests the observed values are compared with calculated 'expected' values. The term expected can be used in three ways:

For some experiments, you have a model that generates expected outcomes for your investigation. This is often referred to as having an *a priori* expectation and arises when you carry out a test of the hypothesis 'Do the data match an expectation?' (this chapter). You can use your own (observed) data to generate 'expected' values given certain rules relating to the statistical test and experimental design to allow you to test the hypothesis 'Is there an association?' (9.1). You may have your own expectations (or predictions) for your experiment. You must never confuse this, your own personal preference or 'expectation', which is usually for significant outcomes or outcomes that match other studies, with the idea of 'expected' values used in hypothesis testing. Having a strong personal preference for a particular outcome can lead to very serious errors in your approach to your work and should therefore be avoided.

In the chi-squared tests we consider in this chapter we use theoretical models to generate the expected values. In the chi-squared tests considered in Chapter 9 the expected values are derived from your own observations. It is therefore important to be sure you have selected the right form of chi-squared test for your type of data and hypothesis.

Expected values
The values calculated using a mathematical or biological model such as the Gaussian equation or a genetic segregation ratio.

 An undergraduate posted a questionnaire on a commonly used site to examine knowledge and attitudes in the general public to genetic testing. She divided the respondents into three discrete age groups: 21–30; 31–40; and over 41, and organized her data into a contingency table with these three age categories. She wished to confirm that the distribution amongst the age groups was random. Which type of 'expectation' does she have?

8.3 Chi-squared goodness-of-fit test: one sample

The chi-squared goodness-of-fit test can be used for two or more categories with a single sample but works most effectively when there are more than three categories and when the sample size and expected values are not 'small'. Solutions to problems of limited designs and small sample sizes are discussed in detail in 9.4 and one solution is illustrated in Box 8.2. As in all chi-squared tests, there are several common steps to be followed. The first of these is to arrange your data into a contingency table.

> ### Example 8.1 The distribution of holly leaf miners on *Ilex aquifolia*
>
> As part of a student project, 200 holly leaves were sampled at random from each of three heights on a holly tree and the number of holly leaf miners was recorded (Table 8.1). If the distribution was random you would expect equal numbers of miners at each height of the tree.

Table 8.1 Contingency table for Example 8.1: the distribution of holly leaf miners on a single *Ilex aquifolia* tree

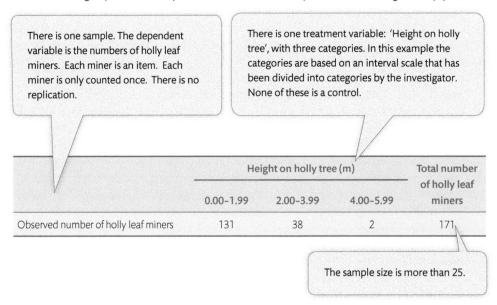

There is one sample. The dependent variable is the numbers of holly leaf miners. Each miner is an item. Each miner is only counted once. There is no replication.

There is one treatment variable: 'Height on holly tree', with three categories. In this example the categories are based on an interval scale that has been divided into categories by the investigator. None of these is a control.

	Height on holly tree (m)			Total number of holly leaf miners
	0.00–1.99	2.00–3.99	4.00–5.99	
Observed number of holly leaf miners	131	38	2	171

The sample size is more than 25.

Table 8.1 is a contingency table for the data from Example 8.1. A contingency table can be the same as a frequency table, as in this example. At this stage, you will only have 'observed' values. The observed data have been organized into classes: 0.00–1.99m, 2.00–3.99m, and 4.00–5.99m. There is no ambiguity about the categories: they are discrete.

8.3.1 Key trends and experimental design

Before undertaking any analysis you should first examine your design and data and note specific design features and the key trends in your data. This will help you to check that you have correctly selected the statistical test and that this test can take into account features such as controls and replicates. For this reason we have annotated Table 8.1 with all the important information about the design. For example none of the height categories can be considered to be a control. In Example 8.1 it is clear that the numbers of holly leaf miners are not evenly spread across the tree. The statistical analysis can confirm if this is by chance or is biologically meaningful.

8.3.2 Using this test

i. To use this test you:

1) Have an *a priori* reason for expecting certain outcomes from your investigation.
2) Wish to examine the effect of one treatment variable and have only one sample.
3) Have data that fall into two* or more discrete categories.
4) Have data that are numbers of items and are not percentages or proportions.
5) Have observations that are independent. Each item only appears once in the results.
6) Have a sample size greater than 25*.
7) Have expected values that are greater than 5*.

* There is an ongoing debate about the effect of limited designs and small sample sizes on the reliability of these tests. We touch on some of these points in 9.4 and one solution is illustrated in Box 8.2.

ii. Does the example meet these criteria?

The annotations on Table 8.1 show that the data meet most of these criteria. In addition the *a priori* expectation is that the holly leaf miners should be randomly distributed if height is not affecting the distribution of the miners across the tree. There should therefore be equal numbers of miners at the different tree heights. The expected values are only worked out as part of the calculation, so cannot easily be checked in advance. If you look at Table 8.2 on the 'expected' row where these values have been calculated, these values are greater than 5. So this criterion is also met. If the criterion is not met, then you should refer to 9.4.

8.3.3 The general calculation

Having organized your observed values in a contingency table, considered for yourself what is happening in the data and checked that the criteria for using this test are met, you can now proceed with the test. In Box 8.1, we show you the general calculation and a specific example of a chi-squared goodness-of-fit test. In addition, you will need to refer to the con-tingency calculation table (Table 8.2). Where steps have been abbreviated, the full calculation is included in the Online Resource Centre.

Table 8.2 Expected values for a goodness-of-fit chi-squared test using the data from Example 8.1: the distribution of holly leaf miners on a single *Ilex aquifolia* tree

Number of holly leaf miners	Height on holly tree (m)			Total number of holly leaf miners
	0.00–1.99	2.00–3.99	4.00–5.99	
Observed	131	38	2	171
Expected	57	57	57	171

BOX 8.1 How to calculate a chi-squared goodness-of-fit test

GENERAL DETAILS	EXAMPLE 8.1
	This calculation is given in full in the Online Resource Centre. For presentation purposes only all values have been rounded to five decimal places.
1. Hypotheses to be tested H_0: There is no difference between the expected and observed values. H_1: There is a difference between the expected and observed values.	1. Hypotheses to be tested H_0: There is no difference between the numbers of holly leaf miners found at various heights (m) on the tree compared with those expected from an equal distribution. H_1: There is a difference between the numbers of holly leaf miners found at various heights (m) on the tree compared with those expected from an equal distribution.
2. Have the criteria for using this test been met?	2. Have the criteria for using this test been met? Yes (8.3.2 ii).
3. How to work out expected values Expected values are calculated using the numerical formula or ratio that you are expecting your data to conform to.	3. How to work out expected values We are expecting a random distribution of the holly leaf miners throughout the tree. This means that we should find equal numbers in each height category, i.e. the ratio we are using is 1:1:1. The number of holly leaf miners we would expect at each height will be 171/3 = 57 (Table 8.2).
4. How to work out $$\chi^2_{calculated} = \Sigma \left[\frac{(observed - expected)^2}{expected} \right]$$	4. How to work out $\chi^2_{calculated}$ $$\chi^2_{calculated} = \frac{(131-57)^2}{57} + \frac{(38-57)^2}{57} + \frac{(2-57)^2}{57}$$ $$= 96.07018 + 6.33333 + 53.07018$$ $$= 155.47369$$
5. How to find $\chi^2_{critical}$ See Appendix D, Table D1. To find the critical value of χ^2, you need to know the degrees of freedom (v). In this goodness-of-fit test, the degrees of freedom are the number of categories (a)-1. Use a chi-squared table of critical values to locate the value at $p = 0.05$.	5. How to find $\chi^2_{critical}$ The categories in this example are 0.00-1.99m, 2.00-3.99m, and 4.00-6.99m, and therefore $v = 3-1 = 2$ and $\chi^2_{critical}$ is 5.99.
6. The rule If $\chi^2_{calculated}$ is greater than $\chi^2_{critical}$, you may reject the null hypothesis.	6. The rule $\chi^2_{calculated}$ (155.47) is greater than $\chi^2_{critical}$ (5.99) at $p = 0.05$ and therefore we reject the null hypothesis. In fact, at $p = 0.001$, the $\chi^2_{critical}$ value is 13.82. Therefore, we can reject the null hypothesis at this higher level of significance. *(continued)*

7. What does this mean in real terms?	7. What does this mean in real terms?
	There is a highly significant difference ($\chi^2_{calculated} = 155.47$, $p < 0.001$) between the numbers of holly leaf miners found at the various levels on the tree compared with those expected, such that the holly leaf miners are not found in equal numbers at all heights.

Further examples relating to the topic including how to use statistical software are included in the Online Resource Centre.

 In an investigation into the visual responses of beetles, a number of coloured pitfall traps were placed at random in grassland. After 24 hours, the pitfall traps were collected and the numbers of beetles recorded. The results for each trap were red: 20 beetles; yellow: 34 beetles; white: 10 beetles; and black: 40 beetles. What is the a *priori* expectation? Calculate the expected values.

8.4 How to check whether your data have a normal distribution using the chi-squared goodness-of-fit test

In Chapter 5, we discussed distributions and in particular the normal distribution. Data with a normal distribution can be analysed using parametric statistics and it is therefore important to be able to check whether your data are normally distributed. Section 5.8 and Box 5.2 explained a number of ways to check your data to see whether they are normally distributed. However, the best support for deciding whether your data are normally distributed is to carry out a test for normality. Tests for normality are best calculated using statistical software and we take you through some of these in the Online Resource Centre. One method is to calculate expected values using the Gaussian equation and then use a chi-squared goodness-of-fit test to check that your observed data are not statistically significant from that predicted by the equation. To do this requires some extra steps in addition to those described in Box 8.1 and a change to how you work out the degrees of freedom. To illustrate the general principles only, we have included a simplified worked example here, where the difference between classes is 1.0. The observed values we will use come from Example 5.2: Length (mm) of two-spot ladybirds (*Adalia bipunctata*). We recommend you refer to the Online Resource Centre statistical walkthroughs and screencast for a fuller explanation.

Although we only include an example relating to the normal distribution, the principles are the same for checking against any known distribution, such as the Poisson distribution and is covered in more detail in the Online Resource Centre.

The contingency data for this example are the same as the frequency table in Chapter 5 (Table 5.9). We reproduce this again here in an amended form so that the whole of this process is represented here for convenience.

Table 5.9 Frequency table of length of two-spot *Adalia bipunctata* (ladybirds)

> There is one treatment variable: 'Size classes for length (mm) of *Adalia bipunctata* (ladybird)', with nine categories. In this example the categories are based on an interval scale that has been divided into categories by the investigator. No category can be said to be a control.

| | Size classes for length (mm) of *Adalia bipunctata* (ladybird) | | | | | | | | | |
	0.0–0.9	1.0–1.9	2.0–2.9	3.0–3.9	4.0–4.9	5.0–5.9	6.0–6.9	7.0–7.9	8.0–8.9	9.0–9.9	10.0–10.9
Mid-point of class (m)	0.45	1.45	2.45	3.45	4.45	5.45	6.45	7.45	8.45	9.45	10.45
Frequency (f)	0	2	3	5	8	12	9	5	4	2	0

> There is one sample and observations are frequencies not percentages or proportions. The total sample size is 50. There is no replication in this design.

8.4.1 Key trends and experimental design

A histogram of these data (Fig. 5.7) shows that this distribution is symmetrical and does appear to be a normal distribution. In Box 5.2 these data have been checked against the general criteria that indicate normally distributed data and are found to comply. At this point we would conjecture that these data are normally distributed and can now use the statistical analysis to test this hypothesis.

8.4.2 Using this test

To use this test the data need to meet the criteria in 8.3.2. The annotations demonstrate that most of the criteria are met. The *a priori* expectation is determined by the Gaussian equation. We take you through the steps for calculating the expected numbers in the next section, but in Table 8.3, you can see that not all the expected values are greater than 5. Therefore, this last criterion is not met. We explain how to overcome this problem of small numbers in general in 9.4 and specifically for this example in Box 8.2.

8.4.3 The general calculation

Having organized your observed values in a contingency table and checked that the criteria for using this test are met, you can now proceed with the test. In Box 8.2, we show you the general calculation and a specific example. In addition, you will need to refer to the contingency calculation table (Table 8.3). Where steps have been abbreviated, the full calculation is included in the Online Resource Centre.

Like the previous example for the chi-squared goodness-of-fit test, you use your *a priori* expectation to determine the expected values. In this case, we use the Gaussian equation, which is the mathematical equation that describes the normal distribution, where:

$$y = \frac{1}{\sqrt{2\pi s^2}} e^{-h}$$

and

$$h = \frac{(x - \bar{x})^2}{2s^2}$$

The terms in this equation are explained in 5.2.1 and Appendix B. The symbols e and π have approximate values of 2.72 and 3.14, respectively. The $\bar{x}$ (mean) and s^2 (variance) were calculated in Box 5.2 and are $\bar{x} = 5.08$mm, $s^2 = 3.74857$mm^2. x is any one observation in your sample. Thus, all these terms have known values apart from y. These y values can be calculated for each x in your sample as follows.

Choose an x value that fits the lowest category of your data. As we have classes, we use the mid-point of that class, so the first $x = 0.45$. By including this x value and all the other known values, you can now work out y. Don't be put off, even though it looks complicated. Break the calculation down into smaller steps.

The first part of the equation is:

$$\frac{1}{\sqrt{2\pi s^2}} = \frac{1}{\sqrt{2 \times 3.14 \times 3.74857}}$$

$$= \frac{1}{\sqrt{23.5529}} = \frac{1}{4.8531} = 0.20605$$

The second part of the equation involves the exponential term e. First work out the value for h:

$$h = \frac{(x - \bar{x})^2}{2s^2} = \frac{(1.45 - 5.08)^2}{2 \times 3.74857} = \frac{13.1769}{7.49714} = 2.85934$$

Use your calculator function buttons to find $e^{-1.75759} = 0.05731$
So

$$y = 0.20605 \times 0.05731 = 0.00118$$

This y value is worked out as a proportion and our final step is to calculate expected numbers from these proportions. As the sample size for our ladybird example is 50, then the expected number for y when $x = 0.45$ is $50 \times 0.00118 = 0.05904$.

Repeat this calculation for all mid-point values from 0.45 to 10.45. These are your expected values (Table 8.3) and you can now carry on with the chi-squared goodness-of-fit test (Box 8.2).

Table 8.3 Contingency table for Example 5.2: The length of two-spot *Adalia bipunctata* (ladybirds) with expected numbers calculated from the Gaussian equation

| | Size classes for length (mm) of *Adalia bipunctata* (ladybird) | | | | | | | | | | |
	0.0–0.9	1.0–1.9	2.0–2.9	3.0–3.9	4.0–4.9	5.0–5.9	6.0–6.9	7.0–7.9	8.0–8.9	9.0–9.9	10.0–10.9
Mid-point of class (*m*)	0.45	1.45	2.45	3.45	4.45	5.45	6.45	7.45	8.45	9.45	10.45
Observed number of ladybirds	0	2	3	5	8	12	9	5	4	2	0
Expected number of ladybirds	0.05904	1.7768	4.0951	7.2283	9.7714	10.1162	8.0209	4.8705	2.2649	0.8066	0.0220

BOX 8.2 To check whether your data are normally distributed using a chi-squared goodness-of-fit test

GENERAL DETAILS	EXAMPLE 5.2
	This calculation is given in full in the Online Resource Centre. For presentation purposes only all values have been rounded to four decimal places.
1. Hypotheses to be tested H_0: There is no difference between the expected and observed values. H_1: There is a difference between the expected and observed values.	1. Hypotheses to be tested H_0: There is no difference between the observed lengths of ladybirds (mm) compared with that expected if the data are normally distributed. H_1: There is a difference between the observed lengths of ladybirds (mm) compared with that expected if the data are normally distributed.
2. Have all the criteria for using this test been met?	2. Have all the criteria for using this test been met? Most of the criteria are met (8.3.2 ii). However, some of the expected values are less than 5 (Table 8.3). We explain how to deal with this in the next step.
3. How to work out expected values Expected values are calculated using the numerical formula you are expecting your data to conform to.	3. How to work out expected values See 8.3.2 ii and Table 8.3 to calculate the expected values. It is clear that five of the nine expected values are less than 5. We discuss problems with small numbers in 9.4. To overcome this, we will add together the expected values for the lower three classes (0.05904 + 1.7768 + 4.0951 = 5.93094) and the expected

(continued)

values in the upper three classes (2.2649 + 0.8066 + 0.022 = 3.0935). Although this last expected value is still less than 5 this adjustment is sufficient to now meet the general requirements to meet this criterion (9.4). These expected values will be compared with observed values that have been combined in the same way.

4. How to work out $\chi^2_{calculated}$

$$\chi^2_{calculated} = \Sigma\left[\frac{(observed - expected)^2}{expected}\right]$$

4. How to work out $\chi^2_{calculated}$

$$\chi^2_{calculated} = \frac{(5 - 5.8719)^2}{5.8719} + \frac{(5 - 7.2283)^2}{7.2283} + \cdots$$
$$+ \frac{(6 - 3.0717)^2}{3.0717} = 4.40299$$

5. How to find $\chi^2_{critical}$

See Appendix D, Table D1. To find the critical value of χ^2 at $p = 0.05$, you need to know the degrees of freedom (v). Usually for a chi-squared goodness-of-fit test, the degrees of freedom would be the number of categories (a)–1. However, because a mathematical formula in which the population value of μ has been estimated using the sample mean ($\bar{x}$), a further degree of freedom is lost. Therefore, for this chi-squared goodness-of-fit test, $v = a–2$.

5. How to find $\chi^2_{critical}$

As we combined the lower three classes and the upper three classes, there were only seven classes in this calculation, so $v = 7–2 = 5$. When $p = 0.05$, $v = 7$, then $\chi^2_{critical}$ is 14.07.

6. The rule

If $\chi^2_{calculated}$ is greater than $\chi^2_{critical}$, you may reject the null hypothesis.

6. The rule

$\chi^2_{calculated}$ (4.34) is less than $\chi^2_{critical}$ (14.07) at $p = 0.05$ and therefore we do not reject the null hypothesis.

7. What does this mean in real terms?

7. What does this mean in real terms?

There is no significant difference ($\chi^2_{calculated}$ = 4.34, $p = 0.05$) between the observed lengths of ladybirds (mm) compared with that expected if the data are normally distributed. The data can be said to be normally distributed.

Further examples relating to the topic including how to use statistical software are included in the Online Resource Centre.

8.5 Replication in a goodness-of-fit test

There are some circumstances where you will have several samples and wish to know whether they can be pooled. For example, you may have carried out an investigation into the genetic inheritance of flower colour where you had several pairs of parent plants. Crossing within these pairs would produce F_1 offspring. If you kept the seed from each cross separate from the others, grew these F_1 plants up, and then crossed these, you could collect the results from several different lineages. These separate lineages are known as accessions. This goodness-of-fit test is known as a goodness-of-fit test for replicates or a test for heterogeneity.

Example 8.2 The genetics of flower colour in *Allium schoenoprasum*

Crosses were carried out in three pairs of plants of *Allium schoenoprasum*. In each cross, one parent had purple flowers and the other was white flowered. Seeds from each cross were collected and grown on. The offspring from each cross were kept as separate accessions. These F_1 plants were all purple flowering and thought to be heterozygous for a single gene that controlled the pigmentation in the flowers, with purple believed to be the dominant allele. These F_1 plants were then crossed among themselves within an accession, and the seed from these crosses grown up and the flower colours recorded (Table 8.4). If the genetic theory is correct, then the F_2 plants should occur in the ratio of three purple-flowering plants: one white-flowering plant.

Table 8.4 Flower colour exhibited by F_2 *Allium schoenoprasum* in four accessions

There is one treatment variable: 'Flower colour in *Allium schoenoprasum*'. There are two nominal categories (purple and white). Neither category can be considered to be a control.

	Flower colour in the F_2 generation		Total number of plants flowering in the F_2
	Purple	White	
Accession 1	127 (75.6%)	41 (24.4%)	168
Accession 2	123 (75.9%)	39 (24.1%)	162
Accession 3	107 (66.9%)	53 (33.1%)	160
Accession 4	130 (75.6%)	42 (24.4%)	172
Total	487 (73.6%)	175 (26.4%)	662

There are four replicates: accessions 1–4. The data are numbers of items. Each plant is only counted once.

The sample size for each replicate is more than 25.

8.5.1 Key trends and experimental design

In this design there are no controls but there are four replicates. However, we are interested in comparing these to see if they differ. The data indicate that accession 3 is not very similar to the other accessions (Table 8.4). Despite this, the total values (487 purple and 175 white) are reasonably close to a 3:1 ratio. When a goodness-of-fit chi-squared test is carried out on just these total values, there is no significant difference ($\chi^2 = 0.73$, $p < 0.05$) between the observed values and that expected for a three purple:one white ratio (Table 8.5). When a chi-squared goodness-of-fit test is carried out on each accession separately (Table 8.5) these analyses confirm that the results from accessions 1, 2, and 4 also conform to a 3:1 ratio at $p = 0.05$. Accession 3, however, is different and the test there indicates that the observed values depart significantly from the predicted 3:1 ratio. Does this difference between accession 3 and accessions 1, 2, and 4 mean that the data should not be pooled? In these circumstances, you may use the chi-squared test to see whether the data are statistically heterogeneous (different).

8.5.2 Using this test

i. To use this test you:

1) Have an *a priori* reason for expecting certain outcomes from your investigation.

2) Wish to test for heterogeneity between samples.

3) Have one treatment variable and have replicates or more than one sample.

4) Have data that fall into two or more discrete categories.

5) Have data that are numbers of items and are not percentages or proportions.

6) Have observations that are independent. Each item only appears once in the results.

7) Have a sample size greater than 25*.

8) Have expected values that are greater than 5*.

* There is an ongoing debate about the effect of limited designs and small sample sizes on the reliability of these tests. We consider these points in 9.4 and illustrate one solution in Box 8.2.

ii. Does the example meet these criteria?

We have checked the data from Example 8.2 against the criteria for using this test and, as the annotations on Table 8.4 show, all the criteria are met. We do wish to test for heterogeneity between samples. All the expected values are greater than 5 (Table 8.5).

8.5.3 The calculation

At this point, you have constructed a contingency table, checked the criteria for using this test, and wish to compare your samples to see whether they are heterogeneous. You can now proceed with the chi-squared goodness-of-fit test with replicates (Box 8.3). Where steps have been abbreviated, the full calculation is included in the Online Resource Centre.

Table 8.5 Contingency table with expected values for a chi-squared goodness-of-fit test with replicates for Example 8.2: Flower colour exhibited by F_2 *Allium schoenoprasum* in four accessions

		Flower colour in the F_2 generation		Total number of plants flowering in the F_2	χ^2 from the goodness-of-fit test for each accession and for the total values
		Purple	White		
Accession 1	Observed	127	41	168	0.03175 (NS)
	Expected	126.0	42.0		
Accession 2	Observed	123	39	162	0.07407 (NS)
	Expected	121.5	40.5		
Accession 3	Observed	107	53	160	5.63333 $(0.05 > p > 0.01)$
	Expected	120.0	40.0		
Accession 4	Observed	130	42	172	0.03101 (NS)
	Expected	129.0	43.0		
Total	Observed	487	175	662	0.72709 (NS)
	Expected	496.5	165.5		

NS, not significant.

BOX 8.3 How to calculate a chi-squared goodness-of-fit test with replicates

GENERAL DETAILS	EXAMPLE 8.2
	This calculation is given in full in the Online Resource Centre. For presentation purposes only all values have been rounded to five decimal places.
1. Hypotheses to be tested H_0: There is no heterogeneity between the samples. H_1: There is heterogeneity between the samples.	**1. Hypotheses to be tested** H_0: There is no heterogeneity between the F_2 accessions of *Allium schoenoprasum*. H_1: There is heterogeneity between the F_2 accessions of *Allium schoenoprasum*.
2. Have all the criteria for using this test been met?	**2. Have all the criteria for using this test been met?** Yes (8.5.2 ii).
3. How to work out expected values These are calculated for each accession and for the total values. The expected values are calculated using the numerical formula you are expecting your data to conform to.	**3. How to work out expected values** We are expecting three purple-flowering plants:one white-flowering plant. As the grand total number of plants examined was 662, then 3/4 should be purple flowering, i.e. 496.5, and 1/4 plants should be white flowering i.e. 165.5. Using the same ratio, we also calculate the expected values for each accession (Table 8.5).

(continued)

4. How to work out $\chi^2_{calculated}$

i. First calculate chi-squared values for each separate sample and for the total, where:

$$\chi^2_{calculated} = \Sigma\left[\frac{(observed - expected)^2}{expected}\right]$$

It is at this point that the process differs from that described for the goodness-of-fit test for one sample and a few further calculations are required.

ii. 'Summed' chi-squared. First sum all the chi-squared values calculated for each accession.

iii. 'Deviation' chi-squared. This is the chi-squared value found when examining the total values.
iv. The 'heterogeneity' chi-squared is found by subtracting the 'deviation' value from the 'summed' value. This is $\chi^2_{calculated}$.

4. How to work out $\chi^2_{calculated}$

i. e.g. Accession 1.

$$\chi^2 = \frac{(127 - 126)^2}{126} + \frac{(41 - 42)^2}{42}$$
$$= 0.00794 + 0.02381$$
$$= 0.03175$$

$$\text{Total } \chi^2 = \frac{(487 - 496.5)^2}{496.5} + \frac{(175 - 165.5)^2}{165.5}$$
$$= 0.18177 + 0.54532$$
$$= 0.72709$$

The other values are shown in Table 8.5.
ii. 'Summed' chi-squared. In our current example, this 'summed' value is: 0.03175 + 0.07407 + 5.63333 + 0.03101 = 5.77016
iii. 'Deviation' chi-squared = 0.72709
iv. $\chi^2_{calculated}$ = 5.77016−0.72709 = 5.04307

5. How to find $\chi^2_{critical}$

See Appendix D, Table D1. First, calculate the degrees of freedom (v). Again, there are several steps.
i. v for the 'deviation' value is the number of categories (columns, a)−1.
ii. v for the 'summed' chi-squared value is the sum of the degrees of freedom from each of the rows (samples) excluding the total.
iii. v for the 'heterogeneity' value is the value from (ii)−the value from (i). The critical value is found at $p = 0.05$ and the heterogeneity degrees of freedom from (iii).

5. How to find $\chi^2_{critical}$

i. There are two categories or columns (purple and white), so $v = 2−1 = 1$.
ii. There are four accessions (rows), each with 1 degree of freedom. Therefore 'summed' $v = 1 + 1 + 1 + 1 = 4$.
iii. For this example, the 'heterogeneity' v is 4−1 = 3. The critical value to test for heterogeneity in the data is found in the chi-squared table at $v = 3$, $p = 0.05$ and is $\chi^2_{critical} = 7.81$.

6. The rule

If $\chi^2_{calculated}$ is greater than $\chi^2_{critical}$, you may reject the null hypothesis.

6. The rule

$\chi^2_{calculated}$ (5.04) is less than $\chi^2_{critical}$ (7.81) at $p = 0.05$ and therefore we do not reject the null hypothesis.

7. What does this mean in real terms?

7. What does this mean in real terms?

There is no statistically significant heterogeneity ($\chi^2_{calculated} = 5.04$, $p = 0.05$) between the accessions of *Allium schoenoprasum* and it is therefore reasonable to sum the data across all accessions and use a goodness-of-fit chi-squared test on the totals.

Further examples relating to the topic, including how to use statistical software, are included in the Online Resource Centre.

In Chapter 7 we outlined an undergraduate investigation which examined an association between a single nucleotide polymorphism (SNP) in the gene TH2 with obsessive compulsive behaviour (Example 7.2). From these data you can calculate the observed frequency of the T and C alleles as 0.59 and 0.41 respectively. A common model used in population genetics is the Hardy–Weinberg model. The observed allele frequencies can be used to estimate the numbers of each genotype that would be expected if the sample was in Hardy–Weinberg equilibrium. These values have been included in Table 8.6. Which of the goodness-of-fit tests outlined in this chapter would be suitable to test whether these data are in the Hardy–Weinberg equilibrium? Are all the criteria for using this test met? If this goodness-of-fit test was calculated what are the degrees of freedom?

Table 8.6 Hardy–Weinberg equilibrium and the T and C allele frequencies of rs4565946 in the TH2 gene in participants with and without obsessive compulsive behaviours (OCD)

	Genotypes of rs4565946 SNP in TH2 gene		
	TT	TC	CC
Total number of participants	10	33	2
Expected number of participants if the sample is in Hardy–Weinberg equilibrium	15.6645	21.7710	7.5645

Summary of Chapter 8

- The statistical tests considered in this chapter are the chi-squared goodness-of-fit tests which test the generic hypothesis 'Do the data match an expected ratio?'
- The chi-squared goodness-of-fit tests may be used if you have count data and wish to examine the effect of one treatment variable, where the data are organized into categories.
- These tests may be used when you have one or more samples or have replication and can test whether your samples are heterogeneous.
- The chi-squared goodness-of-fit tests require you to work out expected values, which are then compared with your observations. In the goodness-of-fit tests, these expected values are derived from *a priori* models.
- There is concern over the reliability of these tests when sample sizes are small (9.4).
- Common errors in the use of these tests are the use of percentage data and organizing the data incorrectly in the contingency table so that the wrong comparisons are made.
- The Online Resource Centre includes interactive exercises that test your understanding of this chapter in combination with other topics, particularly those considered in Chapters 3 and 7–11.

Answers to chapter questions

A1   The student expects the numbers of respondents to be the same in each category. This is the first type of 'expectation' described in 8.1.

A2   The *a priori* expectation is that if the colour of the pitfall trap does not affect beetle behaviour, then there should be equal numbers of beetles in all the pitfall traps. As the total number of beetles observed was $20 + 34 + 10 + 40 = 104$, we would expect $104/4 = 26$ beetles in each trap.

A3   In this example it is clear that we have sampled from a population and have reason to expect a particular genotypic ratio, one that conforms to the Hardy–Weinberg model. You can see from the annotations on Table 8.7 there is only one sample so we exclude the goodness-of-fit test for replicates (8.5). We are not seeing if our data fit a particular distribution where expected values are derived from the mean of the sample, so we can exclude 8.4 and choose the test outlined in 8.3. The annotation on Table 8.7 shows that the criteria for using this test are met. In Box 8.1 it explains that the degrees of freedom are the number of categories (a)–1. In this example the degrees of freedom are therefore $3–1 = 2$.

Table 8.7   Annotated table indicating key features that support the answer to Q3. Are the genotype frequencies for the rs4565946 SNP in the TH2 gene in Hardy–Weinberg equilibrium?

> There is one row of observed values. The data are numbers of items. Each person has only been counted once. In this example the categories are based on a nominal scale.

> There is one treatment variable: the number of individuals with specific genotypes for the rs4565946 SNP in TH2 gene. There are three nominal categories: TT, TC, CC.

	Genotypes of rs4565946 SNP in TH2 gene			
	TT	TC	CC	Total
Total number of participants	10	33	2	45
Expected number of participants if the sample is in Hardy–Weinberg equilibrium	15.6645	21.7710	7.5645	

> The sample size is more than 25.

9 Hypothesis testing:
Associations and relationships

In a nutshell

In this chapter, we consider the second hypothesis that may be tested: a test for an association. An association is where one variable changes in a consistent and similar manner to another variable. There are three groups of statistical tests that we cover that can be used: chi-squared tests for association, correlation analysis, and regression analysis. We discuss what is meant by an association and a relationship, and issues relating to limited experimental designs and small sample sizes. Each test is illustrated with an example from undergraduate research. The statistics tests covered in this chapter are:

In your research, you often investigate the effect of one treatment variable, for example, a comparison of snail shell patterns found in coastal and woodland populations of *Cepaea nemoralis* (Example 9.1). The variable being examined here is 'habitat'. But there are some investigations where you will have more than one treatment variable, such as when comparing leg length and arm length in humans (Example 9.7). In this chapter, therefore, we look at testing hypotheses where you wish to evaluate the possibility of an association between these two variables. This is the second of the three types of hypotheses we discussed in

Chapter 6 (6.1.3). We cover three groups of tests: chi-squared tests for association, correlations, and regressions. If you work through this chapter and the questions, it should take about 2 hours. The answers for these exercises are at the end of the chapter. Most of our examples are based on real undergraduate research projects. If these examples are not in your subject area you will find more in the Online Resource Centre.

9.1 Associations and relationships

It is important to be clear about what we mean by the terms 'association' and 'relationship'. In some text books, these terms are used interchangeably, but we believe it is important to make a distinction. An association is where one treatment variable is found to change in a similar manner to another treatment variable (e.g. Fig. 9.15). A relationship is where one treatment variable is *directly* responsible for causing the change in another treatment variable (e.g. Fig. 9.14).

Association
Where one treatment variable is found to change in a similar manner to another treatment variable.

    Most samples of humans in which arm and leg length are recorded show a significant association between these two variables: if you have relatively long arms, you will also have relatively long legs (Example 9.7). But you would not argue that your arm length directly causes your leg length. This is not sensible. Instead, this association reflects the actions of other phenomena that have not been observed. Therefore, although you may have a significant association, this does not automatically indicate a relationship. You can contrast this example with a significant association observed between the hardness of eggshells and the amount of feed supplement eaten by chickens (Example 9.4). In this case, you could argue that nutritional factors in the feed supplement directly influence eggshell hardness. This is an association for which you can argue that there is a relationship.

Relationship
Where one treatment variable is directly responsible for causing the change in another treatment variable.

    In an association the change in one variable (y) may well be caused by the change in the second variable (x), or the change in the second variable may be caused by the change in the first variable. These are both relationships. In addition to this, a significant association may indicate that the changes in the first and second variables are due to a change in a third, unrecorded variable, as described in our arms and legs example. Alternatively, the statistical association between two variables may be a coincidence. You will need to bear these points in mind when you are interpreting the results from any significant test for association.

    In Chapter 7 we take you through a key to enable you to identify whether you need to use a chi-squared test for association or a correlation or a regression. Your choice tends to be dependent on whether your data are counts and you would plot them on a histogram or bar chart (chi-squared test for association) or whether the data would be plotted on a scatter plot and the distribution tends to be either linear (regressions) or monotonic (correlations).

9.2 Modelling the association

Tests of association can be used for three things: modelling, prediction, and establishing whether there is a significant association. In regressions the line that may be used to describe the trend in the data is a model of this association. For example, in Fig. 9.16, the line that best fits the data from Table 9.8 is shown and indicates a positive association between the amount of feed supplement eaten by pullets and eggshell hardness. This line is described by the equation $y = -2.36 + 0.48x$ (see Box 9.9). This line can also be used to predict a y value for any given x value within the range of your data. For example, if the pullets were given 10g of feed supplement (x), then the hardness of their eggshells would be expected to be about 2.44 units. If these eggshells were hard enough for the eggs to be sold, then the feed supplement given to the chickens could be more economically controlled. For both of these applications (modelling and prediction), you should choose a regression analysis. However, if you wish to establish whether there is a significant association between the two variables, then a correlation analysis or chi-squared test for association are more likely to be appropriate, but you should read on before making your final choice.

9.3 Chi-squared test for association

There are three types of investigations that can generate data suitable for analysis by a chi-squared test for association. These experimental designs differ by how many of the totals are 'fixed' by the investigator. (Note there are different experimental designs or models for regressions and ANOVAs, each are specific for their own group of tests. The models we are describing in this section are not the same as the models we refer to in relation to regressions (9.9) or ANOVAs (Chapter 10).)

- Model 1

  In the first design (Model 1) only the grand total is fixed by the investigator. For example, in a study of 200 anthills the number of active and inactive hills were recorded along with the presence or absence of *Thymus drucei*, a plant commonly associated with anthills. This design is illustrated in Table 9.1 and you can see that the grand total has been fixed by the investigator at $N = 200$.

- Model II

  In a second design the two row totals are fixed by the investigator. For example, in an investigation into an association between hydration and performance 100 volunteers were allocated at random to one of two groups. In one group the participants were then given 2l of water to drink 2 hours before the study, whilst the second group was not given any water to drink. The participants were then tested to see if they could complete a set task within a specific time limit (Table 9.2).

Table 9.1 Association between activity on 200 anthills and the presence or absence of *Thymus drucei*

Presence or absence of T. drucei	Activity of anthills		
	Active	Inactive	Total
Present			
Absent			
Total			$N = 200$

Table 9.2 The effect of hydration on performance

Hydration of participants	Effectiveness at completing the task		
	Completed task	Did not complete task	Total
Hydrated			$n = 50$
Not hydrated			$n = 50$
Total			

> In this design it is the two rows, the numbers of participants who were or were not given 2l of water to drink, that was determined or fixed by the investigator.

- Model III

> The third design is when both row totals and column totals are fixed. Whilst this is theoretically possible, in a biological setting it is difficult to see when this might arise.

There are a number of statistical tests that are designed to test the associations in experiments with these designs. The G test can be used for Models I and II. The Fisher's exact test has been designed for Model III and the chi-squared test is suitable for all of these models. However, in practice all three types of tests are generally used for all models and seem to work reasonably well. The chi-squared test is most commonly in use so this is the test we cover in this chapter. However, chi-squared tests are not thought to be reliable when there are only two rows and two columns or when the sample size is small. Then the Fisher's exact test or the Barnard's test is the preferred option. It is not sensible to calculate either Fisher's exact test or Barnard's test by hand. These tests are therefore demonstrated in the Online Resource Centre. We return to the issues of limited designs and small sample sizes in 9.4.

As in all chi-squared tests, the data are arranged in a contingency table. If the table has more than two rows and two columns this is referred to as a generalized or $r \times c$ test.

Example 9.1 Shell colour in *Cepaea nemoralis* in coastal and hedgerow habitats

An investigation was carried out into the frequency of banding and colour patterns on the shells of the native snail *Cepaea nemoralis* and the habitat the snail was found in (Table 9.3). The association that is under investigation is therefore between the variables 'shell pattern' and 'habitat'.

(continued)

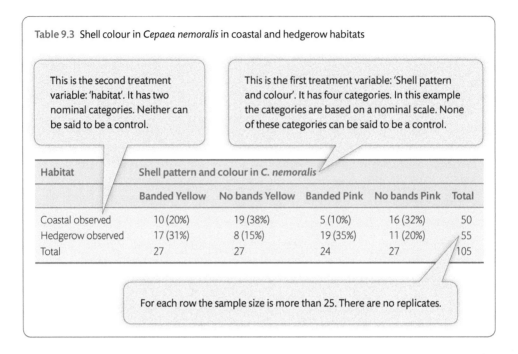

Table 9.3 Shell colour in *Cepaea nemoralis* in coastal and hedgerow habitats

This is the second treatment variable: 'habitat'. It has two nominal categories. Neither can be said to be a control.

This is the first treatment variable: 'Shell pattern and colour'. It has four categories. In this example the categories are based on a nominal scale. None of these categories can be said to be a control.

Habitat	Shell pattern and colour in *C. nemoralis*				
	Banded Yellow	No bands Yellow	Banded Pink	No bands Pink	Total
Coastal observed	10 (20%)	19 (38%)	5 (10%)	16 (32%)	50
Hedgerow observed	17 (31%)	8 (15%)	19 (35%)	11 (20%)	55
Total	27	27	24	27	105

For each row the sample size is more than 25. There are no replicates.

9.3.1 Key trends and experimental design

This is a Model II contingency table where the row totals, the numbers of snails sampled in each habitat, are fixed by the investigator. There are no replicates and no controls. To help you identify the key trends in these data you might plot them on a bar chart. Alternatively you can examine the percentages which we have added to Table 9.3 to make it easier to appreciate what is happening here. It is clear that in the coastal region the most common *Cepaea nemoralis* have no bands on their shell and this contrasts with the snails found in the hedgerows which tend to have banded shells. An analysis will confirm whether this apparent association between habitat and shell pattern is statistically significant or has arisen due to sampling error (1.7).

9.3.2 Using this test

i. To use this test you:

1) Do not have an *a priori* reason for expecting certain outcomes from your investigation but do wish to test for an association between two treatment variables.

2) Have data that are organized into two* or more discrete categories.

3) Have data that are counts or frequencies and are not percentages or proportions.

4) Have observations that are independent of each other.

5) Have a sample size (row totals) greater than 25*.

6) Have expected values that are greater than 5*.

* There is an ongoing debate about the effect of limited designs and small sample sizes on the reliability of chi-squared tests. We consider this in 9.4 and illustrate one solution in Box 8.2.

ii. Does the example meet these criteria?

For Example 9.1 the annotations on Table 9.3 and the expected values on Table 9.4 indicate that all the criteria for using this test for association are met. We can therefore proceed with the calculation.

9.3.3 The calculation

This is the calculation for an $r \times c$ chi-squared test for association. At this point, you will have arranged your data in a contingency table and have checked the criteria for using this test. You can now proceed to test for an association between the two variables (Box 9.1 and Table 9.4). Where steps have been abbreviated, the full calculation is included in the Online Resource Centre.

Table 9.4 Contingency table with expected values for a chi-squared test for association using data from Example 9.1: Shell colour in *Cepaea nemoralis* in coastal and hedgerow habitats

Habitat	Shell pattern and colour in C. *nemoralis*				
	Banded Yellow	No bands Yellow	Banded Pink	No bands Pink	Row Totals
Coastal observed	10	19	5	16	50
Coastal expected	$27/105 \times 50$ $= 12.85714$	$27/105 \times 50$ $= 12.85714$	$24/105 \times 50$ $= 11.42857$	$27/105 \times 50$ $= 12.85714$	
Hedgerow observed	17	8	19	11	55
Hedgerow expected	$27/105 \times 55$ $= 14.14206$	$27/105 \times 55$ $= 14.14206$	$24/105 \times 55$ $= 12.57143$	$27/105 \times 55$ $= 14.14206$	
Column Totals	27	27	24	27	105

BOX 9.1 How to calculate an $r \times c$ chi-squared test for association

GENERAL DETAILS	EXAMPLE 9.1
	This calculation is given in full in the Online Resource Centre. For presentation purposes only all values have been rounded to five decimal places.
1. Hypotheses to be tested H_0: There is no association between the two variables. H_1: There is an association between the two variables.	**1. Hypotheses to be tested** H_0: There is no association between the distribution of shell patterns observed and the habitat (coastal and hedgerow) of *Cepaea nemoralis*. H_1: There is an association between the distribution of shell patterns and the habitat (coastal and hedgerow) of *Cepaea nemoralis*.
2. Have all the criteria for using this test been met?	**2. Have all the criteria for using this test been met?** Yes (9.3.2ii).
3. How to work out expected values In chi-squared tests for association, you have no *a priori* model against which to compare your observed data. Instead, the expected values are calculated from the totals of each column and row.	**3. How to work out expected values** Look first at the expected value in row 1, column 1 in Table 9.4. To calculate this expected value, take the column total for banded yellow (27) ÷ grand total (105) × row total for coastal snails (50) = 12.85714. This calculation is repeated for each row × column combination.
4. How to work out χ^2calculated The rest of the procedure is the same as that described for the goodness-of-fit chi-squared test, where: $$\chi^2_{calculated} = \Sigma \left[\frac{(observed - expected)^2}{expected} \right]$$	**4. How to work out χ^2calculated** $$\chi^2 = \frac{(10-12.85714)^2}{12.85714} + \frac{(19-12.85714)^2}{12.85714}$$ $$+ \frac{(5-11.42857)^2}{11.42857} + \frac{(16-12.85714)^2}{12.85714}$$ $$+ \frac{(17-14.14206)^2}{14.14206} + \frac{(8-14.14206)^2}{14.14206}$$ $$+ \frac{(19-12.57143)^2}{12.57143} + \frac{(11-14.14206)^2}{14.14206}$$ $$\chi^2_{calculated} = 15.18472$$
5. How to find $\chi^2_{critical}$ See Appendix D, Table D1. You now need to calculate the degrees of freedom (v) before looking up the critical value. In this case, v is the (number of rows −1) × (number of columns −1). The critical value is found in the statistical table at $p = 0.05$ and the degrees of freedom just calculated.	**5. How to find $\chi^2_{critical}$** In our example, there are two rows (coastal and hedgerow). Do not include your 'expected' rows as these are part of your calculation. There are four columns (banded yellow, no bands yellow, banded pink, no bands pink). Therefore: $v = (2-1) \times (4-1)$ $= 1 \times 3 = 3$. The critical value where $v = 3$, $p = 0.05$, is $\chi^2_{critical} = 7.81$.

(continued)

6. The rule	6. The rule
If $\chi^2_{calculated}$ is greater than $\chi^2_{critical}$ you may reject the null hypothesis.	$\chi^2_{calculated}$ (15.18) is greater than $\chi^2_{critical}$ (7.81) at $p = 0.05$ and therefore we reject the null hypothesis.
7. What does this mean in real terms?	7. What does this mean in real terms?
	There is a significant association ($p = 0.05$) between the distribution of shell patterns and habitat (coastal and hedgerow) of *Cepaea nemoralis*. In fact, at $p = 0.01$, $\chi^2_{critical} = 11.34$ and $p = 0.001$, $\chi^2_{critical} = 16.27$. Therefore, you may reject the null hypothesis at $p = 0.01$ but not at $p = 0.001$. This can be written as: there is a highly significant association ($\chi^2_{calculated} = 15.18$, $0.01 > p > 0.001$) between the distribution of shell patterns and the habitat of *Cepaea nemoralis*.

Further examples relating to the topic including how to use statistical software are included in the Online Resource Centre.

 Example 9.2 Frequency of *Cepaea nemoralis* and *Cepaea hortensis* in a woodland and a hedgerow

A survey of two species of snail was carried out at two habitats: a woodland and a hedgerow. The numbers of snails at each location were recorded (Table 9.5). The investigators wished to test whether there was an association between the location and the numbers of each species of snail and for this decided to use a chi-squared test for association. Calculate the expected values.

Table 9.5 The distribution of *Cepaea nemoralis* and *Cepaea hortensis* in a woodland and a hedgerow

	Species of snail		Total
	C. nemoralis	*C. hortensis*	
Hedgerow observed	89	59	148
Woodland observed	16	8	24
Total	105	67	172

9.4 The problem with small numbers and limited designs

Chi-squared tests, both those considered in Chapter 8 and those considered here, are believed to be limited in their reliability when: the sample size is small; the number of expected values is less than 5; or you have a 2×2 contingency table with only 1 degree of freedom. We consider each of these points here.

9.4.1 Your sample size is small

When designing an experiment one key consideration is to ensure that your sample size is representative. Small samples are rarely that and therefore a small sample usually indicates a weakness in the design. However, sometimes experiments do not work well or the population itself is small. In these instances, despite your best intentions, the data set you wish to analyse may be limited.

There is considerable debate in the literature as to the minimum size of sample for which chi-squared tests are reliable. Suggestions vary widely from sample sizes of 25 to more than 1000. The value of 25 which we have given in our criteria (e.g. 8.3.2 and 9.3.2) is clearly at the lower end of this range. As it is known that the chi-squared tests do not perform well with a small sample size, unless you have a large sample size of >500 we would encourage you to read the current literature before determining which test to use.

Fisher's exact test
An alternative to any chi-squared test particularly when sample sizes are small or you have a 2 × 2 contingency table. This is best calculated using statistical software.

The Fisher's exact test was developed as a method for analysing the model III design. It is commonly used, however, for all three models and is not susceptible to the same issues of robustness that the chi-squared test suffers from. However, the Fisher's test is mathematically more complex to calculate and is therefore best suited to analysis using statistical software. Fortunately this test is found in most mainstream software packages and we walk you through how to use the Fisher's exact test in the Online Resource Centre.

9.4.2 Your expected values are less than five

One of the criteria for using the chi-squared tests is a requirement for all expected values to be greater than 5. There is some debate about this requirement, but it is generally accepted that, as long as no more than 20% of all expected numbers are less than 5 and none is less than 1, then you may proceed with these tests.

Clearly, therefore, you need to bear this in mind when designing your investigations. Ideally, you need to carry out a preliminary investigation to get a sense of the relative numbers you would expect in each category. If you look at Example 9.2, you can see that not many *Cepaea hortensis* are sampled from the woodland. The sampling strategy used should therefore ensure that adequate numbers of *Cepaea hortensis* are included in the sample for analysis. (In fact this criterion is met in this Example, see A1.) Another example that we frequently come across is when students wish to use a questionnaire as a tool in their honours-year research project. In a study with, for example, 50 participants, if your closed question has six possible answers these 50 responses will be spread across a relatively large number of options. If your closed question only has three possible answers then the responses will only be spread across a smaller number of outcomes. The fewer the categories the less likely you are to have expected values that are less than 5 (3.5).

Even with the best planning, you may still find yourself with expected values less than 5. In these circumstances, you may, if it is sensible, combine categories. For example, in Example 9.3 the results from a survey of the age distribution of oak trees (*Quercus petraea*) at the Wyre Forest and Mortimer Forest, Shropshire, are shown. Initially, the investigator placed the trees in one of four categories: seedling, sapling, immature tree, and mature tree (Table 9.6).

209

Example 9.3 Maturity of *Quercus petraea* in the Wyre Forest and Mortimer's Forest, Shropshire

Quercus petraea was surveyed at random in two woodlands in south Shropshire (the Wyre Forest and Mortimer's Forest). To establish the age distribution of the trees each tree was placed in one of four categories (seedling, sapling, immature tree, and mature tree) (Table 9.6).

Table 9.6 Maturity of *Quercus petraea* at the Wyre Forest and Mortimer Forest: the problem with small numbers in a chi-squared test

Location	Maturity of *Quercus petraea*				
	Seedling	Sapling	Immature tree	Mature tree	Total
Wyre Forest observed	51	21	29	4	105
Wyre Forest expected	51.3	23.3	26.4	3.8	
Mortimer Forest observed	15	9	5	1	30
Mortimer Forest expected	14.6	6.6	7.5	1.1	
Total	66	30	34	5	135

You can see that two of the expected values are less than 5. This is more than 20% of the expected values. In addition, one of the values is about 1. To enable you to analyse these data with a chi-squared test, it is therefore necessary to combine some of these categories and for these data it would appear both biologically acceptable and mathematically useful to combine the data from the immature and mature trees. The revised contingency table and new expected values are shown in Table 9.7.

The expected values are now all greater than 5. This approach has therefore been effective. However, you can now only comment on the seedlings, saplings, and trees in the two woodlands as you have no statistics comparing the immature and mature trees.

This approach may be used for all cases where the expected values are less than 5 with one exception. The exception arises when you wish to carry out a goodness-of-fit test and you

Table 9.7 Revised contingency table with combined classes with data from a survey of the maturity of *Quercus petraea* at the Wyre Forest and Mortimer Forest

Location	Maturity of *Quercus petraea*			
	Seedling	Sapling	Mature tree and immature tree	Total
Wyre Forest observed	51	21	33	105
Wyre Forest expected	51.3	23.3	30.3	
Mortimer Forest observed	15	9	6	30
Mortimer Forest expected	14.6	6.6	8.6	
Total	66	30	39	135

have an *a priori* expectation, but have interval data, so the mid-point is used when calculating the expected values from the predicted ratio (8.4). The relative relationship between the mid-point and predicted ratio becomes disturbed when classes are combined and incorrect expected values will be calculated. In these circumstances, you should work out the expected values for the original classes and then add the expected values together in the classes you are combining (e.g. Box 8.2).

9.4.3 Chi-squared test for a 2 × 2 contingency table

A chi-squared test for association may be carried out on a 2 × 2 design. Here both treatment variables have only two categories (e.g. Table 9.5). When the degrees of freedom are calculated then this value is 'one'. Investigations into the robustness of chi-squared tests have shown that when the degree of freedom is 1 then the chi-squared test will tend to return more significant outcomes than it should. The use of a modification to the test called the Yates correction was a common response to this problem. However, the Yates correction tends to go too far the other way and is too conservative. Common practice now is to not apply the Yates correction but to calculate the chi-squared test anyway. The alternate is to use the Fisher's exact test or the Barnard's test in this situation as both are thought to perform better.

 Three chi-squared models are described in 9.3. Which model is Example 9.3. (Table 9.6.)? Are there any replicates or controls in this design?

9.5 Correlations

If you have data with two treatment variables and you plot this on a scatter plot, it may suggest that there is an association between the two variables. In Example 9.4 the investigator wishes to see if the hardness of eggshells laid by her pullets is affected in a consistent fashion by the amount of a food supplement they have eaten, i.e. is there an association? Unlike the examples we showed for the chi-squared tests this experimental design generates data that are most sensibly plotted on a scatter plot (Fig. 9.1). A correlation analysis examines the data mathematically to see how probable this association is and estimates the degree to which two variables vary together.

Example 9.4 The hardness of eggshells in pullets

A student believes that there is a relationship between the hardness of shells in eggs laid by Maran pullets and the amount of a particular layer's pellet eaten as part of their mixed diet. She selected 13 of her pullets at random and on one day recorded the amount of layer's pellets each consumed and the hardness of the eggs laid, on a scale of 0.0 (softest) to 10.0 (hardest) (Table 9.8).

(continued)

Table 9.8 The hardness of eggshells produced by 13 Maran pullets and their consumption of a food supplement

The items are the pullets.

Two measurements have been recorded for each item: 'amount of food supplement (g)' and 'hardness of eggshell'. Both are interval scales and are random variables.

Pullet	Amount of food supplement (g)	Hardness of shells
1	19.5	7.1
2	11.2	3.4
3	14.0	4.5
4	15.1	5.1
5	9.5	2.1
6	7.0	1.2
7	9.8	2.1
8	11.6	3.4
9	17.5	6.1
10	11.2	3.0
11	8.2	1.7
12	12.4	3.4
13	14.2	4.2

One measure of this association is r, the correlation coefficient. The correlation coefficient may be either positive or negative in sign. A positive sign indicates that there is a positive association between the two variables: as one increases so does the other (Fig. 9.2). A negative sign indicates a negative association, as one variable increases the other decreases (Fig. 9.3).

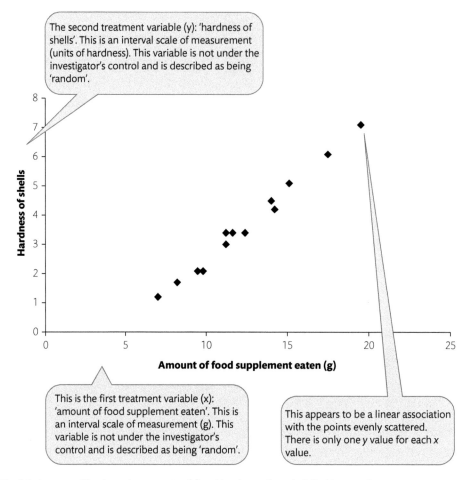

Fig. 9.1 Amount of food supplement eaten (g) and hardness of eggshells in Maran pullets

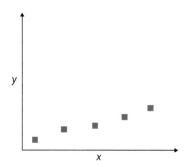

Fig. 9.2 A positive association: as *x* increases, *y* increases

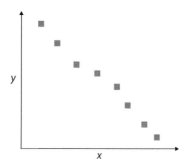

Fig. 9.3 A negative association: as *x* increases, *y* decreases

---

Q3 Is there an apparent association between the two variables illustrated in Fig. 9.4?

A correlation coefficient will lie somewhere between −1 and +1. If you have a value greater than +1 or less than −1, then you have made a mistake somewhere in your calculation. If $r = +1$ or $r = −1$ this indicates a 'perfect' association between the two variables. However, when $r = 0$ the interpretation is less straightforward. An *r* value of zero can mean that there is no association between the two variables. However, it can also mean that the association is not linear or not monotonic (9.6).

A correlation coefficient itself cannot tell you if this association is biologically meaningful. This depends on a range of factors including sample size and the context. However, the coefficient of determination (9.8) which is derived from *r* provides an estimate of the degree to which other factors influence *x* and *y* when all the criteria for its use are met. The correlation coefficient also does not tell you whether this is a statistically significant association. For example, $r = 0.2$ indicates a weak association and $r = 0.8$ indicates a stronger association. Further analysis, however, is required to test whether both or neither of these are statistically significant.

If the analysis indicates a mathematically significant correlation this generally supports the interpretation of a biologically meaningful association between the two factors. However, sample size is an important element in this analysis. For example, a strong association such

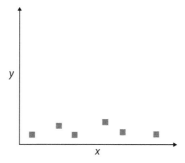

Fig. 9.4 Is there an association between *x* and *y*?

as $r = 0.9$ is likely to be statistically significant, even with a small sample size. However, a small correlation such as $r = 0.2$ may be statistically significant, but only if the sample size is large enough. Therefore a large r and a significant correlation is strong evidence for a biologically important association. The value of a significant association with a small r is less easily either accepted or dismissed.

9.6 Spearman's rank correlation

A Spearman's rank correlation is a ranking test. There are many ranking tests that can be used to test hypotheses. These tests are commonly used for the analysis of non-parametric data where the data do not have to fit a particular type of distribution. These tests are collectively known as ranking tests, because the first step, which they all have in common, requires you to assign ranks to your observations. If you are not familiar with this process, you should first read Box 5.3.

9.6.1 Key trends and experimental design

To illustrate the use of the Spearman's rank test we will use Example 9.4 The hardness of eggshells in pullets. From looking at Fig. 9.1, it is clear that as consumption of the feed supplement increases (for the range 7.0–19.5g) the hardness of the eggshell also increases, i.e. there appears to be a positive association and this increase appears to be linear. A statistical test on this apparent association is needed to confirm that this is mathematically significant or whether it is due to chance.

9.6.2 Using this test

i. To use this test you:

1) Do not have an *a priori* reason for expecting certain outcomes from your investigation but do wish to test for an association between two treatment variables.

2) Need two treatment variables recorded for each item.

3) Have two measures that are ordinal, or one may be ordinal and one interval, or both may be interval scales, and therefore both can be ordered in a consistent manner (rankable). These data are usually represented graphically as a scatter plot.

4) Can use this test if the underlying distribution in the population is not normal.

5) Can use this test if the underlying distribution in the population is normal. However, you would usually then consider using the Pearson's correlation (9.7).

6) Can use this test if the variables are not random but if one or both are fixed and under the control of the investigator. However, if both are fixed it would be more usual to consider a chi-squared test for association (9.3) or an ANOVA (Chapter 10).

7) Have a distribution that appears to be monotonic and the points are reasonably scattered (see Q4).

8) Do not need to have an independent and a dependent variable.

9) Have seven or more pairs of observations. (Having more than 30 pairs of observations does not add to the accuracy of this test.)

10) Should have few tied values, i.e. few of the values in either the x and/or the y data set should be the same as each other. For example if you have 8 x values 2, 2, 2, 2, 2, 4, 3, 2. There are six 2s and these are tied values. The same applies for y values. This can look like an association, but it is not (see Q3).

 Examine Figs 9.2, 9.3, and 9.5(a–d). Which of these fulfil criterion 7? For those distributions that do not meet criterion 7, what can be done about it?

---

ii. Does the example meet these criteria?

The results in Example 9.4 meet the criteria for using Spearman's rank correlation. For each pullet (an item), there are two observations, one for each treatment variable: 'g' and 'hardness'. Observations on these interval scales can be ranked. There are 13 pairs of observations. When plotted on a scatter plot, there appears to be a linear association and the points are reasonably scattered (Fig. 9.1). Both of these variables are random. The highest number of tied values is where there are three 3.4 hardness of eggshells values. There are two 2.1 hardness of shell values and two 11.2 g consumption of feed supplement values. This constitutes few tied ranks and is therefore acceptable. We may therefore proceed with this test.

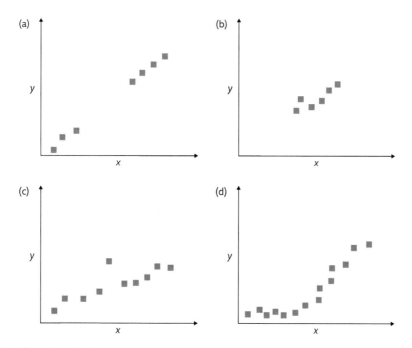

Fig. 9.5 (a–d) Do these distributions meet the criteria for using a correlation analysis?

9.6.3 The calculation

We show you the general calculation and a specific example (Example 9.4) for Spearman's rank correlation in Box 9.2. In addition, you will need to refer to the calculation table (Table 9.9). Full details of all steps in these calculations are given in the Online Resource Centre. To distinguish between the r value calculated using a Spearman's rank test and the Pearson's product moment correlation we refer to this r value as r_s.

BOX 9.2 How to carry out Spearman's rank correlation

GENERAL DETAILS	EXAMPLE 9.4
	This calculation is given in full in the Online Resource Centre. For presentation purposes only all values have been rounded to five decimal places.
1. Hypotheses to be tested	1. Hypotheses to be tested
H_0: There is no correlation between variable 1 and variable 2 in the population from which the samples are taken. H_1: There is a correlation between variable 1 and variable 2 in the population from which the samples are taken.	H_0: There is no correlation between the hardness of shells and the amount of food supplement (g) eaten by 13 Maran pullets. H_1: There is a correlation between the hardness of shells and the amount of food supplement (g) eaten by 13 Maran pullets.
2. Have the criteria for using this test been met?	2. Have the criteria for using this test been met? Yes (9.6.2ii).
3. How to work out $r_{s\ calculated}$ i. For this test it does not matter if you have a dependent and an independent variable; therefore, let the variable on the x axis be variable 1 and the variable on the y axis be variable 2. ii. Arrange the data for variable 1 in numerical order from the smallest to the largest. iii. Assign each observation a rank. Where there is a tie, assign the middle (average) rank. If you are unfamiliar with this process, see Box 5.3. iv. Repeat steps (i) and (ii) for variable 2. v. Return to your original table of data and note down the ranks for each observation. Make sure you do not reorganize the data. The two observations for each item must remain in the same row as each other. vi. Calculate the difference between the ranks for each pair of observations (d). vii. Square each d value and then sum all the d^2 to work out Σd^2. viii. Calculate r_s: $$r_s = 1 - \frac{6\sum d^2}{n(n^2-1)}$$ where n is the number of pairs of observations.	3. How to work out $r_{s\ calculated}$ i. Let the amount of food supplement eaten by the pullets be variable 1 and shell hardness be variable 2 (Table 9.8). ii–v. If you are working this calculation out by hand, it is simplest to use a calculation table. The outcomes from steps (ii)–(vi) are shown in Table 9.9 using the data from Example 9.4. A common error is to reorganize the data in this table to reflect the ranks. This can lead to · the two observations from one item no longer being paired on the same row. If you are not sure, compare Table 9.8 with Table 9.9. The observations remain in the same order in relation to each other. vi, vii. The difference between the ranks and sum of ranks has been added to the calculation table (Table 9.9). viii. In this example, $n = 13$ pairs of observations. So: $$r_s = 1 - \frac{6\times 6.0}{13(13^2-1)} = 0.98352$$

(continued)

4. How to find $r_{s\ critical}$	4. How to find $r_{s\ critical}$
See Appendix D, Table D2. Find $r_{s\ critical}$ in a Spearman's rank r table for a two-tailed test at $p = 0.05$ and where n = the number of pairs of observations.	When $p = 0.05$ and $n = 13$, then $r_{s\ critical}$ is 0.566.
5. The rule	5. The rule
If the absolute*, calculated value of r_s is greater than the critical r_s, we reject H$_0$. (*If you are not familiar with this term please refer to the glossary.)	In our example, r_s was positive, so the fact that we take the absolute value will not change r_s in this example. $r_{s\ calculated}$ (0.984) is greater than $r_{s\ critical}$ (0.566) at $p = 0.05$. In fact, at $p = 0.001$, $r_{s\ critical}$ is 0.824. Therefore, we can reject the null hypothesis at this higher level of significance.
6. What does this mean in real terms?	6. What does this mean in real terms?
	There is a strong and very highly statistically significant positive correlation ($r_s = 0.984$, $p = 0.001$) between the hardness of shells of eggs and the amount of food supplement (g) eaten by Maran pullets.

Further examples relating to the topic including how to use statistical software are included in the Online Resource Centre.

Table 9.9 Calculating Spearman's rank correlation for Example 9.4: The hardness of shells and consumption of food supplement in Maran pullets

Amount of food supplement (g)	Rank	Hardness of shells	Rank	Difference (d)	Difference^2 (d^2)
19.5	13.0	7.1	13.0	0.0	0.00
11.2	5.5	3.4	7.0	−1.5	2.25
14.0	9.0	4.5	10.0	−1.0	1.00
15.1	11.0	5.1	11.0	0.0	0.00
9.5	3.0	2.1	3.5	−0.5	0.25
7.0	1.0	1.2	1.0	0.0	0.00
9.8	4.0	2.1	3.5	0.5	0.25
11.6	7.0	3.4	7.0	0.0	0.00
17.5	12.0	6.1	12.0	0.0	0.00
11.2	5.5	3.0	5.0	0.5	0.25
8.2	2.0	1.7	2.0	0.0	0.00
12.4	8.0	3.4	7.0	1.0	1.00
14.2	10.0	4.2	9.0	1.0	1.00
$n = 13$					$\Sigma d^2 = 6.0$

9.7 Pearson's product moment correlation

This is a parametric test and, unlike the Spearman's rank test, you do not rank the data. However, you do assume that your data have attributes of a normal distribution and that the association is linear, a more stringent requirement than that for the Spearman's rank test.

Like Spearman's rank correlation, the coefficient (r) may indicate the strength and sign of the association that is apparent in your data (9.6) and may be used to test the significance of the association (see Box 9.2), or it may be used to calculate the coefficient of determination (9.8).

9.7.1 Key trends and experimental design

To illustrate the use of the Pearson's product moment correlation we will use Example 9.4 again. The key trends for this were outlined in 9.7 as an apparent positive linear association between the hardness of the eggshells and the amount of food supplement eaten by pullets. The data points are well spread out along the length of the apparent distribution and cover the range 7.0–19.5g.

9.7.2 Using this test

i. To use this test you:

1) Do not have an *a priori* reason for expecting certain outcomes from your investigation but do wish to test for an association between two treatment variables.

2) Have two treatment variables measured for each item. These data are usually represented graphically as a scatter plot.

3) Both variables are 'random' and not fixed and therefore neither is under the control of the investigator.

4) Have a reason to assume that the sample is drawn from a normally distributed population and therefore both variables should be measured on a continuous scale.

5) Have a distribution that appears to be linear when plotted, with points that are reasonably scattered (e.g. Figs 9.2 and 9.3).

6) Have at least three pairs of observations.

7) Have no requirement for there to be an independent or a dependent variable.

8) Have few or no outliers. If you have outliers you should use the Spearman's rank test instead.

ii. Does the example meet these criteria?

The Maran pullet data used in Example 9.4 met the requirements for using a Spearman's rank correlation but they also meet the more stringent criteria for using Pearson's correlation. If you have data that meet the criteria for a parametric test, this is the one you should choose as these tests are more powerful. (If you are not familiar with this term, refer to the glossary and 6.1.6.) The data from Example 9.4 meet the additional and more stringent criteria for the Pearson's correlation in that the distribution is apparently linear, the points are reasonably scattered and both variables are measured on interval scales. There are no outliers.

9.7.3 The calculation

We show you the general calculation and a specific example (Example 9.4) for Pearson's correlation in Box 9.3 and the calculation table (Table 9.10). Where steps have been abbreviated, the full calculation is included in the Online Resource Centre.

BOX 9.3 How to carry out Pearson's product moment correlation

GENERAL DETAILS	EXAMPLE 9.4
	This calculation is given in full in the Online Resource Centre. For presentation purposes all values have been rounded to five decimal places.

1. Hypotheses to be tested

H_0: There is no correlation between variable 1 and variable 2 in the population from which the samples are taken.
H_1: There is a correlation between variable 1 and variable 2 in the population from which the samples are taken.

1. Hypotheses to be tested

H_0: There is no correlation between the hardness of shells and the amount of food supplement (g) eaten by 13 Maran pullets.
H_1: There is a correlation between the hardness of shells and the amount of food supplement (g) eaten by 13 Maran pullets.

2. Have the criteria for using this test been met?

2. Have the criteria for using this test been met?

Yes (9.7.2).

3. How to work out r

i. There is no requirement for there to be a dependent or an independent variable, so you may choose which variable should be x and which y.
ii. Calculate the sum of squares of x (SS (x)):

$$SS(x) = \sum x^2 - \frac{\left(\sum x\right)^2}{n}$$

where n is the number of pairs of observations. This term should be familiar to you as it has figured in the calculation for standard deviation and variances (Box 5.1).
iii. Calculate the sum of squares of y (SS(y)). For SS(y), substitute the 'y' variable data into the same formula as above, i.e.

$$SS(y) = \sum y^2 - \frac{\left(\sum y\right)^2}{n}$$

iv. Calculate the sum of products (SP(xy)). For your data, multiply each x by its related y to produce a column of values (xy). Add these together = $\sum xy$. Using this and values from (ii) and (iii), calculate SP(xy) as follows:

$$SP(xy) = \sum xy - \frac{\left(\sum x\right)\left(\sum y\right)}{n}$$

v. Now you can calculate r.

$$r = \frac{SP(xy)}{\sqrt{[SS(x)SS(y)]}}$$

Remember, r is the correlation coefficient and should fall in the range ±1. The sign indicates whether this is a positive or negative association.

3. How to work out r

i. Let the amount of food supplement eaten be the x variable and hardness of shells be the y variable.

ii. This is an important value that you will come across frequently in this book so we have shown you again in this example how to work this parameter out (Table 9.10).

$$SS(x) = 2153.88 - \frac{25985.44}{13}$$
$$= 155.0$$

iii. For details, see Table 9.10.

$$SS(y) = 208.35 - \frac{2237.29}{13} = 36.25077$$

iv. For details, see Table 9.10.

$$SP(xy) = 661.0 - \frac{161.2 \times 47.3}{13}$$
$$= 74.48$$

v. $r = \frac{74.48}{\sqrt{155.0 \times 36.25077}} = 0.99361$

This indicates a strong positive correlation. But is it statistically significant?

(continued)

4. To test the significance of the association	4. To test the significance of the association
See Appendix D, Table D3. To find the critical value of r, you need to know the degrees of freedom (v). In this Pearson's correlation, the degrees of freedom are the number of pairs of observations (n) – 2. Use an r table of critical values to locate the value at $p = 0.05$.	In this example, there are 13 pullets so $v = 13 - 2 = 11$. At $p = 0.05$, for a two-tailed test, $r_{critical} = 0.553$.
5. The rule	**5. The rule**
If $r_{calculated}$ is greater than $r_{critical}$, you may reject the null hypothesis.	As $r_{calculated}$ (0.994) is greater than $r_{critical}$ (0.553), you may reject the null hypothesis. In fact, at $p = 0.01$, $r_{critical}$ is 0.684. Therefore, we can reject the null hypothesis at this higher level of significance.
6. What does this mean in real terms?	**6. What does this mean in real terms?**
	There is a strong ($r = 0.99$) and highly significant positive correlation ($p < 0.01$) between the amount of food supplement (g) given to the Maran pullets and the hardness of the eggshells that they produce.

Further examples relating to the topic including how to use statistical software are included in the Online Resource Centre.

Table 9.10 Calculating Pearson's product moment correlation for Example 9.4 between the hardness of shells and food consumption in Maran pullets

Amount of food supplement (g) (x)	x^2	Hardness of shells (y)	y^2	$x \times y$
7.00	49.00	1.20	1.44	8.40
8.20	67.24	1.70	2.89	13.94
9.50	90.25	2.10	4.41	19.95
9.80	96.04	2.10	4.41	20.58
11.20	125.44	3.00	9.00	33.60
11.20	125.44	3.40	11.56	38.08
11.60	134.56	3.40	11.56	39.44
12.40	153.76	3.40	11.56	42.16
14.00	196.00	4.50	20.25	63.00
14.20	201.64	4.20	17.64	59.64
15.10	228.01	5.10	26.01	77.01
17.50	306.25	6.10	37.21	106.75
19.50	380.25	7.10	50.41	138.46
$\Sigma x = 161.20$	$\Sigma(x^2) = 2153.88$	$\Sigma y = 47.30$	$\Sigma(y^2) = 208.35$	$\Sigma xy = 661.00$
$(\Sigma x)^2 = 25985.44$		$(\Sigma y)^2 = 2237.29$		
$\bar{x} = 12.40$		$\bar{y} = 3.63846$		
$n = 13$		$n = 13$		

9.8 Coefficient of determination

$r^2 \times 100$ is a useful measure called the coefficient of determination (%). It is a measure of the proportion of variability in one treatment variable that is accounted for by variation in the other treatment variable and therefore indicates to what extent other factors are influencing x and y. The use of the coefficient of determination is dependent on whether your data satisfy the criteria for Pearson's product moment correlation and whether an r value can be calculated. From Box 9.3, we know that for Example 9.4. $r = 0.99361$. Therefore, $r^2 \times 100 = 98.7\%$. This indicates that nearly all the variation observed in the hardness of the eggshells is due to the variation in the amount of food supplement eaten.

9.9 Regressions

If a regression line is drawn through the data and the means of x and y, the observations may fall on the line or on one side of the line and/or the other. The distances between the observations and the hypothetical line are called residuals. In effect regressions minimize the residuals so that the line sits at the point where the positive and negative residuals are cancelled out (Fig. 9.6). The primary role of a regression analysis is to mathematically fit a line to your data using approaches like the one described.

A regression analysis therefore differs from a correlation analysis in that it calculates the mathematical equation that describes or models your data. This model may be used as a description of any significant association between the two variables and/or used to predict values. We describe three methods for modelling a linear regression. However, there are also ways to fit a curvilinear line. We have not been able to include these here, but we refer you to other texts (e.g. Legendre & Legendre, 2012; Sokal & Rohlf, 2011).

Curvilinear When data are plotted and where part or all of the distribution is found to curve rather than follow a straight line.

Regression analyses fall into two groups. The first (Model I) may be appropriate when you, the investigator, determine one of the variables. For example, you may set up an experiment to examine the number of pollen grains falling on agar plates set at specific distances from a field of oilseed rape. In this investigation, the location of the agar plates will be under your control and is not therefore subject to sampling error (2.2.5). If neither variable is under your control, then you should consider using a Model II regression.

In the Model I regression (the simple linear regression) the two important values that are calculated are the slope or gradient of the line (b) and the point at which the line cuts the

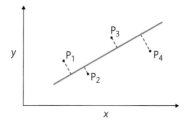

Fig. 9.6 Minimization of differences from a regression line. Differences from P_1 and P_3 to the line are positive; differences from P_2 and P_4 are negative.

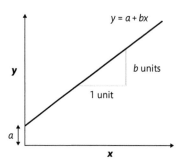

Fig. 9.7 The general line described by a Model I simple linear regression

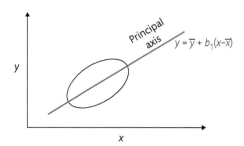

Fig. 9.8 The general line described by the Model II principal axis and ranged principal axis regressions

y axis (a). The general mathematical formula that describes this line is $y = a + bx$ (Fig. 9.7). In the Model II regressions (the principal axis regression and the ranged version of the principal axis regression) the line is determined by 'drawing' an ellipse around the data. The line that passes along the longest axis of the ellipse is called the principal axis, and this can also be found by calculating the slope b_1. In these Model II regressions, the general mathematical formula for the line (the principal axis) is $y = \bar{y} + b_1(x - \bar{x})$ (Fig. 9.8). The slope of the line, b or b_1, can be either positive or negative. These b values are referred to as regression coefficients.

You could fit a line through any data on a scatter plot. For example, you could draw a line on Fig. 9.9, although clearly this would be meaningless. Just drawing a line is not enough to indicate an association. It is therefore necessary to show whether this line represents a statistically significant trend in the data. For a simple linear regression, a modified t-test (see Box 9.5) or

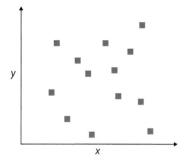

Fig. 9.9 Distribution of data with no obvious linear association

ANOVA (analysis of variance) (see Box 9.7) may be used. However, for the principal axis regression and ranged principal axis regression, a correlation should be carried out to test the significance of the association before making use of any regression analysis to then model the data.

Reporting correlations and regressions is usually achieved using a scatter plot. For a correlation, this is sufficient if the results from the correlation, including the correlation coefficient and p value, are included in the figure legend. If a regression analysis has been used, then it is common practice to add the regression line on the scatter plot and label this with the regression equation. We consider the topic of drawing lines on figures in more detail in Chapter 12, but while we are on the subject of regressions, we would like to make two points:

- A trend may not reflect an association or a relationship. Lines (we refer to them as trend lines) are often drawn on figures, especially by the common software packages. These lines indicate the apparent trend but have no statistical meaning. There are many common errors associated with these trend lines (12.1.10). One way to overcome these errors, if you have appropriate data, is to carry out a regression analysis to model a significant association. A significant regression line is a reasonable mathematical description of the trend in your data. You should show how important the line is by adding the regression equation (e.g. Fig. 9.16 compared with Fig. 9.1). A regression line that is shown not to be significant should not be used.

- Do not extend the regression line beyond your data. A common error is to extend the significant regression line beyond the data set. Q5 illustrates why this is not appropriate.

In the following sections, we describe the methods for drawing and testing a linear regression when you have two variables. However, regressions can be modified in a number of ways, for example if you have three or more variables and if your data are measured on a continuous scale. Here you might consider using a general linear model or generalized linear model, which is outside the scope of this book.

General linear and generalized linear model

Statistical approaches that may be taken when you have three or more treatment variables with data that are measured on a continuous scale and some of the apparent associations appear to be linear.

9.10 Model I: simple linear regression: only one y for each x

The first of the linear regressions we consider in this chapter is the simple linear regression where the line is determined by mathematically minimizing the distance between the observations and the line (Fig. 9.6). Clearly, in this process some distances will be negative and some positive and would cancel each other out, so these values are squared. This has the effect of making all the values positive. The regression 'fits' a line that minimizes these deviations from the line. This regression is also therefore known as a least-squares regression.

A simple linear regression analysis is the most commonly used type of regression. In this analysis, first the coefficients a and b are calculated that allow you to mathematically describe and to draw the regression line. To test the significance of this line, a modified t value is calculated. One of the criteria that must be met to use a simple linear regression is that one of the variables is taken without sampling error and is therefore usually under the investigator's control. This is always denoted the x variable, and as regressing y on x is not the same as regressing x on y, you must take care to identify which is the variable taken without sampling error (x) and which is not (y). Although a change in x may not directly cause a change in y, y is often referred to as the dependent variable and x as the independent variable.

Example 9.5 Heavy metal contamination of soil under electricity pylons

An undergraduate investigated heavy metal tolerance in plants growing under electricity pylons. As part of her study, she recorded the concentration of zinc in soil samples taken at regular intervals moving away from the pylons (Table 9.11). One of the objectives of her investigation was to see whether there was an association between the distance from the pylon and the concentration of zinc in the soil.

Table 9.11 Zinc concentrations in soil at specific distances from an electricity pylon

> There are five items: the points at which the measurements were taken. There are no replicates and no one category can be said to be a control.

> Two measurements have been recorded for each item: distance (m) and zinc concentration (μg Zn/g soil). Both are interval scales. The distance at which these measurements are taken has been fixed by the investigator.

Distance from pylon (m)	Zinc concentration (μg Zn/g soil)
1.0	648
1.5	610
2.0	534
2.5	500
3.0	472

> There is an apparently linear distribution with the points evenly scattered along the line.

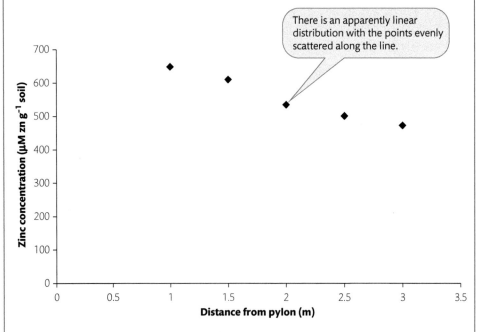

Fig. 9.10 Zinc concentration in soil (μg Zn/g soil) at specific distances from an electricity pylon

9.10.1 **Key trends and experimental design**

The data for Example 9.5 appear to have a negative association, so that the further away from the pylon the soil sample has been taken, the lower the concentration of zinc. There are only five points and these are regularly spaced on the x axis. These are fixed points determined by the investigator and as such there was no sampling error in these measurements. The association appears to be linear and in Fig. 9.10 it has a rather flat gradient. It would be interesting to calculate the equation of the line that best describes these data and then to test if the association is significant. If there is a significant association this would allow the investigator to predict results at points between those measured.

9.10.2 **Using this test**

i. To use this test you:

1) Do not have an *a priori* reason for expecting certain outcomes from your investigation but do wish to test for an association between two treatment variables.

2) Have two treatment variables measured for each item. These data are usually represented graphically as a scatter plot.

3) Measured both treatment variables on an interval scale.

4) Have only one y value for each value of x.

5) Have an association that appears to be linear, with the points reasonably scattered (see Q4).

6) Know that one variable (x) has been taken without sampling error and as such is usually under the investigator's control.

7) Should know that each value of y varies normally with each value of x and they should have a similar variance. (This sounds complicated but you can tell whether your data probably meet this criterion if the points on your scatter plot are fairly evenly distributed along the whole line (see Fig. 9.11 compared with Fig. 9.12).)

8) Have three or more pairs of observations.

9) Should note that, if you are using the modified t-test to test the significance of the association, then the data should be parametric.

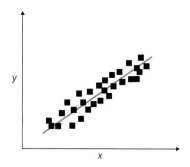

Fig. 9.11 Points evenly distributed about the line, indicating that y varies normally with x

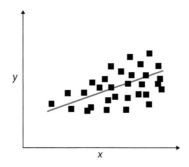

Fig. 9.12 Points becoming more spread out as *x* increases, indicating that *y* does not vary normally with *x*

ii. Does the example meet these criteria?

The annotations on Table 9.11 and Fig. 9.10 show that criteria 1–8 are met. However, if we wished to test the significance of this association we should test that these data are normally distributed, as this is usually taken to indicate that the population from which this sample is derived is normally distributed. It is very difficult to do this as there is so little data; this is a common problem with undergraduate studies in particular. Clearly the larger the study the more representative of the population your estimate will be. However, given the limits of time and inexperience undergraduates do find themselves in this position. Statistical software may carry out tests for normality even on small sample sizes such as this. Alternatively you can accept that as both variables are measured on interval scales, then the data *may* be parametric (see Box 5.2). If you have reason to doubt that these are parametric data then you should use the Spearman's rank correlation instead.

9.10.3 The calculation

There are two parts to this regression analysis. In the first part, you calculate the coefficients that allow you to draw the regression line through your data. Box 9.4 explains how to calculate the regression line for a simple linear regression. Full details of the calculation are included in the Online Resource Centre. The second step of the simple linear regression is to see whether there is a statistically significant association that is described by this line. Box 9.5 shows you this part of the calculation.

BOX 9.4 How to carry out a Model I simple linear regression (one *y* value for each *x* value): drawing a regression line

GENERAL DETAILS	EXAMPLE 9.5
	This calculation is given in full in the Online Resource Centre. For presentation purposes all values have been rounded to five decimal places.

(continued)

1. How to work out the regression coefficient b

i. The independent variable is that plotted on the x axis. The dependent variable is plotted on the y axis.

ii. Calculate the following:
Σx Add all x values together.
$(\Sigma x)^2$ Square the Σx value.
$\Sigma(x^2)$ Square all the x values and sum.
Σy Add all the y values together.
$\Sigma(xy)$ Multiply each x by its corresponding y. Add all these xy values together.

iii. $b = \dfrac{n\sum xy - \sum x \sum y}{n\sum x^2 - \left(\sum x\right)^2}$

1. How to work out the regression coefficient b

i. Variable x, the distance from the pylon is under the investigator's control and is the independent variable. Variable y is the concentration of zinc in the soil and is the dependent variable.

ii. Table 9.12 summarizes these calculations.

iii. In this example.

$$b = \frac{(5 \times 5297) - (10 \times 2764)}{(5 \times 22.5) - 100} = -92.4$$

2. How to work out a

i. Examine your data and record the means for each variable.

ii. a can be calculated using the general formula for this straight line:
$a = \bar{y} - b\bar{x}$

2. How to work out a

i. The means for the data are shown in Table 9.12.

ii. Therefore:
$a = 552.8 - (-92.4 \times 2)$

$= 552.8 + 184.8 = 737.6$

3. What is the regression equation?

For a Model I regression, the general equation of the line is:
$y = a + bx$

3. What is the regression equation?

$y = a + bx$

$y = 737.6 + (-92.4)x$

$= 737.6 - 92.4x$

4. How to draw the line

The regression line indicates the apparent trend in your data. If you want to draw this line on your scatter plot you will need to calculate three pairs of data points as follows:
Take any x value from your observed data and place this and the values for a and b into the general equation $y = a + bx$. The only unknown value will be y. Work out the y value and plot the x you selected and this a value on your graph. Repeat for at least two more points and draw a line through them within the range of your data set. This is the regression line and should be labelled with the regression equation.

4. How to draw the line

When $x = 1m$,
$y = 737.6 - (92.4 \times 1)$

$= 645.2\ \mu g\ Zn/g\ soil.$

When $x = 2m$,
$y = 737.6 - (92.4 \times 2)$

$= 552.8\ \mu g\ Zn/g\ soil.$

When $x = 3m$
$y = 737.6 - (92.4 \times 3)$
$= 460.4\ \mu g\ Zn/g\ soil$

Fig. 9.13 illustrates the scatter plot with the regression line added. But you need to confirm that this is a statistically significant association: how to do this is shown in Box 9.5.

GO TO BOX 9.5.

Table 9.12 Calculating a Model I simple linear regression (one *y* value for each *x* value) using the data from Example 9.5: Heavy metal contamination of soil under electricity pylons

Distance from pylon (m) (*x*)	Zinc concentration (µg Zn/g soil) (*y*)	$x \times y$
1.0	648	648
1.5	610	915
2.0	534	1068
2.5	500	1250
3.0	472	1416
$n = 5$	$n = 5$	$n = 5$
$\Sigma x = 10$	$\Sigma y = 2764$	$\Sigma(xy) = 5297$
$(\Sigma x)^2 = 100$	$(\Sigma y)^2 = 7639696$	
$\Sigma(x^2) = 22.5$	$\Sigma(y^2) = 1549944$	
$\bar{x} = 2.0$	$\bar{y} = 552.8$	

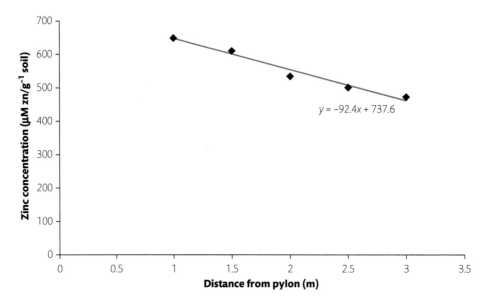

Fig. 9.13 Highly significant association ($0.01 > p > 0.001$) between the concentration of zinc in soil and the distance from the pylon so that $y = -92.4x + 737.6$

BOX 9.5 How to carry out a Model I simple linear regression (one *y* value for each *x* value): testing the significance of the association

GENERAL DETAILS	EXAMPLE 9.5
	This calculation is given in full in the Online Resource Centre. For presentation purposes all values have been rounded to five decimal places.

(continued)

1. Hypotheses to be tested	**1. Hypotheses to be tested**
H_0: There is no linear association between x and y. H_1: There is a linear association between x and y, described by $y = a + bx$.	H_0: There is no linear association between the distance from the pylons (m) and concentrations of zinc in the soil (μg Zn/g soil). H_1: There is a linear association between the distance from the pylons (m) and concentrations of zinc in the soil (μg Zn/g soil).
2. Have the criteria for using this test been met?	**2. Have the criteria for using this test been met?** Sort of (9.10.2ii).
3. How to work out $t_{calculated}$ i. First calculate the basic terms SS(x), SS(y) and SP(xy). $$SS(x) = \sum\left(x^2\right) - \frac{\left(\sum x\right)^2}{n}$$ $$SS(y) = \sum\left(y^2\right) - \frac{\left(\sum y\right)^2}{n}$$ $$SP(xy) = \sum xy - \frac{\left(\sum x\right)\left(\sum y\right)}{n}$$ More details about how to calculate these terms are included in Box 9.4. ii. Calculate the residual variance $\left(s_r^2\right)$ $$s_r^2 = \frac{1}{n-2}\left[SS(y) - \frac{(SP(xy))^2}{SS(x)}\right]$$ iii. Calculate the standard error of b (SE(b)). $$SE(b) = \sqrt{\frac{s_r^2}{SS(x)}}$$ $$t_{calculated} = \left|\frac{b}{SE(b)}\right|$$ The brackets in this last step indicate that you need the absolute value and therefore any negative sign for $t_{calculated}$ can be ignored.	**3. How to work out $t_{calculated}$** i. See Table 9.10. $$SS(x) = 22.5 - \frac{100}{5} = 2.5$$ $$SS(y) = 1549944 - \frac{7639696}{5} = 22004.8$$ $$SP(xy) = 5297 - \frac{10 \times 2764}{5} = 231.0$$ ii. The residual variance is: $$s_r^2 = \frac{1}{5-2}\left[22004.8 - \frac{(231.0)^2}{2.5}\right]$$ $$s_r^2 = 220.13333$$ iii. $$SE(b) = \sqrt{\frac{220.13333}{2.5}}$$ $$= 9.38367$$ $$t_{calculated} = \left|\frac{-92.4}{9.38367}\right| = |-9.84689| = 9.84689$$
4. How to find $t_{critical}$ See Appendix D, Table D6. Use a t table for a two-tailed test where $p = 0.05$. The degrees of freedom (v) are $n - 2$.	**4. How to find $t_{critical}$** When $v = 5 - 2 = 3$ and $p = 0.05$, $t_{critical} = 3.182$.
5. The rule If the absolute value of $t_{calculated}$ is greater than the value of $t_{critical}$, then you may reject H_0.	**5. The rule** As $t_{calculated}$ (9.847) is more than $t_{critical}$ (3.182) at $p = 0.05$, you may reject the null hypothesis. In fact, at $p = 0.01$, $t_{critical} = 5.841$, and at $p = 0.001$, $t_{critical} = 12.941$, so you may reject the null hypothesis at $p = 0.01$ but not at $p = 0.001$.

(continued)

6. What does this mean in real terms?	6. What does this mean in real terms?
	There is a highly significant ($0.01 > p > 0.001$) negative linear association between the distance from the pylons (m) and concentrations of zinc in the soil (µg Zn/g soil) described by $y = 737.6 - 92.4x$.

Further examples relating to the topic including how to use statistical software are included in the Online Resource Centre.

9.11 Model I: linear regression: more than one y for each value of x, with equal replicates

In some experiments, you may have more than one y for each value of x. For example, in the experiment described above (Example 9.5), the student may have taken four soil samples along transects placed at each leg of the pylon rather than just one. This would provide her with four replicates of zinc concentration in soil (y values) for each distance from the pylon (x). This is still a Model I regression, as one of the variables (distance) is under the control of the investigator and so taken without sampling error. The test we outline in this section is for experiments like this.

Example 9.6   The weight (g) of juvenile hamsters (4–14 weeks)

An undergraduate studying the effect of a number of commercially available hamster foods on reproduction examined, as one objective, the weight of baby hamsters produced to mothers on one particular diet. The hamster babies were sampled at random from a number of similarly aged babies and weighed at age 4 to 14 weeks of age (Table 9.13).

Table 9.13   The weight (g) (y) of hamsters aged 4–14 weeks

The items in this experiment are the baby hamsters (A–E). These are also the replicates.

Two measurements have been recorded for each item: age (weeks; x) and weight (g; y). Both are interval scales. The time at which the hamsters were weighed is fixed by the investigator. None of these weeks can be said to be a control.

Hamsters	Age of hamsters (weeks) (x)					
	4	6	8	10	12	14
A	18.1	27.0	26.3	35.0	35.5	41.0
B	19.5	26.9	29.0	34.0	37.5	42.5
C	21.0	25.4	29.7	31.1	38.5	41.5
D	23.0	24.6	31.0	33.1	39.7	43.4
E	23.0	22.1	31.2	35.2	39.6	42.1
$\bar{y}$	20.92	25.20	29.44	33.68	38.16	42.10

(continued)

Table 9.13 Continued

Hamsters	Age of hamsters (weeks) (x)					
	4	6	8	10	12	14
s^2	4.6570	4.0350	3.9130	2.7870	3.0180	0.8550
s	2.15800	2.00873	1.97813	1.66943	1.73724	0.92466
n	5	5	5	5	5	5
Σy	104.6	126.0	147.2	168.4	190.8	210.5
$\Sigma(y^2)$	2206.86	3191.34	4349.22	5682.86	7293.00	8865.47

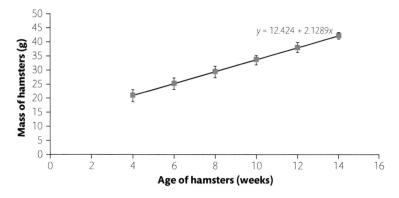

Fig. 9.14 An association ($y = 12.42362 + 2.12886x$, $p < 0.001$) between the increase in weight of baby hamsters over a 10-week period.

To test the significance of this regression, you use a parametric ANOVA. We cover these types of tests primarily in the next chapter (Chapter 10) and you may find it helpful to read the introduction there. Important points to note at this stage are that the ANOVA we describe is for parametric data and therefore you need to confirm that your data meet this criterion before proceeding. Secondly, the ANOVAs are usually used to test the hypothesis 'There is no difference between samples'. This is very different from the tests we have been looking at in this chapter, which examine the hypothesis 'There is no association between two variables'. This difference in hypothesis is important. Tests for association examine the data to see whether, as one variable changes, the data relating to the second variable change in a similar, consistent way. (For most of the tests we describe here, this change is assumed to be linear.) However, the hypothesis that tests for 'differences between samples' is more open and is not based on the concept of any consistency in changes seen in the two variables. The ANOVA we outline here allows you to test both for a linear association and for a difference between samples. Due to the important difference in these two hypotheses, it is possible to have a statistically significant difference between samples and a non-significant association. This can arise when the samples are different but in a non-linear way.

There are many types of ANOVA. They are known as 'one-way', 'two-way', or 'three-way' depending on the number of treatment variables you wish to test the effect of. In Example 9.6, there are two treatment variables: weight of hamsters (g) and age of hamsters (weeks). In the tests for an association, we would say we are testing the significance of an *association* between these two variables. However, when testing for a difference between samples, we would say we wish to examine the *effect* of age on the weight of hamsters. Therefore, we are testing an effect of one variable (age) only. This seems odd but is more convincing if you try to reverse the statement and say you wish to test the significance of the effect of weight on age. This is not sensible: the weight of the hamster cannot alter the age of the hamster. Therefore, although we have two treatment variables, there is only one independent variable (age) and so will use a one-way ANOVA.

In our example there are equal replicates but this test can easily be extended for unequal replicates and we include an example illustrating this in the Online Resource Centre.

9.11.1 Key trends and experimental design

In Example 9.6 it appears that in the first few weeks of life the increase in weight in baby hamsters is linear and there is very little variation in weight gain between the hamsters in this study (Fig. 9.14). It would be interesting to model this association; to examine whether age has a significant effect on the weight gained by these hamsters and whether these are consistent increments?

9.11.2 Using this test

i. To use this test you:

1) Do not have an *a priori* reason for expecting certain outcomes from your investigation but do wish to test for an association between two treatment variables.

2) Have two treatment variables measured for each item. These data are usually represented graphically as a scatter plot.

3) Measured both treatment variables on an interval scale.

4) Have more than one y value for each value of x with equal numbers of y values (replicates) for each x.

5) Have an association that appears to be linear, with the points reasonably scattered (see Q4).

6) Know that one variable (x) has been taken without sampling error and as such is usually under the investigator's control.

7) Should know that each value of y varies normally with each value of x and they should have a similar variance. (This sounds complicated but you can tell whether your data probably meet this criterion if the points on your scatter plot are fairly evenly distributed along the whole line (see Fig. 9.11 compared with Fig. 9.12) and by using an F_{max} test.)

8) Have three or more pairs of observations.

9) Can confirm that the observations for both treatment variables are parametric.

ii. Does the example meet these criteria?

The data from Example 9.6 meet all the criteria for using this one-way parametric ANOVA for linear regressions with equal replicates. We do wish to test for an association between two variables (age and weight). Both variables are measured on an interval scale and the age at which the hamsters were measured was determined by the investigator as specific points in their development. This association appears to be linear with the points reasonably scattered (Fig. 9.14). There are six x values (weeks) and equal replicates (hamsters). It is not possible to be confident that the y values are parametric as there are so few observations at each age point. However, the data are measured on an interval scale and we will therefore have to make an assumption that this criterion is met (Box 5.2). To check whether the variances are similar, an F_{max} test is carried out (Box 9.6). For Example 9.6, the F_{max} test confirms that the variances are homogeneous and we may proceed with the ANOVA (Box 9.7).

BOX 9.6 How to carry out an F_{max} test to check for homogeneous variances before carrying out an ANOVA for linear regressions

GENERAL DETAILS	EXAMPLE 9.6
	This calculation is given in full in the Online Resource Centre. For presentation purposes all values have been rounded to five decimal places.
1. Hypotheses to be tested	**1. Hypotheses to be tested**
H_0: There is no difference between the variances of the two samples. The variances are homogeneous. H_1: There is a difference between the variances of the two samples.	H_0: There is no difference between the variances of the weight of hamsters (g) at different ages. H_1: There is a difference between the variances of the weight of hamsters (g) at different ages.
2. How to work out $F_{max\ calculated}$ i. For each sample, calculate the variance (s^2) using the method described in Box 5.1. ii. $F_{max\ calculated}$ is the ratio found by dividing the largest variance by the smallest variance. $$F_{max\ calculated} = \frac{\text{largest sample variance}(s_1^2)}{\text{smallest sample variance}(s_2^2)}$$	**2. How to work out $F_{max\ calculated}$** i. See Table 9.13. ii. $F_{max\ calculated} = \dfrac{4.6570}{0.8550} = 5.44678$
3. How to find $F_{max\ critical}$ See Appendix D, Table D7. To look up the value for an F_{max} test that precedes an ANOVA, you will need to know the number of samples (a) and the degrees of freedom ($v = n_s - 1$). n_s is the number of observations in each sample or category.	**3. How to find $F_{max\ critical}$** For our example, measurements were taken at six different ages, so $a = 6$. There are five hamsters in each category so the degrees of freedom are $v = 5 - 1 = 4$. So at $p = 0.05$, $F_{max\ critical} = 29.5$.
4. The rule If $F_{max\ calculated}$ is less than $F_{max\ critical}$, then you may accept the null hypothesis and proceed with the ANOVA.	**4. The rule** In this example, $F_{max\ calculated}$ (5.45) is less than $F_{max\ critical}$ (29.5). Therefore, you do not reject the null hypothesis.

(continued)

5. What does this mean in real terms?	5. What does this mean in real terms?
	There is no significant difference ($F_{max} = 5.45$, $p = 0.05$) between the variances of the weight of hamsters (g) at different ages. You may proceed with the ANOVA. If the null hypothesis is not accepted, then you should consider transforming your data (5.10).

9.11.3 The calculation

We have now satisfied ourselves that our data meet all the criteria for using this test. In Box 9.7, we have organized the calculation for the ANOVA to test the significance of a linear regression with equal replicates under a number of subheadings:

- Calculate general terms.
- Calculate the sums of squares.
- Construct and complete an ANOVA calculation table.
- Test hypotheses.

At each point, we show how these steps can be applied to Example 9.6. Some of the calculation is included in Tables 9.13 and 9.14. Where steps have been abbreviated, the full calculation is included in the Online Resource Centre.

BOX 9.7 How to carry out a one-way parametric ANOVA for a linear regression with equal replicates of y for each value of x

This calculation is given in full in the Online Resource Centre. For presentation purposes all values have been rounded to five decimal places. The calculation is illustrated using data from Example 9.6.

1. **General hypotheses to be tested.** There are three pairs of hypotheses that may be tested. These are explained at the end of the box in the context of the calculation itself.

2. **Have the criteria for using this test been met?** Yes. The criteria for using this test on the data from Example 9.6 have been met (9.11.2ii).

3. **How to work out $F_{calculated}$.**

A. Calculate general terms

1. Add together all the observations in all the categories (grand total): Σy.

2. Square each observation in all the categories and add these together: $\Sigma(y^2)$.

3. For each category add all the observations together (Σy_s). Square this value ($\Sigma y_s)^2$. Add all the squared values together $\Sigma(\Sigma y_s)^2$. Divide this number by the number of observations in any one category (n), i.e.

$$\frac{\Sigma\left(\Sigma y_s\right)^2}{n}$$

(continued)

4. Take the result from step **1**, square it, and divide by N, where N is the total number of observations in all the samples combined:

$$\frac{\left(\sum y\right)^2}{N}$$

5. Add all the x values together (Σx) and multiply by n: $n(\Sigma x)$.

6. Square each x value (x^2) and add these together ($\Sigma(x^2)$). Multiply by n.

7. For each x value, there are a number of y values, which have been summed (Σy_s). For each x value, multiply it by its own sum of y: ($x(\Sigma y_s)$). Add these together:

$$\sum\left(x\left(\sum y_s\right)\right)$$

8. Square the result from step **5** and divide it by the number of categories (a) multiplied by the number of observations in each category (n):

$$\frac{(\text{result from step }\mathbf{5})^2}{a \times n}$$

9. Subtract the result from step **8** from the result from step **6**.

10. First multiply together the results from steps **5** and **1**. Divide this value by $a \times n$. Subtract the result from step **7**.

In our example, the calculations are shown in the ten steps here and with reference to Table 9.12:

1. $\Sigma y = 18.1 + 27.0 + \dots 39.6 + 42.1 = 947.50$

2. $\Sigma y^2 = (18.1)^2 + (27.0)^2 + \dots (39.6)^2 + (42.1)^2$
   $= 2206.86 + 3191.34 + 4349.22 + 5682.86 + 7293.0 + 8865.47$
   $= 31588.75$

3.
$$\frac{(104.6)^2 + (126.0)^2 + (147.2)^2 + (168.4)^2 + (190.8)^2 + (210.5)^2}{5}$$
   $= (10941.16 + 15876.0 + 21667.84 + 28358.56 + 36404.64 + 44310.25)/5$
   $= 157558.45/5 = 31511.69$

4. $(947.50)^2/30 = 897756.25/30 = 29925.2083$

5. $5(4+6+8+10+12+14) = 5 \times 54 = 270.0$

6. $((4)^2 + (6)^2 + (8)^2 + (10)^2 + (12)^2 + (14)^2) \times 5$
   $= 556.0 \times 5 = 2780.00$

7. $(4 \times 104.6) + (6 \times 126.0) + (8 \times 147.2) + (10 \times 168.4) + (12 \times 190.8) + (14 \times 210.5)$
   $= 418.4 + 756.0 + 1177.6 + 1684.0 + 2289.6 + 2947.0 = 9272.6$

8. $(270.0)^2/(6 \times 5) = 72900.0/30 = 2430.0$

9. $2780.00 - 2430.0 = 350.0$

10. $9272.6 - \dfrac{270.0 \times 947.5}{6 \times 5} = 9272.6 - (255825.0/30)$
    $= 9272.6 - 8527.5 = 745.1$

B. Calculate the sums of squares (SS)

11. SS_{total} = result from step **2** – result from step **4**.

12. $SS_{between*}$ = result from step **3** – result from step **4** (*between* categories).

13. $SS_{within**}$ = result from step **11** – result from step **12** (**within* categories).

14. $SS_{regression}$ = square result from step **10** and divide by result from step **9**.

15. $SS_{deviation from regression}$ = result from step **12** – result from step **14**.

In our example, the calculations are:

11. $SS_{total} = 31588.75 - 29925.2083 = 1663.5417$

12. $SS_{between} = 31511.69 - 29925.2083 = 1586.4817$

13. $SS_{within} = 1663.5417 - 1586.4817 = 77.06$

14. $SS_{regression} = (745.1)2/350.0 = 555174.01/350.0 = 1586.21146$

15. $SS_{deviation from regression} = 1586.4817 - 1586.21146 = 0.27024$

C. Construct and complete an ANOVA calculation table

i. Draw an ANOVA table as illustrated here.

Source of variation	SS	v	Mean squares	F
Between categories	Step 12			
Linear regression	Step 14			
Deviation from linear regression	Step 15			
Within categories	Step 13			
Total variation	Step 11			

(continued)

ii. Transfer the results from the calculations for the sums of squares into the ANOVA table in column 1 (Table 9.14).

iii. Calculate the degrees of freedom (v).
$v_{between}$ = number of categories − 1 = $a - 1$
$v_{regression}$ = 1
$v_{deviations\ from\ regression}$ = $a - 2$
v_{within} = total number of observations − number of categories = $N - a$
v_{total} = total number of observations − 1 = $N - 1$
Enter these results in column 2 (Table 9.14).

iv. Calculate the mean squares (MS) for between and within categories, regression, and deviation from regression by dividing the relevant sum of squares (e.g. SS$_{within}$) by the associated degrees of freedom (Table 9.14).

D. Calculation of *F* values, finding critical *F* values, and testing hypotheses

There are three pairs of hypotheses we may test. The hypotheses we test are written below in relation to Example 9.6.

To test for a deviation from linearity
These hypotheses are tested in step **16**.

H$_0$: There is no deviation from linearity between the variables 'weight of hamsters' (g) and 'age of hamsters' (weeks).

H$_1$: There is a deviation from linearity between the variables 'weight of hamsters' (g) and 'age of hamsters' (weeks).

Testing the significance of this null hypothesis is your first step. A significant decision at this point suggests there is significant deviation from linearity and either an association is not linear (e.g. curvilinear) or there is significant variation in the data that is masking any linearity. If you have a significant outcome in this step, you may either consider transforming your data to produce a linear distribution and/or proceed to step **18**.

To test for significant linearity in the data
These hypotheses are tested in step **17**.

H$_0$: There is no significant linear association between the weight of hamsters (g) and their age (weeks).

H$_1$: There is a significant linear association between the weight of hamsters (g) and their age (weeks).

There is a debate about how the appropriate *F* value is calculated to test this hypothesis. We describe the ratio that is correct most of the time and for the method we have described. A significant result here would indicate a significant regression and you may then proceed to step **19**, which shows you how to calculate the regression equation. A non-significant result here would suggest either that any association is not linear or that there is too much variation in the data, which is masking any linear trend. You should consider transforming your data and/or proceed to step **18**.

To test for a difference between samples of hamsters at different ages
These hypotheses are tested in step **18**.

H$_0$: There is no difference between the hamsters at different ages (weeks) in relation to their weights (g).

H$_1$: There is a difference between the hamsters at different ages (weeks) in relation to their weights (g).

This is a typical one-way ANOVA that tests the hypotheses described here. The test does not consider the linearity or any other specific type of association. The hypothesis testing instead focuses on any significant differences between samples. You may have a non-significant outcome when testing for linearity but a significant outcome here. If you wish to find out more about the nature of the difference, you may then proceed to a Tukey's test (10.8).

16. To test for a deviation from linearity.

$F_{calculated}$ = MS$_{deviations}$ divided by MS$_{within}$.

In our example, this $F_{calculated}$ value = 0.06756/3.21083 = 0.02104. The $F_{critical}$ value is found in Appendix D, Table D8, where the degrees of freedom are those for the MS$_{deviations}$ (v_1) and the MS$_{within}$ (v_2). In our example, $v_1 = 4$ and $v_2 = 24$. When $p = 0.05$, $F_{critical} = 2.78$.

The rule for this test is that if $F_{calculated}$ is more than $F_{critical}$, you may reject the null hypothesis. In our example, $F_{calculated}$ (0.02) is less than $F_{critical}$ (2.78) so we may not reject the null hypothesis. There is therefore no significant deviation from linearity between the variables 'weight of hamsters' (g) and 'age of hamsters' (weeks). We may now test for the significance of the regression by proceeding to

(continued)

the next step. If this were not the case, we would skip the next step and proceed to step **18**.

17. To test for significant linearity in the data.

$F_{calculated} = MS_{regression}$ divided by $MS_{deviation}$

In our example, this $F_{calculated}$ value = 1586.21146/0.06756 = 23 478.55921. The $F_{critical}$ value is found in Appendix D, Table D8, where the degrees of freedom are those for $MS_{regression}$ (v_1) and $MS_{deviation}$ (v_2). In our example, $v_1 = 1$ and $v_2 = 4$. When $p = 0.05$, $F_{critical} = 7.71$. The rule for this test is that if $F_{calculated}$ is more than $F_{critical}$, you may reject the null hypothesis. In our example, $F_{calculated}$ (23478.56) is more than $F_{critical}$ (7.71), so we may reject the null hypothesis. In fact, at $p = 0.001$ (Appendix D, Table D10), $F_{critical} = 74.13$ so we may reject the null hypothesis at this higher level of significance. There is therefore a very highly significant linear association between weight (g) and age (weeks) of juvenile hamsters. We may now proceed to step **19** to calculate the equation for the significant regression.

18. To test for a difference between samples of hamsters at different ages.

If the hypothesis testing at step **16** is significant or step **17** is not significant, it is worth carrying out step **18** to test for a significant difference between the samples.

$F_{calculated} = MS_{between}$ divided by MS_{within}

In our example, this $F_{calculated}$ value = 317.29634/3.21083 = 98.82066. The $F_{critical}$ value

is found in Appendix D, Table D8, where the degrees of freedom are those for the $MS_{between}$ (v_1) and MS_{within} (v_2). In our example, $v_1 = 5$ and $v_2 = 24$. When $p = 0.05$, $F_{critical} = 2.62$. The rule for this test is that if $F_{calculated}$ is more than $F_{critical}$, you may reject the null hypothesis. In our example, $F_{calculated}$ (98.82) is more than $F_{critical}$ (2.62) so we may reject the null hypothesis. In fact, at $p = 0.001$ (Appendix D, Table D8), $F_{critical} = 5.98$ so we may reject the null hypothesis at this higher level of significance. There is therefore a very highly significant difference between the weights (g) of hamster juveniles in relation to their age (weeks).

19. The regression equation.

The general equation for this regression line is $y = a + bx$. The regression coefficient is the gradient of the line (b) (Fig. 9.14) and can be calculated as b = result from step **10**/result from step **9**. The point where the line crosses the y axis is a, where $a = \bar{y} - b\bar{x}$ = result from step **1**/N – [b (result from step **5**/N)], where N is the total number of observations.

In this example:

$b = \dfrac{745.1}{350.0} = 2.12886$

$a = \dfrac{947.50}{30} - 2.12886\left(\dfrac{270.0}{30}\right)$

$= 31.58333 - 19.15974 = 12.42359$

The full regression equation is $y = 12.42359 + 2.12886x$.

Table 9.14 ANOVA table for testing the significance of the linear regression where there are replicates of y for each value of x, and for the testing of the hypothesis 'There is no difference between samples' for data from Example 9.6

Source of variation	SS	v	Mean squares	$F_{calculated}$
Between categories	1586.4817	6 – 1 = 5	317.29634	$MS_{between}/MS_{within}$ 317.29634/3.21083 = 98.82066
Linear regression	1586.21146	1	1586.21146	$MS_{regression}/MS_{deviation}$ 1586.21146/0.06756 = 23478.55920
Deviation from linear regression	0.27024	6 – 2 = 4	0.06756	$MS_{deviations}/MS_{within}$ 0.06756/3.21083 = 0.02104
Within categories	77.06	30 – 6 = 24	3.21083	
Total variation	1663.5417	29		

9.12 Model II: principal axis regression

Unlike the simple linear regression model, the principal axis regression model can be used when both variables are subject to sampling error. Usually this means that neither variable is under the investigator's control. However, the principal axis regression cannot be used to test the significance of an association, and in this instance, a correlation should be used. The principal axis regression and ranged principal axis regression (9.13) are part of a group of tests called principal component analyses. In these Model II regressions, the line is determined by placing an ellipse around the data and drawing a line through the longest axis, known as the principal axis. The general formula for this principal axis is $y = \bar{y} + b_1(x - \bar{x})$. As in the simple linear regression, there is a regression coefficient (b), but to denote the difference in the method used to calculate the Model I and Model II values of b, we call the regression coefficient calculated for the principal axis regression b_1.

Example 9.7 The lower arm and lower leg length (cm) of a small cohort of female undergraduates

In an investigation into human morphology, the lower arm and lower leg lengths of 12 female undergraduates were recorded (Table 9.15) on one particular day.

Table 9.15 Lower arm and lower leg length (cm) in a small cohort of female undergraduates

Each student is the item. There are therefore 12 items. These can also be described as replicates.

Two measurements have been recorded for each item: lower arm length and lower leg length. Both are measured on the same interval scale (cm). Neither measure has been fixed by the investigator and is subject to sampling error.

Student	Lower arm length (cm)	Lower leg length (cm)	Student	Lower arm length (cm)	Lower leg length (cm)
1	24.0	39.0	7	26.0	41.0
2	25.0	40.0	8	25.5	40.0
3	23.0	36.0	9	24.2	36.7
4	26.0	43.0	10	25.0	38.0
5	24.0	38.0	11	24.0	41.0
6	23.0	35.0	12	24.5	40.0

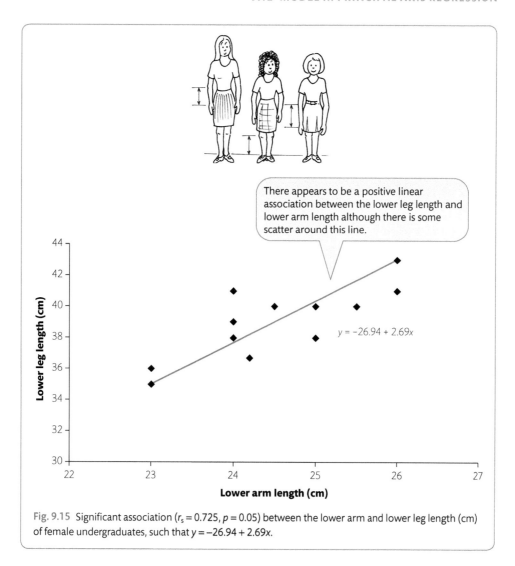

Fig. 9.15 Significant association ($r_s = 0.725$, $p = 0.05$) between the lower arm and lower leg length (cm) of female undergraduates, such that $y = -26.94 + 2.69x$.

9.12.1 Key trends and experimental design

Figure 9.15 does indicate a positive linear association between the lower leg length and lower arm length in these female undergraduates. There is some variation especially towards the mid x distribution. We need to carry out a regression analysis to allow us to model the association and so plot a line of best fit and use a correlation analysis to see if the association is significant.

9.12.2 Using this test

i. To use this test you:

1) Do not have an *a priori* reason for expecting certain outcomes from your investigation but do wish to test for an association between two treatment variables.

2) Have two treatment variables measured for each item. These data are usually represented graphically as a scatter plot (see Q4).

3) Have an association that appears to be linear, with the points reasonably scattered (see Q4).

4) Have confirmed using a correlation analysis that the association is significant.

5) Have two treatment variables that are measured on the same interval scale.

6) Can confirm that neither treatment variable is under the investigator's control; therefore, both treatment variables are subject to sampling error.

7) Have three or more pairs of measurements.

ii. Does the example meet these criteria?

As the annotations on Table 9.15 and Fig. 9.15 show, the data from Example 9.7 meet all the criteria for a principal axis regression apart from point 4 for which further analysis is needed. When a Spearman's rank correlation is carried out (see Online Resource Centre, Chapter 9, interactive exercise 1) this confirms that there is a significant association ($r_s = 0.725, p = 0.05$) between lower leg length and lower arm length. Therefore this criterion is also met.

9.12.3 The calculation

We show you the general process and a specific example of a Model II principal axis regression in Box 9.8. In addition, you will need to refer to the calculations in Table 9.16.

BOX 9.8 How to carry out a Model II principal axis regression: drawing a regression line

GENERAL DETAILS	EXAMPLE 9.7
	This calculation is given in full in the Online Resource Centre. For presentation purposes all values have been rounded to five decimal places.
1. Have the criteria for using this test been met?	1. Have the criteria for using this test been met?
	Yes (9.12.2ii).
2. How to work out b_1	2. How to work out b_1
i. First decide which of the variables to call x and which to call y. ii. Calculate the basic terms. To carry out this regression analysis, you first need to calculate the following for the two samples: Means ($\bar{x}$ and $\bar{y}$) Sum of x values (Σx), sum of y values (Σy) The variances (s_x^2 and s_y^2) The number of pairs of observations n The method for calculating these is given in Box 5.1.	i. Let the measurements of arm length (cm) be x and the measurements of leg length (cm) be y (Fig. 9.15). ii. The general methods for calculating such terms as Σx and s_x^2 are described in Chapter 5 (Box 5.1) and therefore are not included here. The basic terms are: $\bar{x} = 24.51666, \bar{y} = 38.975$ $\sum x = 294.2, \sum y = 467.7$ $s_x^2 = 1.03061, s_y^2 = 5.38932$ $n = 12$

(continued)

iii. Calculate the variance for products xy, $\left(s_{xy}^2\right)$, where

$$s_{xy}^2 = \frac{SP(xy)}{n-1}$$

The term $SP(xy)$ has figured already in this chapter (Box 9.3) and is

$$SP(xy) = \sum xy - \frac{\left(\sum x\right)\left(\sum y\right)}{n}$$

iv. The slope of the principal axis (b_1) is calculated as:

$$b_1 = \frac{s_y^2 - s_x^2 + \sqrt{\left[\left(s_y^2 - s_x^2\right)^2 + 4\left(s_{xy}^2\right)^2\right]}}{2s_{xy}^2}$$

iii. To calculate the variance for the products $\left(s_{xy}^2\right)$, first note or calculate each of the terms you need. Σx, Σy, and n are above (ii). To calculate Σxy, see Table 9.15: $\Sigma xy = 11487.14$

$$SP(xy) = \sum xy - \frac{\left(\sum x\right)\left(\sum y\right)}{n}$$

$$= 11487.14 - \frac{(294.2)(467.7)}{12}$$

$$= 20.695$$

$$s_{xy}^2 = \frac{SP(xy)}{n-1} = \frac{20.695}{11}$$

$$= 1.88136$$

iv. To calculate the slope of the principal axis b_1, first calculate the components of this equation:

$$s_y^2 - s_x^2 = 5.38932 - 1.03061$$

$$= 4.35871$$

$$\left(s_y^2 - s_x^2\right)^2 = (4.35871)^2$$

$$= 18.99837$$

$$4\left(s_{xy}^2\right)^2 = 4 \times (1.88136)^2$$

$$= 14.15812$$

$$2s_{xy}^2 = 2 \times 1.88136$$

$$= 3.76273$$

Therefore:

$$b_1 = \frac{4.35871 + \sqrt{(18.99837 + 14.15812)}}{3.76227}$$

$$= 2.68871$$

3. What is the regression equation?

The equation for the principal axis is:
$$y = \bar{y} + b_1(x - \bar{x})$$
Substitute the known numerical values where possible.

3. What is the regression equation?

Using the values you have already calculated:
$$y = \bar{y} + b_1(x - \bar{x})$$
$$= 38.975 + 2.68871\,(x - 24.51667)$$
$$= 38.975 + 2.68874x - 65.91822$$
$$= -26.94319 + 2.68871x$$

4. How to draw the line

In the same way as other regressions, you need to calculate the x and y values for three points using x values that lie within your data set, and your regression equation. Your line then passes through these points within the range of your data set. This is the regression line and if you add it to your figure it should be labelled with the regression equation.

4. How to draw the line

The full regression equation for this example is:
$$y = -26.94319 + 2.68871x$$
Therefore if $x = 23$cm,
$$y = -26.94319 + (2.68871 \times 23)$$
$$= 34.9\text{cm}$$
and if $x = 24.5$cm, then $y = 38.9$cm
and if $x = 26$cm, then $y = 43$cm.
The data from Table 9.15 and the regression line are shown in Fig. 9.15.

(continued)

5. Testing the significance of the association	5. Testing the significance of the association
The principal axis regression allows you to identify a regression line that reflects the linear nature of the association between your two variables. However, unlike the simple linear regression, there is no simple associated process for testing a hypothesis relating to the significance of the association. Therefore, you should also use a correlation analysis to confirm the strength and significance of the association between the two variables.	An examination of the data indicates that they are non-parametric; therefore, a two-tailed Spearman's rank correlation was carried out. From this, it was concluded that there is a moderate ($r_s = 0.725$), statistically significant ($p = 0.05$), positive correlation between the lower arm and lower leg length in a cohort of female students, which can be described by the equation: $y = -26.94319 + 2.68871x$ (Fig. 9.15).

Further examples relating to the topic including how to use statistical software are included in the Online Resource Centre.

Table 9.16 Calculating a Model II principal axis regression using the data from Example 9.7: Arm and leg length (cm) in a small cohort of female students

Length of arm (cm) (x)	Length of leg (cm) (y)	$x \times y$
24.00	39.00	936.00
25.00	40.00	1000.00
23.00	36.00	828.00
26.00	43.00	1118.00
24.00	38.00	912.00
23.00	35.00	805.00
26.00	41.00	1066.00
25.50	40.00	1020.00
24.20	36.70	888.14
25.00	38.00	950.00
24.00	41.00	984.00
24.50	40.00	980.00
$n = 12$	$n = 12$	$n = 12$
$\Sigma x = 294.2$	$\Sigma y = 467.70$	$\Sigma(xy) = 11487.14$
$\bar{x} = 24.51667$	$\bar{y} = 38.97500$	

9.13 Model II: ranged principal axis regression

Many investigators wish to determine whether there is an association between two variables, neither of which has been manipulated by the experimenter. In Example 9.4, neither hardness of eggshell nor the amount of food supplement eaten was under the investigator's control; therefore, a Model II regression is appropriate. Where the units of measurement for the two variables differ, as in this example, a Model II ranged principal axis regression may be used (Legendre & Legendre, 2012). Unlike the principal axis regression (9.12), the first step in the calculation is to transform the data so that each scale of measurement is revised to one that only extends from 0 to 1. These ranged data then satisfy the criteria for using a principal axis regression. The principal axis regression is carried out on this ranged data and the slope b′ is then transformed back to the original units of measurement before the regression line is calculated. Like the previous principal axis regression, the line is the one that passes longitudinally through an ellipse which encompasses the data. This ranged principal axis regression cannot easily be used as the basis for a test for association so a correlation should be carried out to examine the strength and significance of any association before using this test.

9.13.1 Key trends and experimental design

The example we use to illustrate this method of a Model II regression analysis is Example 9.4 on the association between eggshell hardness and food supplement consumed by pullets. As these data were used to illustrate the Spearman's rank correlation and the Pearson's product moment correlation, the trends that are seen have been outlined in 9.6.1.

9.13.2 Using this test

i. To use this test you:

1) Do not have an *a priori* reason for expecting certain outcomes from your investigation but do wish to test for an association between two treatment variables.

2) Have two treatment variables measured for each item. These data are usually represented graphically as a scatter plot.

3) Have an association that appears to be linear, with the points reasonably scattered (see Q4).

4) Have confirmed using a correlation analysis that the association is significant.

5) Have two treatment variables that are measured on different interval scales.

6) Can confirm that neither treatment variable is under the investigator's control; therefore, both treatment variables are 'random' and subject to sampling error.

7) Have three or more pairs of measurements.

ii. Does the example meet these criteria?

The data from Example 9.4 satisfy the criteria for using a ranged principal axis regression because there are two treatment variables (hardness of eggshell and amount of food supplement eaten). The association appears to be linear and the points reasonably scattered (Fig. 9.1). Neither the hardness of eggshell nor the amount of food additive eaten was manipulated by the investigator. There are 13 pairs of observations. The hardness of eggshell is an arbitrary unit of measure and the amount of food additive eaten is measured in grams. Both are interval scales. Using Pearson's product moment correlation (Box 9.3), we have confirmed that there is a strong ($r = 0.99$) and highly significant positive correlation ($p < 0.001$) between these two variables.

9.13.3 The calculation

As discussed earlier, there are more steps in this ranged principal axis regression than in the principal axis regression. However, the regression itself is the same. As in the principal axis regression it is necessary to carry out a correlation analysis to test the significance of the association. The ranged principal axis regression is shown in Box 9.9 with some calculations included in Table 9.17.

BOX 9.9 How to carry out a Model II ranged principal axis regression: drawing a regression line

GENERAL DETAILS	EXAMPLE 9.4
	This calculation is given in full in the Online Resource Centre. For presentation purposes all values have been rounded to five decimal places.
1. Have the criteria for using this test been met?	1. Have the criteria for using this test been met? Yes (9.13.2ii).
2. How to work out b' i. First decide which of the variables to call x and which to call y.	2. How to work out b' i. Example 9.4 has been used to illustrate Pearson's and Spearman's rank correlations (Box 9.3 and Box 9.2). Therefore, in line with these earlier calculations, we will let the amount of food supplement eaten (g) be the x variable and hardness of shells be the y variable.
ii. Transform each x and y value into x' and y' as follows: $$x' = \frac{x - \text{lowest value of } x}{\text{highest value of } x - \text{lowest value of } x}$$ $$y' = \frac{y - \text{lowest value of } y}{\text{highest value of } y - \text{lowest value of } y}$$	ii. The transformed data are given in Table 9.17. The first original x observation was 19.5g. The lowest x value was 7.0g and the highest value was 19.5g. Therefore: $$x' = \frac{19.5 - 7.0}{19.5 - 7.0} = 1.0$$ The second original x observation was 11.20g. Therefore: $$x' = \frac{11.2 - 7.0}{19.5 - 7.0} = 0.336$$ The first original y observation was 7.1 units, the lowest y value was 1.2 units and the highest was 7.1 units. Therefore: $$y' = \frac{7.1 - 1.2}{7.1 - 1.2} = 1.0$$ etc.
iii. Calculate basic terms To carry out this regression analysis you need to first calculate the following for the two samples: Means ($\bar{x}'$ and $\bar{y}'$) Sum of x' values ($\Sigma x'$) Sum of y' values ($\Sigma y'$) The variances $\left(s_{x'}^2 \text{ and } s_{y'}^2\right)$ Number of pairs of observations n The method for calculating these is given in Box 5.1.	iii. Calculate basic terms As this method has been demonstrated elsewhere (Boxes 5.1 and 9.3), we will not repeat it here. The basic terms are: $\bar{x}' = 0.43662, \bar{y}' = 0.41330$ $\sum x' = 5.676, \sum y' = 5.37288$ $s_{x'}^2 = 0.08183, s_{y'}^2 = 0.08678$ $n = 13$

(continued)

iv. Calculate the variance $\left(s^2_{x'y'}\right)$ for the products $x'y'$, where:

$$s^2_{x'y'} = \frac{SP(x'y')}{n-1}$$

The term $SP(x'y')$ has figured already in this chapter and is:

$$SP(x'y') = \sum x'y' - \frac{\left(\sum x'\right)\left(\sum y'\right)}{n}$$

v. The slope of the principal axis (b') for the ranged data is:

$$b' = \frac{s^2_{y'} - s^2_{x'} + \sqrt{\left[\left(s^2_{y'} - s^2_{x'}\right)^2 + 4\left(s^2_{x'y'}\right)^2\right]}}{2s^2_{x'y'}}$$

vi. Transform b' to original units to obtain b_1.

$$b_1 = b' \times \frac{\text{highest } y \text{ value} - \text{lowest } y \text{ value}}{\text{highest } x \text{ value} - \text{lowest } x \text{ value}}$$

3. What is the regression equation?

The equation for the principal axis is:

$$y = \bar{y} + b_1(x - \bar{x})$$

You need first to calculate the means for the original x and y values, then substitute these and b_1 in the equation.

---

iv. To calculate the variance $\left(s^2_{x'y'}\right)$ for the products $x'y'$, first note or calculate each of the terms you need. $\sum x'$, $\sum y'$, and n are above (iii). To calculate $\sum x'y'$ see Table 9.17. $\sum x'y' = 3.35132$

$$SP(x'y') = \sum x'y' - \frac{\left(\sum x'\right)\left(\sum y'\right)}{n}$$

$$= 3.35132 - \frac{(5.676)(5.37288)}{13}$$

$$= 1.00544$$

$$s^2_{x'y'} = \frac{SP(x'y')}{n-1} = \frac{1.00544}{13-1} = 0.08901$$

v. To find the slope of the principal axis (b') for the ranged data.
First calculate the components of this equation.

$$s^2_{y'} - s^2_{x'} = 0.08678 - 0.08183 = 0.00495$$

$$\left(s^2_{y'} - s^2_{x'}\right)^2 = (0.00495)^2$$

$$= 2.45025 \times 10^{-5}$$

$$4\left(s^2_{x'y'}\right)^2 = 4 \times (0.089011)^2$$

$$= 0.03169$$

$$\sqrt{\left[(s^2_{y'} - s^2_{x'})^2 + 4(s^2_{x'y'})^2\right]}$$

$$= \sqrt{(2.45025 \times 10^{-5} + 0.03169)}$$

$$= 0.17809$$

$$2s^2_{x'y'} = 2 \times 0.08901$$

$$= 0.17802$$

Therefore:

$$b' = \frac{0.00495 + 0.17809}{0.17802}$$

$$= 1.02820$$

vi. Transform b' to original units to obtain b_1.

$$b_1 = 1.02820 \times \frac{7.1 - 1.2}{19.5 - 7.0}$$

$$= 0.48531$$

3. What is the regression equation?

Calculate the means for the original observations (Table 9.8). Then:

$$y = \bar{y} + b_1(x - \bar{x})$$

$$= 3.63846 + 0.48531(x - 12.4)$$

$$= 3.63846 + 0.48531x - 6.01784$$

$$= -2.37938 + 0.48531x$$

(continued)

4. How to draw the line

In the same way as other regressions, you need to calculate the x and y values for three points using x values that lie within your data set, using your regression equation. Your line then passes through these points within the range of your data set. This is the regression line and should be labelled with the regression equation.

4. How to draw the line

The regression equation for this example is
$y = -2.37938 + 0.48531x$
Therefore, if:
$x = 8.2g$
$y = -2.37938 + (0.48531 \times 8.2)$
  $= 1.6$ units
and when $x = 19.50g$, $y = 7.1$ units and
if $x = 15.10g$, $y = 4.9$ units.

5. Testing the significance of the association

The ranged principal axis regression allows you to identify a regression line that reflects the linear nature of the association between your two variables. However, unlike the simple linear regression, there is no simple associated process for testing a hypothesis relating to this association. Therefore, you should also use a correlation analysis to confirm the strength and significance of the association between the two variables.

5. Testing the significance of the association

These data were used to illustrate both Spearman's rank correlation (Box 9.2) and Pearson's correlation (Box 9.3). If we use the more powerful parametric Pearson's correlation, it can be seen that there is a strong ($r = 0.99$) and highly significant positive correlation ($p < 0.001$) between the amount of food supplement (g) given to Maran pullets and the hardness of the eggshells they produce. This linear association can be described by the regression equation: $y = -2.37938 + 0.48531x$ (Fig. 9.16).

Further examples relating to the topic including how to use statistical software are included in the Online Resource Centre.

Table 9.17 Calculation of ranged values for a ranged principal axis regression using data from Example 9.4: The association between the amount of food supplement eaten and eggshell hardness in Maran pullets

Pullet	Amount of food supplement (g) (x)	Ranged x' values	Hardness of shells (y)	Ranged y' values	$x' \times y'$
1	19.5	1.000	7.1	1.00000	1.0
2	11.2	0.366	3.4	0.37288	0.13647
3	14.0	0.560	4.5	0.55932	0.31322
4	15.1	0.648	5.1	0.66102	0.42834
5	9.5	0.200	2.1	0.15254	0.03051
6	7.0	0.000	1.2	0.00000	0.00000
7	9.8	0.224	2.1	0.15254	0.03417
8	11.6	0.368	3.4	0.37288	0.13722
9	17.5	0.840	6.1	0.83051	0.69763
10	11.2	0.366	3.0	0.30508	0.11166
11	8.2	0.096	1.7	0.08475	0.00814
12	12.4	0.432	3.4	0.37288	0.16108
13	14.2	0.576	4.2	0.50847	0.29288
		$\Sigma x' = 5.676$		$\Sigma y' = 5.37288$	$\Sigma x'y' = 3.35132$

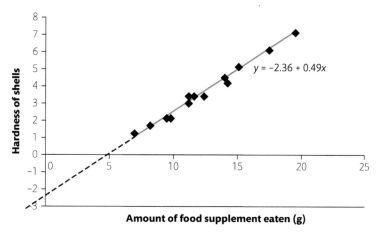

Fig. 9.16 Significant association ($r = 0.99$; $p < 0.001$) between the amount of food supplement eaten and the hardness of eggshells in Maran pullets, described by $y = -2.36 + 0.49x$. The figure also illustrates why you should not extrapolate beyond your known data (see Q5).

 The regression equation that we calculated for Example 9.4 in Box 9.9 between the amount of food supplement eaten by Maran pullets and the hardness of the eggshells is $y = -2.36 + 0.49x$ (Fig. 9.16). If the farmer wishes to economize and only gives the pullets 1g of food supplement, how hard would you predict the eggshells will be?

Summary of Chapter 9

- In this chapter, we have considered how to test whether there is an association between two treatment variables. The tests have fallen into three groups: the chi-squared tests; correlations; and regressions.

- The term association indicates a statistical and not necessarily a biological relationship between the two variables.

- The chi-squared test for association is distribution-free, which differs from the requirements of the correlations and regression, but this test struggles when the design is limited or the sample sizes are small.

- The chi-squared tests for association are used when you have counts or frequencies, and categories can be based on nominal, ordinal, or interval scales. In the correlations and regressions, neither variable can be measured on a nominal scale and in some cases both must be measured on interval scales.

- A correlation analysis will generate a coefficient that indicates the strength of association. This coefficient may then be used to test the significance of the association. The coefficient is affected by sample size; therefore, a weak correlation may not be significant if the sample size is small but may be significant if the sample size is large. We considered two correlation analyses: Spearman's rank test for non-parametric data and Pearson's product moment test for parametric data.

- Regression analyses are primarily used to determine the mathematical equation for the line that may be drawn through the data. The significance of this line as a line of best fit can be determined only in the simple linear regression with no replication using a modified t test. Where there is replication, an extended one-way ANOVA may be used. In a principal axis and ranged principal axis regression, the significance of the association can be determined using a correlation analysis.

- The regression analyses fall into two groups: Model I and Model II. In a Model I regression (simple linear regression), one variable is fixed—the observations are recorded without sampling error, and are therefore usually under the control of the investigator. In Model II regressions (principal axis and ranged principal axis regressions), both variables may be measured with some sampling error, either using the same scale (principal axis) or different scales (ranged principal axis).

- Correlation and regression analysis may be extended to testing or describing the association in non-linear distributions, or more than two variables. We are not able to include these tests here and refer you to other texts, such as Legendre & Legendre (2012) and Sokal & Rohlf (2011).

- The Online Resource Centre includes interactive exercises that test your understanding of this chapter with other topics.

Answers to chapter questions

A1 The expected values for the data from Example 9.2 have been calculated in Table 9.18. All these values are greater than 5 so meet the criteria for using a chi-squared test for association.

A2 The main design features of Example 9.3 can be seen in the annotations that have been included in Table 9.19. In this example the row totals are determined by the investigator; these are fixed. So this is a Model II design. There are no replicates or controls.

A3 No. As the x values increase, the value of y stays the same. Therefore, there is no association between these two variables.

Table 9.18 The distribution of *Cepaea nemoralis* and *Cepaea hortensis* in a woodland and a hedgerow with calculations of the expected values

	Species of snail		Total
	C. nemoralis	C. hortensis	
Hedgerow observed	89	59	148
Hedgerow expected	(105/172) × 148 = 90.34883	(67/172) × 148 = 57.65116	
Woodland observed	16	8	24
Woodland expected	(105/172) × 24 = 14.65116	(67/172) × 24 = 9.34884	
Total	105	67	172

Table 9.19 Maturity of *Quercus petraea* at the Wyre Forest and Mortimer Forest: the problem with small numbers in a chi-squared test

> This is the second treatment variable: 'woodland location'. There are two nominal categories. Neither of these is a control.

> This is the first treatment variable: 'age class of the trees'. There are four categories that have been chosen by the investigator, so are 'fixed'. These are nominal categories. None of these is a control.

Number of trees	Maturity of *Quercus petraea*				
	Seedling	Sapling	Immature tree	Mature tree	Total
Wyre Forest observed	51	21	29	4	105
Mortimer Forest observed	15	9	5	1	30
Total	66	30	34	5	135

> The data are counts and there are no replicates.

> The size of the two samples has been determined by the investigator. So this is a Model II design.

A4 Only Figs 9.2 and 9.3 meet criterion 7. Figure 9.2 illustrates a positive linear distribution, whereas Fig. 9.3 illustrates a negative linear distribution.

Figure 9.5(a) has a gap in the middle of the distribution. To resolve this, you would need to repeat the experiment, ensuring that observations are recorded for the middle part of the distribution.

In Fig. 9.5(b), the data are bunched and again you would gain a better understanding of your topic under investigation by repeating the experiment and expanding the range of measurements taken.

In Fig. 9.5(c), there is one outlying observation. You cannot ignore this datum point. If you have data like this, you should first carry out your data analyses with this datum point included. You can then repeat the analysis with this datum point excluded. You should report on the outcomes of both analyses.

Figure 9.5(d) shows an apparent curvilinear association. Curvilinear data can be analysed using a Spearman's rank correlation. For the other correlations and regressions in this chapter, one or both sets of observations should be transformed (5.10) so that a linear distribution results, or you would use methods not described in this book.

A5 In our original investigation, the minimum feed supplement eaten was 7.0g, but in theory you could use the regression equation to calculate this unknown y value. In this example, if $x = 1.0g$, then $y = -2.37938 + (0.48531 \times 1.0) = -1.89407$ *units*. But what would an egg with a negative shell hardness look like? This illustrates the folly of working outside the boundaries of your known data set. For all we know, the pullets may die unless they eat a minimum of food supplement, or they may do brilliantly without the supplement and produce really hard eggshells. You must therefore never extrapolate outside the range of your existing observations.

10 Hypothesis testing:
Do my samples come from the same population? Parametric data

 In a nutshell

Parametric tests of hypotheses are powerful and flexible but should only be used when the population from which the samples are obtained is normally distributed. In this chapter, we have included the tests most often used by our undergraduates to test the hypothesis 'Do the samples come from the same statistical population?' for one to three variables with or without replicates. There are many key principles that need to be understood in relation to the use and interpretation of these tests, including how to tell if your data are parametric, absolute values, fixed and random treatment variables, nested designs, and the interaction between treatment variables. These points are covered at the point when they first become relevant to the statistics tests being considered.

The statistics tests covered in this chapter are:

10.1 Two sample z-test for unmatched data

10.2 Two sample Student's t-test for unmatched data

10.3 Two sample unequal variance t-test for unmatched data

10.4 Two sample z- and t-tests for matched data

10.5 One sample t-test

10.7 One-way parametric ANOVA with equal numbers of replicates

10.8 Tukey's test following a parametric one-way ANOVA with equal replicates

10.9 One-way parametric ANOVA with unequal replicates

10.10 Tukey–Kramer test following a parametric one-way ANOVA with unequal replicates

10.12 Two-way parametric ANOVA with equal replicates

10.13 Tukey's test following a parametric two-way ANOVA with equal replicates

10.14 Two-way parametric ANOVA with unequal replicates

10.15 Two-way parametric ANOVA with no replicates

10.16 Two-way nested parametric ANOVA with equal replicates

10.17 Factorial three-way parametric ANOVA with no replicates

This chapter tells you how you may analyse parametric data to test the hypothesis that two or more samples come from the same statistical population. These data are usually presented in tables or summary histograms rather than a scatter plot and the treatment variables are usually measured on an ordinal and/or nominal scale or occasionally measured on an interval scale and arranged into discrete categories. Other (dependent) variables are measured on an interval scale.

All these tests use the underlying mathematical relationships that come from the data having a normal distribution. This is discussed in more detail in 5.8. If you are not sure whether your data are normally distributed, you can use the criteria in Box 5.2 to check. There are a number of tests for normality which are carried out as a standard within the statistical software commonly available to you. We explain this in the Online Resource Centre statistical walkthroughs. As these tests are derived from the Gaussian equation they use terms relating to this equation. Two of these are calculations for the sum of squares and the variance. If you are not familiar with these terms, you can refer to Box 5.1.

In many of the statistical tests described in this chapter, a variance ratio or F-test is calculated. For example, an F-test is used to check one of the criteria for using a z-test. Similarly, an F-test is used before an analysis of variance (ANOVA) to check that the data are suitable for using this test. An F-test is also used at the end of the ANOVA enabling you to test your hypotheses. These F-tests differ in the tables of critical values that are used to compare the calculated F values to. You must therefore always check to make sure you are using the correct table for the particular F-test.

The tests outlined in this chapter cover a wide range of experimental designs with one or more treatment variables, with or without replicates. Particular design features are covered in the relevant sections; however, whether your observations are matched or unmatched is a central point common to all these tests.

One simple design considered in this chapter is for one treatment variable which has a number of categories and there is more than one observation in each category. In Example 10.1 the height of periwinkle shells from the mid- and lower shore at Aberystwyth is measured once and so appears only once in the data table. This design can be represented in a general way (Table 10.1).

By comparison the fencers (the items) in Example 10.3 are measured before and after the match. In this example the treatment variable is weight (kg) and each item has two observations measured for the same treatment variable. This general design is illustrated in Table 10.2.

Table 10.1 Experimental design with one treatment variable and more than two categories

Treatment variable	
Category 1	Category 2
Observation 1_1	Observation 1_2
Observation 2_1	Observation 2_2
Observation 3_1, etc.	Observation 3_2, etc.

Table 10.2 Experimental design with one treatment variable and for matched data

Item	Treatment variable	
	Category 1	Category 2
1	Observation 1_1	Observation 2_1
2	Observation 1_2	Observation 2_2
3	Observation 1_3	Observation 2_3
4, etc.	Observation 1_4	Observation 2_4

In the second of these designs the data are said to be matched or to be repeated measures. In this chapter we consider tests for matched data or repeated measures in 10.4 (one treatment variable) and 10.15 (two treatment variables).

To identify the correct test for your experimental design and/or data you can follow the key in Chapter 7 or refer to the quick guide to choosing your statistics tests in the appendices. However, each test has additional criteria that need to be met. The full criteria including these additional requirements are included in this chapter so that you can check these against your design and/or data. Fifteen tests are covered in this chapter and worked examples are given for each. If this is the first time you have used a test, you should work through the example and check your answers before using the test on your own data. If your answer differs considerably from that given, you can check your calculation by going to the Online Resource Centre. If you work through these examples, we would expect it to take you 8 hours. If you prefer to work on parts of this chapter, we recommend you look at the t- and z-tests first, then the one- and two-way ANOVAs, and finally the three-way ANOVA and two-way nested ANOVA. To study these, we would expect you to take 2 hours, 3 hours, and 3 hours respectively. The answers for these exercises are at the end of the chapter. Our examples are usually based on real undergraduate research projects. If these examples are not in your subject area, you will find more in the Online Resource Centre.

10.1 Two sample z-test for unmatched data

In this section and in 10.2 we cover z- and t-tests that can be used for very similar sets of data with one treatment variable and two columns of data. The underlying basis for these tests requires the use of the population parameters of the mean (μ) and variance (σ^2) but these are rarely known as we do not often measure every item in a population. Instead we derive estimates of these population parameters from our samples as the mean ($\bar{x}$) and variance (s^2). The sample therefore has to be large to provide a reliable estimate of the population values. The boundary for 'large' in this context is generally accepted as more than 30 observations in each sample. When the sample sizes are more than 30 then the t- and z-tests give very similar outcomes and both can be used although it is common practice to use the z-test. When sample sizes are less than 30 the sample parameters are poor indicators of the population values so an allowance has to made in the hypothesis test. In this case the t-test is more robust than the z-test.

A second effect of sample size is that as the sample size increases the need to have a population that has a normal distribution lessens as the distribution of the means of samples derived from this population will still be normally distributed and it is on this that the analysis is based. In theory therefore it is possible to use these tests on non-parametric data if your sample size is large enough. However, the literature is not clear on how large a sample needs to be before it can be considered to be large enough and in most undergraduate studies sample sizes would not be sufficient to meet this criterion. For simplicity we therefore retain just the one sample size criteria to delineate between the use of the t-and z-test and refer to both these tests as parametric tests.

The z-test relies on the known relationship between the mean and standard deviation in normally distributed data and between the mean and standard error of the mean (Fig 6.3). If samples are from the same population, then you would expect the difference between the sample means to be close to zero. The greater the difference between the sample means, the more unlikely it is that the means are from the same population. The z-test estimates, in terms of standard errors, how far apart the means are. This then indicates the probability that the means are from the same population.

There are two forms of this z-test: one for unmatched data (10.1) and one for matched data (10.4). These two tests are quite different from each other. If you are not clear of the distinction in your experiment you should refer to the key in Chapter 7.

10.1.1 Key trends and experimental design

In Example 10.1 it is clear that the mean height of the shells of the periwinkles on the lower shore ($\bar{x}_1 = 5.55$mm) is less than that on the mid shore ($\bar{x}_2 = 7.47$mm). The variation about both sample means as indicated by the variance (s^2) is not substantial ($s_1^2 = 3.59$ and $s_2^2 = 4.1$), is reasonably similar between the two samples although the coefficient of variation is greater in the sample from the mid-shore compared to the lower shore (5.6). There are no outliers. It would be interesting to see if the difference in height between these two samples is significant and therefore supporting the conjecture that this species is evolving into two distinct groups.

Example 10.1 The evolution of *Littorina littoralis* at Aberystwyth, 2002

Several years ago it was suggested that a species of periwinkle, *Littorina littoralis*, was evolving sympatrically into two species, *Littorina obtusata* and *Littorina mariae*, through niche partitioning. *Littorina obtusata* apparently grazes on the brown alga *Ascophyllum nodosum* on the mid-shore, whilst putative *Littorina mariae* feeds on the epiphytes growing on *Fucus serratus* on the lower shore. In a study of the sympatric evolution of *Littorina littoralis*, a representative sample of the two groups of individuals was collected from Aberystwyth in 2002 and their shell height (mm) recorded (Table 10.3). The investigators wished to test the hypothesis that there is no difference between shell height of the two groups of periwinkles from the mid- and lower shore.

(continued)

Table 10.3 Height of shells of two putative species of periwinkles from the mid- and lower shore at Aberystwyth, 2002

> There is one treatment variable: 'location on the shore' and two categories. The two categories are nominal. The dependent variable is shell height which is measured on an interval scale. The two locations were determined by the investigator.

The height of the periwinkle shells (mm) at two locations on the shore					
Periwinkles from the lower shore			Periwinkles from the mid-shore		
5.3	4.3	6.5	11.7	9.3	11.3
8.7	8.0	6.8	4.1	7.7	7.0
5.3	10.2	5.0	8.8	6.6	6.8
5.3	5.3	5.1	9.7	9.4	8.3
5.9	7.8	4.9	5.0	8.8	4.5
2.8	7.0	3.7	6.7	5.8	6.0
8.7	6.1	2.8	6.6	5.6	7.0
2.0	5.0	3.8	7.8	7.5	6.5
5.3	5.4	3.0	7.0	6.3	6.6
6.5	5.7	4.2	7.6	12.5	5.6

$$n_1 = 30$$

$$\sum x_1 = 166.4$$

$$\left(\sum x_1\right)^2 = 27688.96$$

$$\sum\left(x_1^2\right) = 1027.08$$

$$\bar{x}_1 = 5.54667$$

$$s_1^2 = 3.59016$$

$$n_2 = 30$$

$$\sum x_2 = 224.1$$

$$\left(\sum x_2\right)^2 = 50220.81$$

$$\sum\left(x_2^2\right) = 1792.85$$

$$\bar{x}_2 = 7.47$$

$$s_2^2 = 4.09734$$

> There are 30 observations in each category. These are replicates. There are no controls in this investigation.

> Each periwinkle has only been included once in a sample. These observations are therefore independent.

10.1.2 Using this test

i. To use this test you:

1) Do not have an *a priori* reason for expecting certain outcomes from your investigation.

2) Usually present these data in tables or summary histograms rather than a scatter plot.

3) Wish to examine the effect of one treatment variable with two categories. These categories are usually nominal or ordinal.

4) Have observations (the dependent variable) which are independent and unmatched and measured on an interval scale.

5) Have data which are derived from a normally distributed population, they are parametric.

6) Have more than 30 observations in each sample but the sample sizes need not be the same in the two samples.

7) Have two samples where the variances are similar (homogeneous).

ii. Does the example meet these criteria?

At this point we can tell that all but criterion 7 are met. There is one treatment variable (location on the shore) and two nominal categories (mid- and lower shore). There are 30 observations for each sample. Each periwinkle has been measured only once so the data are unmatched. The data have been examined to see if they are probably parametric and they do meet these criteria (see interactive exercise 10 in the Online Resource Centre).

One of the common major assumptions underlying many parametric tests of hypotheses is that the variances of the different samples are homogeneous (similar). This requirement is flagged up in criterion 7. A variance ratio or F-test may be used to examine whether this is probably true (Box 10.1). This check needs to be carried out before using the z-test.

BOX 10.1 How to carry out an F-test to check for homogeneous variances before carrying out a two-sample z-test for unmatched data

GENERAL DETAILS	EXAMPLE 10.1
	This calculation is given in full in the Online Resource Centre. For presentation purposes only all values have been rounded to five decimal places.
1. Hypotheses to be tested H_0: There is no difference between the variances of the two samples. The variances are homogeneous. H_1: There is a difference between the variances of the two samples.	1. Hypotheses to be tested H_0: There is no difference between the variances of the shell height of periwinkles from the lower and mid-shore. H_1: There is a difference between the variances of the shell height of periwinkles from the lower and the mid-shore.
2. Have the criteria for using this test been met?	2. Have the criteria for using this test been met? Yes (10.1.2ii) apart from criterion 7 which is being checked here.
3. How to work out $F_{calculated}$ i. Decide which is sample 1 and which is sample 2.	3. How to work out $F_{calculated}$ i. Let the periwinkles from the lower shore be sample 1 and the periwinkles from the mid-shore be sample 2.

(continued)

ii. For each sample calculate the variance s_1^2 and s_2^2 using the method described in Box 5.1, where the sum of squares of x is:

$$SS(x) = (\Sigma x^2) - \frac{(\Sigma x)^2}{n}$$

The variance is:

$$s^2 = \frac{SS(x)}{n-1}$$

ii. For the periwinkles from the lower shore,

$$\Sigma x_1 = 166.4, n = 30$$

$$\frac{(\Sigma x_1)^2}{n} = 922.96533$$

$$\Sigma(x_1^2) = 1027.08$$

$$SS(x_1) = 1027.08 - 922.96533$$
$$= 104.11467$$

$$s_1^2 = 3.59016$$

For periwinkles from the mid-shore

$$\Sigma x_2 = 224.1, n = 30$$

$$\frac{(\Sigma x_2)^2}{n} = 1674.027,$$

$$\Sigma(x_2^2) = 1792.85$$

$$SS(x_2) = 118.823, s_2^2 = 4.09734$$

iii. $F_{calculated}$ is the ratio between the two variances.

$$F_{calculated} = \frac{\text{larger sample variance}}{\text{smaller sample variance}}$$

iii.

$$F_{calculated} = \frac{4.09734}{3.59016}$$
$$= 1.14127$$

---

4. How to find $F_{critical}$

You will come across three F-tests. Each uses a different table to find $F_{critical}$. You must take great care to use the correct table.

See Appendix D, Table D4. To look up the value in the F table, you will need to know the degrees of freedom (v) for each sample.

Sample 1 $v_1 = n_1 - 1$

Sample 2 $v_2 = n_2 - 1$

where n is the number of observations in that category or sample.

4. How to find $F_{critical}$

For our example, $v_1 = 30 - 1 = 29$ and $v_2 = 30 - 1 = 29$

Interpolating (6.4.4) from the F table:

$F_{critical} = 2.10$ at $p = 0.05$

---

5. The rule

If $F_{calculated}$ is less than $F_{critical}$, you do not reject the null hypothesis and may proceed with the z-test.

5. The rule

In this example, $F_{calculated}$ (1.14) is less than $F_{critical}$ (2.10). Therefore, you do not reject the null hypothesis.

---

6. What does this mean in real terms?

6. What does this mean in real terms?

There is no significant difference ($p = 0.05$) between the variances of shell height of periwinkles from the lower and mid-shores. The variances are homogeneous. You may proceed with the z-test. If the null hypothesis is not accepted, then you should consider transforming your data (5.10) or using a non-parametric test, e.g. Mann–Whitney U test (11.1).

Further examples relating to the topic including how to use statistical software are included in the Online Resource Centre.

10.1.3 The calculation

Having satisfied ourselves that our data meet all the criteria for using this z-test for unmatched data, we can then proceed (Box 10.2).

BOX 10.2 How to carry out a two-sample z-test for unmatched data

GENERAL DETAILS	EXAMPLE 10.1
	This calculation is given in full on the Online Resource Centre. For presentation purposes only all values have been rounded to five decimal places.
1. Hypotheses to be tested H_0: There is no difference between the mean values of the populations from which the samples are taken. H_1: There is a difference between the mean values of the populations from which the two samples are taken.	**1. Hypotheses to be tested** H_0: There is no difference between the mean shell height (mm) of the periwinkles from the lower and mid-shore at Aberystwyth in 2002. H_1: There is a difference between the mean shell height (mm) of the periwinkles from the lower and mid-shore at Aberystwyth in 2002.
2. Have the criteria for using this test been met?	**2. Have the criteria for using this test been met?** Yes (10.1.2ii and Box 10.1).
3. How to work out $z_{calculated}$ i. Decide which is sample 1 and which is sample 2. ii. Use the variances (s^2) and n from Box 10.1 calculated for the F-test. iii. Calculate the means ($\bar{x}$) for each sample. iv. The standard error of the difference of the means (SE$_D$) is calculated as: $$SE_D = \sqrt{\left(\frac{s_1^2}{n_1} + \frac{s_2^2}{n_2}\right)}$$ v. $z_{calculated} = \dfrac{\bar{x}_1 - \bar{x}_2}{SE_D}$	**3. How to work out $z_{calculated}$** i. Let the periwinkles from the lower shore be sample 1 and the periwinkles from the mid-shore be sample 2. ii. $s_1^2 = 3.59016$, $s_2^2 = 4.09734$ $n_1 = n_2 = 30$ iii. $\bar{x}_1 = 5.54667$, $\bar{x}_2 = 7.47$ iv. Standard error of the difference of the means (SE$_D$): $$SE_D = \sqrt{\left(\frac{3.59016}{30} + \frac{4.09734}{30}\right)}$$ $$= 0.50621$$ v. $z_{calculated} = \dfrac{5.54667 - 7.47}{0.50621}$ $$= -3.79947$$

(continued)

4. How to find $z_{critical}$	4. How to find $z_{critical}$
See Appendix D, Table D5. The z distribution is unusual in that it is independent of sample size. For any one p value, there is only one z value.	For a two-tailed test at $p = 0.05$, $z_{critical} = 1.96$
5. The rule	5. The rule
If the absolute value (i.e. ignore the sign) of $z_{calculated}$ is more than or equal to $z_{critical}$, then you may reject the null hypothesis (H_0). Samples are not likely to have come from the same statistical population.	The absolute value of $z_{calculated}$ (3.8) is more than $z_{critical}$ (1.96). Therefore, you may reject the null hypothesis. In fact, at $p = 0.001$, $z_{critical} = 3.29$. You may therefore reject the null hypothesis at this higher level of significance.
6. What does this mean in real terms?	6. What does this mean in real terms?
	There is a very highly significant difference ($z = 3.8$, $p < 0.001$) between the mean shell heights (mm) of the periwinkles from the lower and mid-shore at Aberystwyth. *Littorina mariae* is the smaller of the two apparently distinct species.

Further examples relating to the topic including how to use statistical software are included in the Online Resource Centre.

10.2 Two-sample Student's *t*-test for unmatched data

The z-test compares two samples of parametric data. However, when the sample sizes are relatively small, then the distribution on which the z statistic is based changes with the sample size. This can be taken into account by using a t-test. With a t-test, the sample sizes are used both in the estimate of the standard error and when looking up the critical value of t in tables. There are a number of t-tests. The most commonly used is Student's t-test after the pen name (Student) of W. Gosset who introduced this test (Student, 1908). This test requires that the two samples have similar (homogeneous) variances and we therefore carry out an F-test first to check that this criterion is met (Box 10.1). An alternative t-test (also known as the Welch test or Smith–Welch–Satterwaite test) does not require the variances to be homogeneous, although both sets of data still need to be normally distributed (Ruxton, 2006). When variances are similar, this test performs as well as the more commonly used Student's t-test. Both tests are commonly found in statistical software packages so in this edition we include both the Student's t-test and the unequal variances t-test. These two-sample t-tests are the parametric equivalent of the Mann–Whitney U test (11.1).

Example 10.2 The evolution of *Littorina littoralis* at Porthcawl, 2002

The study at Aberystwyth (2002) indicated that a species of periwinkle, *Littorina littoralis*, was evolving sympatrically into two species (Example 10.1, Box 10.2). To extend this study, the researchers also sampled from Porthcawl in the same year. Here, however, they found far fewer periwinkles (Table 10.4).

(continued)

Table 10.4 Height of shells in two putative species of periwinkles from the mid- and lower shore at Porthcawl, 2002

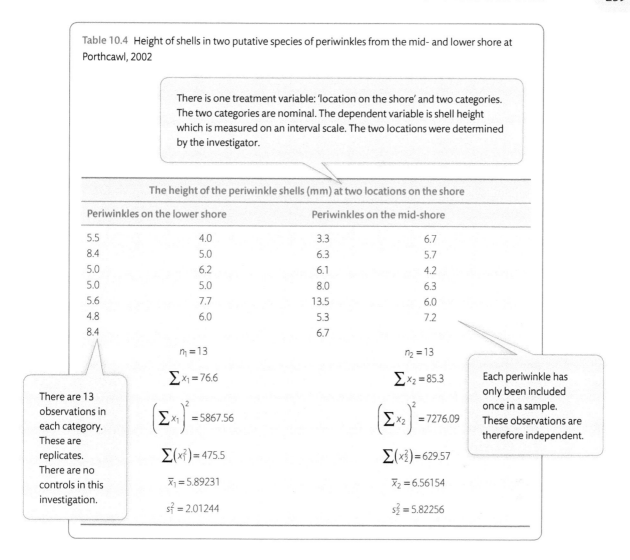

There is one treatment variable: 'location on the shore' and two categories. The two categories are nominal. The dependent variable is shell height which is measured on an interval scale. The two locations were determined by the investigator.

The height of the periwinkle shells (mm) at two locations on the shore

Periwinkles on the lower shore		Periwinkles on the mid-shore	
5.5	4.0	3.3	6.7
8.4	5.0	6.3	5.7
5.0	6.2	6.1	4.2
5.0	5.0	8.0	6.3
5.6	7.7	13.5	6.0
4.8	6.0	5.3	7.2
8.4		6.7	

$$n_1 = 13 \qquad\qquad n_2 = 13$$

$$\sum x_1 = 76.6 \qquad\qquad \sum x_2 = 85.3$$

$$\left(\sum x_1\right)^2 = 5867.56 \qquad\qquad \left(\sum x_2\right)^2 = 7276.09$$

$$\sum\left(x_1^2\right) = 475.5 \qquad\qquad \sum\left(x_2^2\right) = 629.57$$

$$\bar{x}_1 = 5.89231 \qquad\qquad \bar{x}_2 = 6.56154$$

$$s_1^2 = 2.01244 \qquad\qquad s_2^2 = 5.82256$$

There are 13 observations in each category. These are replicates. There are no controls in this investigation.

Each periwinkle has only been included once in a sample. These observations are therefore independent.

10.2.1 Key trends and experimental design

Like Example 10.1 the mean height of periwinkle shells on the lower shore ($\bar{x}_1 = 5.89$mm) is less than that on the upper shore ($\bar{x}_2 = 6.56$mm). There is a smaller difference between the two locations than in Example 10.1 and there is more variation in the mid-shore sample ($s_2^2 = 5.8$ compared to $s_1^2 = 3.52.01$). This is largely as a result of an outlier in the mid-shore sample which has a value of 13.5mm. As there is no reason for excluding this outlier from a biological perspective then we will continue to include this. However, usual practice would be to complete the calculation with the outlier present and then again without this value and comment then on the effect this single individual is having on the outcome of the analysis.

10.2.2 Using this test

i. To use this test you:

1) Do not have an *a priori* reason for expecting certain outcomes from your investigation.

2) Usually present these data in tables or summary histograms rather than a scatter plot.

3) Wish to examine the effect of one treatment variable with two categories. These categories are usually nominal or ordinal.

4) Have observations (the dependent variable) which are independent and unmatched and measured on an interval scale.

5) Have data which are derived from a normally distributed population, they are parametric.

6) Can use this test if you have less than 30 observations in each sample. The sample sizes need not be the same in the two samples.

7) Have two samples where the variances are similar (homogeneous).

ii. Does the example meet these criteria?

The annotation on Table 10.4 confirms that criteria 1–4 and 6 are met. With such a small sample size it is very difficult to confirm that these data themselves are parametric; however, given the previous study (Example 10.1) it appears likely that the population from which these samples are derived should be normally distributed. In addition the data are measured on an interval scale. We will therefore conclude based on these points that criterion 5 is also met. The F-test described in Box 10.1 can be used to confirm that the two variances are similar. For this example $s_1^2 = 2.01$ and $s_2^2 = 5.82$, $F_{calculated} = 2.9$ with $v_1 = 12$, $v_2 = 12$ and $F_{critical} = 3.28$ at $p = 0.05$. Therefore since $F_{calculated}$ (2.9) is less than $F_{critical}$ (3.28) there is no significant difference between the variances of the two periwinkle samples and we may proceed with the t-test.

10.2.3 The calculation

Having satisfied ourselves that our data meet all the criteria for using this t-test, we can then proceed (Box 10.3).

BOX 10.3 How to carry out a two-sample Student's t-test for unmatched data

GENERAL DETAILS	EXAMPLE 10.2
	This calculation is given in full in the Online Resource Centre. For presentation purposes only all values have been rounded to five decimal places.

(continued)

1. Hypotheses to be tested	**1. Hypotheses to be tested**
H_0: There is no difference between the means for the two samples. H_1: There is a difference between the means for the two samples.	H_0: There is no difference between the mean shell height (mm) of the periwinkles from the lower and mid-shore at Porthcawl in 2002. H_1: There is a difference between the mean shell height (mm) of the periwinkles from the lower and the mid-shore at Porthcawl in 2002.
2. Have the criteria for using this test been met?	**2. Have the criteria for using this test been met?** Yes (10.2.2ii).
3. How to work out $t_{calculated}$ i. Decide which is sample 1 and which is sample 2.	**3. How to work out $t_{calculated}$** i. Let the periwinkles from the lower shore be sample 1 and the periwinkles from the mid-shore be sample 2.
ii. Use the variances (s_1^2 and s_2^2) calculated for the *F*-test, calculate the means ($\bar{x}_1$ and $\bar{x}_2$), and record the number of observations (*n*) for each sample.	ii. Table 10.4. and (10.2.2.ii.) outline these calculations. $\bar{x}_1 = 5.89231$, $s_1^2 = 2.01244$, $n_1 = 13$ $\bar{x}_2 = 6.56154$, $s_2^2 = 5.82256$, $n_2 = 13$
iii. Determine the common variance as: $$s_c^2 = \frac{(n_1-1)s_1^2 + (n_2-1)s_2^2}{(n_1-1)+(n_2-1)}$$	iii. Determine the common variance (s_c^2) as: $$s_c^2 = \frac{(13-1)2.01244 + (13-1)5.82256}{(13-1)+(13-1)}$$ $$= 3.9175$$
iii. Use the common variance to calculate the standard error of the difference between the means (SE_D) and then $t_{calculated}$. $$SE_D = \sqrt{\left(\frac{s_c^2}{n_1} + \frac{s_c^2}{n_2}\right)}$$	iii. $$SE_D = \sqrt{\left(\frac{3.9175}{13} + \frac{3.9175}{13}\right)}$$ $$= 0.77633$$
iv. $t_{calculated} = \dfrac{\bar{x}_1 - \bar{x}_2}{SE_D}$	iv. $t_{calculated} = \dfrac{5.89231 - 6.56154}{0.77633}$ $$= -0.86204$$
4. How to find $t_{critical}$ i. Calculate the degrees of freedom where $v = (n_1-1)+(n_2-1)$ ii. $t_{critical}$ can be found in Appendix D in Table D6 at this degrees of freedom and for $p = 0.05$ for a two-tailed test.	**4. How to find $t_{critical}$** i. $v = (13-1)+(13-1) = 24$ ii. For a two-tailed test $t_{critical} = 2.064$ at $p = 0.05$
5. The rule If the absolute value of $t_{calculated}$ is more than or equal to $t_{critical}$ you may reject the null hypothesis.	**5. The rule** When you use an absolute value, the negative sign is ignored. In this example, $t_{calculated}$ is therefore 0.862 and you can conclude that the absolute value of $t_{calculated}$ (|-0.862|) is less than $t_{critical}$ (2.064). Therefore, you do not reject the null hypothesis.

(continued)

6. What does this mean in real terms?	6. What does this mean in real terms?
	There is no significant difference ($t = 0.862$, $p = 0.05$) in the mean shell height (mm) between the periwinkles on the lower and mid-shores of Porthcawl. Therefore, these data do not support the notion that there are two distinct species.

Further examples relating to the topic including how to use statistical software are included in the Online Resource Centre.

10.3 Two-sample unequal variance *t*-test for unmatched data

An explanation about sample sizes and the *t*- and *z*-tests is given in 10.1. In 10.2 we explained the choice you can make between the Student's *t*-test and this unequal variances *t*-test. The *t*-test we explain in this section does not have the same requirement for homogeneous variances as the Student's *t*-test and therefore there is no need to calculate the *F*-test. However, the estimate of the degrees of freedom is more involved. We illustrate this test using Example 10.2.

10.3.1 Key trends and experimental design

The main aspects of the design for Example 10.2 were outlined in 10.2.1. It is clear from the means and variances (Table 10.4) that there appears to be a difference between these two locations in the shell height. Confirming this is statistically meaningful and not due to sampling error would benefit the study of this possible example of sympatric evolution.

10.3.2 Using this test

i. To use this test you:

1) Do not have an *a priori* reason for expecting certain outcomes from your investigation.

2) Usually present these data in tables or summary histograms rather than a scatter plot.

3) Wish to examine the effect of one treatment variable with two categories. These categories are usually nominal or ordinal.

4) Have observations (the dependent variable) which are independent and unmatched and measured on an interval scale.

5) Have data which are derived from a normally distributed population, they are parametric.

6) Have less than 30 observations in each sample. The sample sizes need not be the same in the two samples.

ii. Does the example meet these criteria?

We have confirmed in 10.2.2ii that the data from Example 10.2 meet these criteria.

10.3.3 **The calculation**

Having satisfied ourselves that the data from Example 10.2 meet all the criteria for using this *t*-test, we can then proceed (Box 10.4). To distinguish between the $t_{calculated}$ for Student's *t*-test, we use the statistic t' for the $t_{calculated}$ using this unequal variances *t*-test.

BOX 10.4 **How to carry out an unequal variance *t*-test for unmatched data**

GENERAL DETAILS	EXAMPLE 10.2
	This calculation is given in full in the Online Resource Centre. For presentation purposes only all values have been rounded to five decimal places.
1. Hypotheses to be tested H_0: There is no difference between the means for the two samples. H_1: There is a difference between the means for the two samples.	1. Hypotheses to be tested H_0: There is no difference between the mean shell height (mm) of the periwinkles from the lower and mid-shore at Porthcawl in 2002. H_1: There is a difference between the mean shell height (mm) of the periwinkles from the lower and the mid-shore at Porthcawl in 2002.
2. Have the criteria for using this test been met?	2. Have the criteria for using this test been met? Yes (10.3.2ii).
3. How to work out $t'_{calculated}$ i. Decide which is sample 1 and which is sample 2. ii. Calculate the variances (s_1^2 and s_2^2) and means ($\bar{x}_1$ and $\bar{x}_2$) and record the number of observations (*n*) for each sample. If you are not sure how to calculate the variances see Box 5.1. iii. Calculate the standard error of the difference between the means (SE_D). $$SE_D = \sqrt{\left(\frac{s_1^2}{n_1} + \frac{s_2^2}{n_2}\right)}$$ iv. $t'_{calculated} = \dfrac{\bar{x}_1 - \bar{x}_2}{SE_D}$	3. How to work out $t'_{calculated}$ i. Let the periwinkles from the lower shore be sample 1 and the periwinkles from the mid-shore be sample 2. ii. $\bar{x}_1 = 5.89231$ $s_1^2 = 2.01244$ $n_1 = 13$ $\bar{x}_2 = 6.56154$ $s_2^2 = 5.82256$ $n_2 = 13$ iii. $SE_D = \sqrt{\left(\dfrac{2.01244}{13} + \dfrac{5.82256}{13}\right)}$ $= \sqrt{(0.15480 + 0.44789)}$ $= \sqrt{0.60269} = 0.77633$ iv. $t'_{calculated} = \dfrac{5.89231 - 6.56154}{0.77633}$ $= -0.86204$

(continued)

4. How to find $t_{critical}$

See Appendix D, Table D6. The critical value of t' is found using a t table and the degrees of freedom (v).

$$v = \frac{\left(\frac{1}{n_1}+\frac{u}{n_2}\right)^2}{\left(\frac{1}{n_1^2(n_1-1)}\right)+\left(\frac{u^2}{n_2^2(n_2-1)}\right)}$$

This is a more complex calculation than is usual for the degrees of freedom so we have broken this down into a number of steps.

i. First calculate u

$$u = \frac{s_2^2}{s_1^2}$$

ii. Calculate the numerator

$$\left(\frac{1}{n_1}+\frac{u}{n_2}\right)^2$$

iii. Calculate the first part of the denominator

$$\left(\frac{1}{n_1^2(n_1-1)}\right)$$

iv. Calculate the second part of the denominator

$$\left(\frac{u^2}{n_2^2(n_2-1)}\right)$$

v. Using these terms, now calculate the degrees of freedom. $t'_{critical}$ can be found in Appendix D Table D6 at this degrees of freedom and for $p = 0.05$.

4. How to find $t_{critical}$

i. Calculating the degrees of freedom for Example 10.2.

$$u = \frac{s_2^2}{s_1^2} = \frac{5.82256}{2.01244} = 2.89328$$

ii. $\left(\frac{1}{n_1}+\frac{u}{n_2}\right)^2 = \left(\frac{1}{13}+\frac{2.89328}{13}\right)^2$

$$= (0.07692+0.22256)^2$$
$$= (0.29948)^2 = 0.08969$$

iii. $\left(\frac{1}{n_1^2(n_1-1)}\right) = \frac{1}{(13)^2(13-1)}$

$$= \frac{1}{169\times12} = \frac{1}{2028}$$
$$= 0.00049$$

iv. $\left(\frac{u^2}{n_2^2(n_2-1)}\right) = \frac{(2.89328)^2}{(13)^2(13-1)}$

$$= \frac{8.37109}{169\times12} = 0.00413$$

v. $v = \frac{0.08969}{(0.00049)+(0.00413)}$

$v = 19.41342$

So $t'_{critical}$ is 2.093

5. The rule

If the absolute value of $t_{calculated}$ is more than or equal to $t_{critical}$ you may reject the null hypothesis. The term 'absolute' is explained in the glossary and Appendix E.

5. The rule

When you use an absolute value, the negative sign is ignored. In this example, $t_{calculated}$ is therefore 0.862. You can therefore conclude that the absolute value of $t_{calculated}$ ($|-0.862|$) is less than $t_{critical}$ (2.093). Therefore, you do not reject the null hypothesis.

(continued)

6. What does this mean in real terms?	6. What does this mean in real terms?
	There is no significant difference ($t = 0.862$, $p = 0.05$) in the mean shell height (mm) between the periwinkles on the lower and mid-shores of Porthcawl. Therefore, these data do not support the notion that there are two distinct species.

Further examples relating to the topic including how to use statistical software are included in the Online Resource Centre.

Q1 Examine the results from Boxes 10.2 and 10.4. Suggest reasons why the outcomes from these two investigations differ.

10.4 Two-sample z- and t-tests for matched data

If you are designing an investigation in which you will measure an item before a particular event or treatment and then measure the same item after the event or treatment, this will generate matched data. Matched observations are also sometimes known as 'repeated measures'. We have illustrated this in Table 10.2. The two observations for each item are not independent of each other as they relate to a single item. This needs to be allowed for in the analysis of the data. Unlike unmatched z- and t-tests, the comparison of matched samples is based on the difference between each pair of measurements. If the samples are similar, then the differences between each pair of observations will be small. The mean and standard error of these differences are used to generate $z_{calculated}$.

Matched z- and t-tests are virtually identical; the main difference again arises in that in the t-test there is an allowance for the relatively small sample sizes being investigated. As these two tests are so similar, we have amalgamated the information in one box (Box 10.5). If you have taken more than two repeated measurements from one item you can consider using the two-way ANOVA with no replicates (10.15).

Example 10.3 Weight loss by members of a fencing club during a 1-day competition

The members of a fencing club were aware that on days when they competed, they lost weight (Table 10.5). The weight loss was thought to be the result of dehydration. However, as not all members lost weight, the team wished to know whether the weight loss was significant (T. Richards, personal communication, 9.1.05).

(continued)

Table 10.5 Weight loss during a 1-day fencing competition by members of a fencing club

There are 13 competitors. These are replicates and items.

There is one treatment variable: 'the competition'. There are two observations for this one treatment variable for each item. The observations are in nominal categories 'before the competition' and 'after the competition'. The dependent variable (weight) is measured on an interval scale.

Competitor	Weight before competition (kg)	Weight after competition (kg)	Difference between weights (kg)
1	60.00	59.55	0.45
2	59.15	58.70	0.45
3	60.20	59.80	0.40
4	62.40	61.90	0.50
5	57.20	57.20	0.00
6	60.35	59.85	0.50
7	59.80	59.40	0.40
8	60.10	60.10	0.00
9	60.20	59.90	0.30
10	59.90	59.90	0.00
11	60.00	60.00	0.00
12	61.20	60.75	0.45
13	58.50	58.05	0.45
$\bar{x}$	59.92308	59.62308	0.30
s^2	1.51656	1.35942	0.04583

10.4.1 Key trends and experimental design

It is clear that most of the competitors lost weight as a result of taking part in the fencing competition (before the competition $\bar{x} = 59.92$kg; after the competition $\bar{x} = 59.62$kg). The within-sample variance of the difference in weight loss ($0.05kg^2$) suggests the competitors were fairly consistent in how much weight they lost during the competition. The weight loss unsurprisingly is not that great but is it significant?

10.4.2 Using this test

i. To use this test you:

1) Do not have an *a priori* reason for expecting certain outcomes from your investigation.

2) Usually present these data in tables or summary histograms rather than a scatter plot.

3) Wish to examine the effect of one treatment variable with two categories. These categories are usually nominal or ordinal.

4) Have two observations (the dependent variable) for each item measured for the one treatment variable. These data are therefore matched. Each pair must be independent of any other pair.

5) Have data which are derived from a normally distributed population, they are parametric and measured on an interval scale.

6) Have fewer than 30 pairs of observations. With 30 or more observations in each sample common practice is to use the z-test.

7) Have two samples where the variances are similar (homogeneous).

ii. Does the example meet these criteria?

The data for Example 10.3 will be used to illustrate the *t*-test as there are fewer than 30 observations in each sample. This example meets criteria 1–4 and 6 for the *t*-test (Table 10.5). The data are matched as two observations have been recorded for each person for this one treatment variable. To check criterion 5 it is common practice to check if the samples are normally distributed and take this as an indicator that the population from which the samples are derived is the same. In this example the sample size is small and it is realistically only possible to check the first criterion in Box 5.2. and confirm that the observations (kg) are measured on an interval scale. However, weight in humans is known to be usually a normally distributed characteristic. Therefore we will assume that this criterion is also met.

To check criterion 7 we use the test for homogeneity of variances (Box 10.1). For Example 10.3, $F_{calculated} = 1.51656/1.35942 = 1.11559$, $v = 13 - 1 = 12$ for both samples, and at $p = 0.05$, $F_{critical} = 3.28$. As $F_{calculated}$ is less than $F_{critical}$, there is no significant difference between the two sample variances and we may proceed with the *t*-test.

10.4.3 The calculation

Having satisfied ourselves that the data for Example 10.3 meet all the criteria for using this *t*-test, we can then proceed (Box 10.5). If you are carrying out a z-test for matched data, the procedure is exactly the same until finding the critical values. Some of the calculations are included in Table 10.5. The full calculation is given in the Online Resource Centre.

BOX 10.5 How to carry out a two-sample z- and t-test for matched data

GENERAL DETAILS	EXAMPLE 10.3
	This calculation is given in full in the Online Resource Centre. For presentation purposes only all values have been rounded to five decimal places.
1. Hypotheses to be tested H_0: There is no difference between the means of the populations from which sample 1 and sample 2 are taken. H_1: There is a difference between the means of the populations from which sample 1 and sample 2 are taken.	1. Hypotheses to be tested H_0: There is no difference between the mean weights (kg) of individuals before and after a fencing competition. H_1: There is a difference between the mean weights (kg) of individuals before and after a fencing competition.
2. Have the criteria for using this test been met?	2. Have the criteria for using this test been met? Yes (10.4.2ii).
3. How to work out $z_{calculated}$ or $t_{calculated}$ i. Work out the difference (D) for each pair of observations. ii. Calculate the mean $(\bar{D})$ and variance $\left(s_D^2\right)$ for these differences, where n is the number of pairs of observations. iii. The standard error of the difference of the means $SE_D = \sqrt{\dfrac{s_D^2}{n}}$	3. How to work out $t_{calculated}$* (*In this example we are working out t; the steps would be the same here if you were working out z.) i. The difference between each pair of observations is included in Table 10.5. ii. The method for calculating a variance is given in Box 5.1. The results for this example are included in Table 10.5. iii. $SE_D = \sqrt{\dfrac{0.04583}{13}} = 0.05938$
iv. $z_{calculated}$ or $t_{calculated} = \dfrac{\bar{D}}{SE_D}$	iv. For this example, we are using a t-test, therefore $t_{calculated} = \dfrac{0.3}{0.05938} = 5.05245$
4. How to find $z_{critical}$ or $t_{critical}$ If this is a t-test for matched pairs, then $t_{critical}$ is found for a two-tailed t-test where $v = n - 1$ at $p = 0.05$ (Appendix D, Table D6) If this is a z-test for matched pairs, then $z_{critical}$ is found from the two-tailed table of z values at $p = 0.05$ (Appendix D, Table D5).	4. How to find $t_{critical}$ In our example, $v = 13 - 1 = 12$ and for a two-tailed t-test at $p = 0.05$, $t_{critical} = 2.179$.

(continued)

5. The rule	5. The rule
If the absolute value of $z_{calculated}$ or $t_{calculated}$ is more than or equal to $z_{critical}$ or $t_{critical}$, respectively, then you may reject the null hypothesis.	$t_{calculated}$ (5.053) is greater than $t_{critical}$ (2.179), so we reject the null hypothesis. In fact, $t_{critical}$ at $p = 0.001$ is 4.318, so we may reject the null hypothesis at this higher level of significance.
6. What does this mean in real terms?	6. What does this mean in real terms?
	There is a very highly significant difference ($t = 5.053$, $p = 0.001$) between the mean weights (kg) of individuals before and after a fencing competition.

Further examples relating to the topic including how to use statistical software are included in the Online Resource Centre.

10.5 One-sample *t*-test

Some of our undergraduates find themselves wishing to compare a single sample to a single value. For example measurements from a sample of living individuals of a species compared to a measurement from one museum specimen or measures of nutritional intake such as amount of calcium ingested each day by a group of adults compared to the reference nutritional intake (RNI) or daily recommended value (DRV). In this instance a modified *t*-test may be used. The test in essence asks 'what is the probability that the single value is derived from the same population as the sample?' As explained in 10.1, the *t*-test functions like *z* when the sample size increases so this one test is suitable for all sample sizes. However, the issues about 'representativeness' (1.4.1) should be taken into account when interpreting the findings if your sample size is small. A non-significant result where the sample size is small could indicate that you do not have sufficient data to reject the null hypothesis.

There are a number of consequences in having one value in one of the samples. One of these is that unlike the Student's *t*-test and matched *t*-test you have to assume that the variances are homogenous as there is no variance estimate for the single value.

> **Example 10.4** The change in nutrition of international students when coming to a university in the UK
>
> An undergraduate investigated how nutrition in international students from the Balkans changed on coming to the UK to attend a university. She collected food diaries from the students when they were at home and whilst at university and analysed the nutritional content for a number of different components including folate. The data for males and females were analysed separately and the results for the male students compared to the estimated average UK intake are included in Table 10.6.
>
> *(continued)*

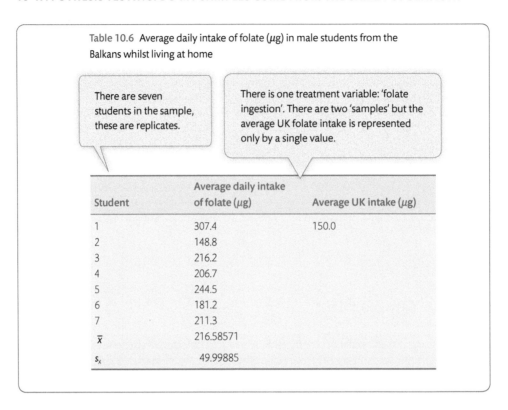

Table 10.6 Average daily intake of folate (μg) in male students from the Balkans whilst living at home

There are seven students in the sample, these are replicates.

There is one treatment variable: 'folate ingestion'. There are two 'samples' but the average UK folate intake is represented only by a single value.

Student	Average daily intake of folate (μg)	Average UK intake (μg)
1	307.4	150.0
2	148.8	
3	216.2	
4	206.7	
5	244.5	
6	181.2	
7	211.3	
$\bar{x}$	216.58571	
s_x	49.99885	

10.5.1 Key trends and experimental design

It is clear that there is some variation between the students. The first student's average daily intake of folate whilst living at home is double that of the second. Nonetheless the mean from the sample of students is close to the average UK intake of folate.

10.5.2 Using this test

i. To use this test you:

1) Do not have an *a priori* reason for expecting certain outcomes from your investigation.

2) Usually present these data in tables or summary histograms rather than a scatter plot.

3) Wish to examine the effect of one treatment variable with two categories. These categories are usually nominal or ordinal.

4) Have a single value only in one category and more than one observation in the second category.

5) Have data which are derived from a normally distributed population, they are parametric and measured on an interval scale.

ii. Does the example meet these criteria?

We can confirm that the data meet criteria 1–4 and that the data are measured on an interval scale. With this small sample size it is difficult to use the sample to judge whether the population is normally distributed. However, a check of the criteria in Box 5.2 with these data suggests that this sample appears to be normally distributed so we will assume that this criterion is also met.

10.5.3 The calculation

Having satisfied ourselves that the data for Example 10.4 meet all the criteria for using this *t*-test, we can then proceed (Box 10.6). Some of the calculations are included in Table 10.6. The full calculation is given in the Online Resource Centre.

BOX 10.6 How to carry out a one-sample *t*-test

GENERAL DETAILS	EXAMPLE 10.4
	This calculation is given in full in the Online Resource Centre. For presentation purposes only all values have been rounded to five decimal places.
1. Hypotheses to be tested	1. Hypotheses to be tested
H_0: There is no difference between the means for the two samples. H_1: There is a difference between the means for the two samples.	H_0: There is no difference between the average daily intake of folate (μg) in male students from the Balkans whilst living at home compared to average UK intake of folate. H_1: There is a difference between the average daily intake of folate (μg) in male students from the Balkans whilst living at home compared to the average UK intake of folate.
2. Have the criteria for using this test been met?	2. Have the criteria for using this test been met? Yes (10.5.2ii).
3. How to work out $t_{calculated}$ i. From the single sample calculate the mean ($\bar{x}$), standard deviation (s), and record n. The method for calculating the standard deviation is outlined in Box 5.1 where the sum of squares of x is: $$SS(x)=\left(\Sigma x^2\right)-\frac{\left(\Sigma x\right)^2}{n}$$ The standard deviation is: $$s_x=\sqrt{\frac{SS(x)}{n-1}}$$ ii. Use these values to determine $t_{calculated}=\dfrac{\bar{x}-y}{s_x/\sqrt{n}}$	3. How to work out $t_{calculated}$ i. $n=7$ $\bar{x}=216.58571$ $s_x=49.99885$ ii. In this example y = 200 $$t_{calculated}=\frac{216.58571-150}{49.99885/\sqrt{7}}=\frac{66.58571}{18.89779}$$ $$=3.52347$$

(continued)

4. How to find $t_{critical}$	4. How to find $t_{critical}$
Calculate the degrees of freedom where $v = n_x - 1$ $t_{critical}$ can be found in Appendix D in Table D6 at this degrees of freedom and for $p = 0.05$ for a two-tailed test.	$v = n_x - 1 = 7-1 = 6$ For a two-tailed test $t_{critical} = 2.447$ at $p = 0.05$
5. The rule	5. The rule
If the absolute value of $t_{calculated}$ is more than or equal to $t_{critical}$ you may reject the null hypothesis.	When you use an absolute value, the negative sign is ignored. In this example $t_{calculated}$ is a positive value anyway. You can conclude that the absolute value of $t_{calculated}$ ($|3.523|$) is more than $t_{critical}$ (2.447). Therefore, you reject the null hypothesis.
6. What does this mean in real terms?	6. What does this mean in real terms?
	There is a significant difference ($t = 3.523$, $p = 0.05$) in the average daily folate intake of male students from the Balkans whilst living at home and average UK folate intake.

Further examples relating to the topic including how to use statistical software are included in the Online Resource Centre.

10.6 Introduction to parametric ANOVAs

The acronym ANOVA stands for analysis of variance. Parametric ANOVAs are flexible and powerful. There are many versions of the ANOVA that allow you to handle data with two or more samples, with or without replicates, with one or more treatment variables, and with or without incomplete data sets. In this chapter, we examine the ANOVAs our undergraduates most often use. There are many versions of parametric ANOVAs that we do not cover. If you think that an ANOVA is likely to be the best test for your experimental design but you cannot find the one you need in this chapter, then we recommend Sokal & Rohlf (2011) or Zar (2009).

ANOVAs are used to test the hypothesis 'There is no difference between the mean values for a given number of samples'. These samples may come from different variables or treatments or from blocks or replicates within variables. We consider experimental designs relating to ANOVAs at the start of the one-way ANOVA section (10.6) and the two-way section (10.11). Although the hypotheses refer to sample means, the ANOVA 'looks' at this by comparing variation in the data. In Fig. 10.1, data from two samples are plotted. The means for both samples are similar and the variances are similar and relatively small. When the two samples are combined there is little change in either the mean or the total variance. Compare this with Fig. 10.2. Here the two samples have very different means, although the variances for each sample are relatively small and similar to each other. If these two samples are

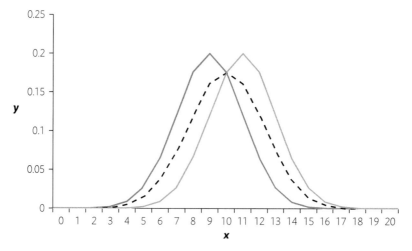

Fig. 10.1 Two samples with similar means and variances (––), which when combined (--) produce a distribution with a mean and variance similar to those of the two original samples

combined, the mean for this total data set is different and the combined variance is relatively large. In an analysis of the data from Fig. 10.1, you would probably not expect a significant difference between the two samples and this can be detected by comparing the total variance with the variance for the two samples. However, in the data from Fig. 10.2, the total variance is relatively large because the means of the two samples are so different from each other. In

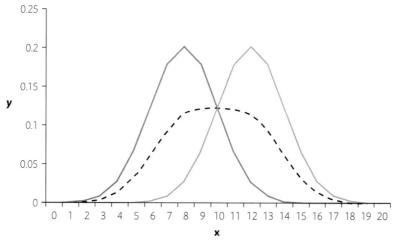

Fig. 10.2 Two samples with very different means (––), which when combined (--) produce a distribution with a variance much greater than the variance of either sample 1 or sample 2

this case, a comparison of the two sample variances with the total variance would suggest that these two samples will be significantly different from each other. It is this comparison of variances in the ANOVA that detects the relative differences in the means of the samples.

Calculating ANOVAs by hand is straightforward but tedious. However, there are so many ANOVAs that a common mistake when using computer software is to select the wrong one. For this reason, we include the ANOVAs used most frequently by our undergraduates. We hope that this introduction will provide you with sufficient familiarity to be confident when using statistical packages.

The ANOVAs we include show you how to analyse data from experiments with one (one-way ANOVA, 10.7) to three (three-way ANOVA, 10.17) treatment variables. Most of these designs require equal replicates and this should be considered when planning your experiment. In the two-way ANOVAs, we have a two-way ANOVA and a nested two-way ANOVA. These reflect a Model I and mixed model experimental design (10.11).

The method for calculating ANOVAs does not lend itself to the format we have followed in the boxes elsewhere in the book. However, we include both general instructions and a specific example to help you understand the calculations. In addition, there are calculation tables and ANOVA tables that illustrate both how to carry out the analysis and how to report these results.

10.7 One-way parametric ANOVA with equal numbers of replicates

The first and simplest ANOVA we consider is one that allows you to examine the effect of one treatment variable. For example if you wished to examine the organic content of soil in four different woodlands the variable you are looking at the effect of (the treatment variable) is 'woodlands'. Usually when there is one treatment variable under investigation, the ANOVA is called a one-way ANOVA. An exception to this can arise if you have matched data or repeated measures (10.11.5). This parametric one-way ANOVA can examine Model I or Model II designs with data that have an equal number of replicates (10.7) or with a small adjustment can analyse data with unequal replicates (10.9).

Example 10.5 The effectiveness of weaning plantlets of *Lobelia* 'Hannah' from tissue culture onto one of four composts

An undergraduate studying horticulture was interested in how plants propagated by tissue culture were then weaned onto other growing media, such as soil-based compost. She designed a series of trials that examined aspects of this weaning process. In one trial, she transferred *Lobelia* 'Hannah' plantlets at random into plugs containing one of four different composts (A–D). After 8 weeks, the plantlets were examined and their fresh weight recorded (Table 10.7).

(continued)

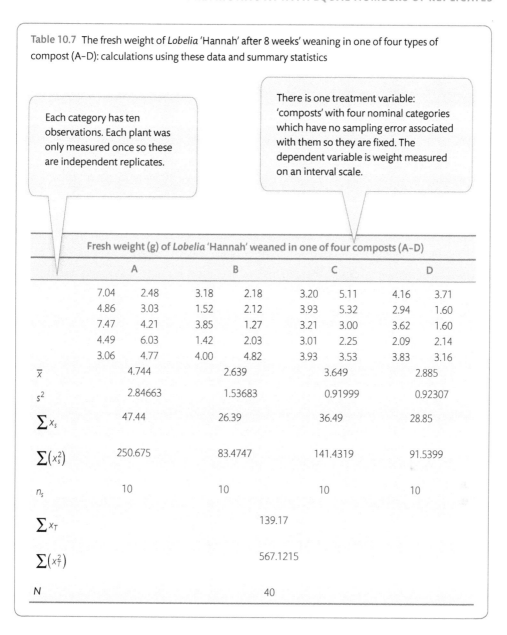

Table 10.7 The fresh weight of *Lobelia* 'Hannah' after 8 weeks' weaning in one of four types of compost (A–D): calculations using these data and summary statistics

Each category has ten observations. Each plant was only measured once so these are independent replicates.

There is one treatment variable: 'composts' with four nominal categories which have no sampling error associated with them so they are fixed. The dependent variable is weight measured on an interval scale.

Fresh weight (g) of *Lobelia* 'Hannah' weaned in one of four composts (A–D)

	A		B		C		D	
	7.04	2.48	3.18	2.18	3.20	5.11	4.16	3.71
	4.86	3.03	1.52	2.12	3.93	5.32	2.94	1.60
	7.47	4.21	3.85	1.27	3.21	3.00	3.62	1.60
	4.49	6.03	1.42	2.03	3.01	2.25	2.09	2.14
	3.06	4.77	4.00	4.82	3.93	3.53	3.83	3.16
$\bar{x}$	4.744		2.639		3.649		2.885	
s^2	2.84663		1.53683		0.91999		0.92307	
$\sum x_s$	47.44		26.39		36.49		28.85	
$\sum(x_s^2)$	250.675		83.4747		141.4319		91.5399	
n_s	10		10		10		10	
$\sum x_T$			139.17					
$\sum(x_T^2)$			567.1215					
N			40					

10.7.1 Key trends and experimental design

If we examine the mean values for the fresh weight of the *Lobelia* plants compost A ($\bar{x}$ = 4.74g) appears to be the most effective transplant media followed by compost C ($\bar{x}$ = 3.65g). The variation within compost A, however, is by far the greatest (s^2 = 2.85g^2). It may be that this variation indicates that other factors are affecting the fresh weight in this species and therefore a test of significance relating to the compost as a treatment will clarify whether the compost is an important factor.

10.7.2 **Using this test**

i. To use this test you:

1) Have no *a priori* reason for expecting certain outcomes from your investigation.

2) Have data which are usually presented in tables or summary histograms rather than a scatter plot.

3) Wish to examine the effect of one treatment variable with more than two categories.

4) Can confirm that the treatment variable is usually measured on an ordinal and/or nominal scale but could be measured on an interval scale and arranged into discrete categories.

5) Have a dependent variable which is measured on an interval scale and where the data are parametric.

6) Have the same number of observations (replicates) in each category.

7) Have an experimental design that means that each item is assigned at random to the categories and each item is only measured once.

8) Have samples where the variation is similar (homogeneous).

ii. **Does the example meet these criteria?**

The annotations on Table 10.7 show that Example 10.5 meet criteria 1–4, 6, and 7. With a sample size of only 10 it is again difficult to numerically confirm that these data are normally distributed. The data are measured on an interval scale and in most studies weight is usually a normally distributed variable. We will therefore assume that these data are parametric. Clearly, however, this uncertainty is something that must be considered when discussing the results. To check whether the variances are similar (criterion 8) an F_{max} test is carried out (Box 10.7). This test is very similar to the F-test outlined in Box 10.1; however, the table of critical values is not the same. You must therefore use the correct F_{max} table of critical values.

BOX 10.7 **How to carry out an F_{max} test to check for homogeneous variances, before carrying out an ANOVA**

GENERAL DETAILS	EXAMPLE 10.5
	This calculation is given in full in the Online Resource Centre. For presentation purposes only all values have been rounded to five decimal places.
1. Hypotheses to be tested	1. Hypotheses to be tested
H_0: There is no difference between the variances of the two samples. The variances are homogeneous.	H_0: There is no difference between the variances of the fresh weight (g) of plantlets grown in the four different composts.

(continued)

H_1: There is a difference between the variances of the two samples.	H_1: There is a difference between the variances of the fresh weight (g) of plantlets grown in the four different composts.
2. Have the criteria for using this test been met?	2. Have the criteria for using this test been met? Yes (10.7.2ii).
3. How to work out $F_{max\ calculated}$ i. For each sample, calculate the variance (s^2) using the method described in Box 5.1. ii. $F_{max\ calculated}$ is the ratio between two values found by dividing the largest variance by the smallest variance. $$F_{max\ calculated} = \frac{\text{largest sample variance}\left(s^2_1\right)}{\text{smallest sample variance}\left(s^2_2\right)}$$	3. How to work out $F_{max\ calculated}$ i. See Table 10.7. ii. $F_{max\ calculated} = \dfrac{2.84663}{0.91999} = 3.09420$
4. How to find $F_{max\ critical}$ See Appendix D, Table D7. To look up the value for an F-test that precedes an ANOVA, you will need to know the number of categories or samples (a) and the degrees of freedom ($v = n_s - 1$). n_s is the number of observations in each category or sample.	4. How to find $F_{max\ critical}$ For our example, there are four samples ($a = 4$). There are ten observations in each sample so the degrees of freedom are $v = 10 - 1 = 9$ So at $p = 0.05$, $F_{max\ critical} = 6.31$.
5. The rule If $F_{max\ calculated}$ is less than $F_{max\ critical}$, then you may accept the null hypothesis and proceed with the ANOVA.	5. The rule In this example, $F_{max\ calculated}$ (3.09) is less than $F_{max\ critical}$ (6.31). Therefore, you do not reject the null hypothesis.
6. What does this mean in real terms?	6. What does this mean in real terms? There is no significant difference ($F_{max} = 3.09$, $p = 0.05$) between the variances of the fresh weight of the plantlets weaned on different composts. You may proceed with the ANOVA. If the null hypothesis is not accepted, then you should consider transforming your data (5.10) or using a non-parametric test, e.g. Kruskal–Wallis (11.3).

Further examples relating to the topic including how to use statistical software are included in the Online Resource Centre.

10.7.3 The calculation

Having satisfied ourselves that our data meet all the criteria for using this ANOVA, we can then proceed. In Box 10.8, we have organized the calculation of $F_{calculated}$ under a number of subheadings: (A) Calculate general terms; (B) Calculate the sums of squares; and (C) Construct and complete an ANOVA calculation table. At each point, we show how these steps can be applied to Example 10.5. Some of the calculations are included in Tables 10.7 and 10.8.

Table 10.8 ANOVA table for one-way parametric ANOVA of data from Example 10.5

| Source of variation | Fresh weight of *Lobelia* 'Hannah' after 8 weeks' weaning in one of four types of compost (A–D) | | | |
	SS	*v*	MS	F
Between composts	26.87561	3	26.87561/3 = 8.95854	5.75509
Within composts	56.03867	36	56.03867/36 = 1.55663	
Total variation	82.91428	39		

BOX 10.8 How to carry out a one-way parametric ANOVA with equal replicates

This calculation is given in full in the Online Resource Centre. For presentation purposes only all values have been rounded to five decimal places. The calculation is illustrated using data from Example 10.5.

1. **General hypotheses to be tested**

    H_0: There is no difference between the means of the populations from which the samples are taken.

    H_1: There is a difference between the means of the populations from which the samples are taken.

    In our example the hypotheses we are testing are:

    H_0: There is no difference between the mean fresh weight (g) of plantlets of *Lobelia* 'Hannah' weaned on the four composts after 8 weeks.

    H_1: There is a difference (g) between the mean fresh weight of plantlets of *Lobelia* 'Hannah' weaned on the four composts after 8 weeks.

2. **Have the criteria for using this test been met?**

    In this example, the criteria have been met (10.7.2ii).

3. **How to work out $F_{calculated}$**

A. Calculate general terms

1. Add together all the observations in all the categories or samples (grand total): $\sum x_T$

2. Square each observation in all the categories and add these together: $\sum \left(x_T^2\right)$

3. Add all the observations in a category or sample: $\sum x_s$. This is the sample total. Do this for all samples. Square each sample total: $\left(\sum x_s\right)^2$. Add these together. Divide this total by the number of observations in each sample (n_s).

4. Take the result from step 1 and square it: $\left(\sum x_T\right)^2$. Divide by N, where N is the total number of observations in all the categories or samples combined.

In our example, the calculations are:

1. $\sum x_T = 139.17$

2. $\sum \left(x_T^2\right) = (7.04)^2 + (4.86)^2 + \ldots + (2.14)^2 + (3.16)^2$
    $= 567.1215$

3. $\dfrac{(47.44)^2 + (26.39)^2 + (36.49)^2 + (28.85)^2}{10}$
    $= 511.8283$

4. $(139.17)^2 / 40 = 484.20722$

B. Calculate the sums of squares (SS)

5. SS_{total} = result from step 2 – result from step 4

6. $SS_{between*}$ = result from step 3 – result from step 4 (*between samples*)

7. $SS_{within**}$ = result from step 5 – result from step 6 (** *within samples*)

    Optional check on calculations. *If you calculate the SS for each sample and add these together, this figure should be the same as SS_{within}. The SS for each sample is calculated as:*

    $$SS_{sample} = \Sigma \left(x^2\right) - \frac{\left(\Sigma x\right)^2}{n}$$

In our example (Table 10.7) the calculations are:

5. $SS_{total} = 567.1215 - 484.20722 = 82.91428$

6. $SS_{between} = 511.8283 - 484.20722 = 26.87561$

7. $SS_{within} = 82.91428 - 26.87561 = 56.03867$

(continued)

C. Construct and complete an ANOVA calculation table

i. Draw an ANOVA table as illustrated here.

Source of variation	SS	v	MS	F
Between samples	Step **6**			
Within samples	Step **7**			
Total variation	Step **5**			

ii. Transfer the results from the calculations for $SS_{between}$, SS_{within}, and SS_{total} into the ANOVA table in column 2 (see Table 10.8).

iii. Calculate the degrees of freedom (v):

v_{total} = total number of observations - 1 = N - 1

$v_{between}$ = number of samples - 1 = a - 1

v_{within} = total number of observations - number of samples = $N - a$

Enter these results in column 3.

iv. Calculate the mean squares (MS) for between and within samples.

Note: Variances in this context are also known as mean squares (MS).

$$s^2 = MS = \frac{SS}{v}$$

In the 'between samples' row, take the value for SS and divide it by its value for v.

In the 'within samples' row, take the value for SS and divide it by its value for v.

Put the results from these calculations in column 4 (Table 10.8).

v. Finally calculate the variance ratio $F_{calculated}$ where:

$$F_{calculated} = \frac{\text{'between samples' MS}}{\text{'within samples' MS}}$$

It is usual to place the results from this calculation in column 5 in the first row (Table 10.8).

4. To find $F_{critical}$

See Appendix D, Table D7. Identify the critical F value using the F tables for an ANOVA. You may need to interpolate (6.4.4) these values. To find the critical value, you need the degrees of freedom relating to the variances 'between samples' (v_1) and 'within samples' (v_2).

In our example, 'between samples' $v_1 = 3$ and 'within samples' $v_2 = 36$.

Interpolating from the F table for $p = 0.05$, $F_{critical}$ = 2.872.

5. The rule

If $F_{calculated}$ is greater than or equal to $F_{critical}$, then you may reject the null hypothesis.

In our example, $F_{calculated}$ (5.76) is greater than $F_{critical}$ (2.87) so you may reject the null hypothesis.

In fact, at $p = 0.01$ (Table D9), $F_{critical} = 4.39$, so you may reject the null hypothesis at this higher level of significance.

6. What does this mean in real terms?

There is a highly significant difference ($F = 5.76$, $p = 0.01$) between the mean fresh weight (g) of *Lobelia* 'Hannah' when weaned on four different composts.

Further examples relating to the topic including how to use statistical software are included in the Online Resource Centre.

10.8 Tukey's test following a parametric one-way ANOVA with equal replicates

The one-way ANOVA described in 10.7 has allowed us to test the general hypothesis that there was no difference between the success of weaning as indicated by the mean fresh weight of *Lobelia* 'Hannah' on four composts. The analysis showed that there is a significant difference (Box 10.8). It would be helpful to be able to be more exact in our interpretation of the data. Looking at the mean values in Table 10.7, compost A appears to be most effective in this regard, but is this significant? To make a specific evaluation of this sort, you could decide to make a number of comparisons using a *t*-test, comparing compost A to compost B, compost A to compost C, and so on. By taking this approach you would carry out six different pairwise comparisons. In 6.1.6 we discuss the Type I and Type II errors that can arise in statistics. If a Type I error occurs there is no real biological effect but the statistics test, by chance, returns a significant result. In a Type II error there is a real biological effect of the factor you are investigating but the statistics does not detect this. The more tests you carry out the more likely you are to be affected by these errors. Therefore if you wish to carry out further more specific tests, having already identified using a general test such as an ANOVA that there is a difference, you should use a *post hoc* test. *Post hoc* tests, of which there are many, are specifically designed to reduce the Type I error rate allowing you to make many comparisons with more confidence in the outcome.

A clear advantage of following a general ANOVA test with a more specific *post hoc* test is that you are able to compare any controls explicitly to each treatment as well as comparing each treatment to every other treatment.

Commercial statistical software invariably carries out a number of different *post hoc* tests. Before using any of these you should read about each one and determine which is right for your data as these make different assumptions about the equality of variance, group sizes, and the number of comparisons you wish to make. We cover the Tukey's test which is reasonably conservative, powerful, and balances the risk of a Type I error occurring against the risk of a Type II error occurring. By being conservative it means it does not detect as many significant differences as other *post hoc* tests and is therefore more stringent. The Tukey's test is used only when the ANOVA demonstrates a general significant difference. You are therefore not embarking on a fishing expedition looking to see if despite your overarching experimental design involving numbers of categories any one pair is different.

The Tukey's test we describe is suitable for use following an ANOVA where the number of replicates is the same for all samples, as in our example. If you wish to use a *post hoc* test following a one-way ANOVA with unequal replicates, refer to the Tukey–Kramer test (10.10).

10.8.1 Key trends and experimental design

To illustrate this Tukey's test we are using Example 10.5. In a Tukey's test we compare each mean with every other mean. Clearly there is quite a big difference between the mean fresh weight for the *Lobelia* grown on compost A ($\bar{x} = 4.744$) compared to compost B ($\bar{x} = 2.639$). There is also a difference between these mean fresh weights for compost B ($\bar{x} = 2.639$) and compost D ($\bar{x} = 2.885$) but the difference is less. To identify in our analysis which of these is significantly different will be invaluable from a biological perspective as it will enable us to have a more in depth understanding of this system.

10.8.2 Using this test

i. To use this test you:

1) Wish to test more specific hypotheses following a significant outcome in a parametric one-way ANOVA.

2) Should have equal numbers of observations in each category or sample.

ii. Does the example meet these criteria?

The data in Example 10.5 have been anaysed using a one-way parametric ANOVA in Box 10.8. Therefore, we are using the Tukey's test having obtained a significant difference in the ANOVA (Table 10.8) and there are ten observations in each category.

10.8.3 The calculation

In a Tukey's test, each mean value for a sample is compared with every other mean value in the investigation. This difference between the pairs of means is the calculated T value, which you then compare with a critical T value, which you also work out. We demonstrate this in Box 10.9 with supporting calculations in Table 10.9.

BOX 10.9 How to carry out Tukey's test after a significant one-way parametric ANOVA with equal replicates

GENERAL DETAILS	EXAMPLE 10.5
	This calculation is given in full in the Online Resource Centre. For presentation purposes only all values have been rounded to five decimal places.
1. Hypotheses to be tested	1. Hypotheses to be tested
You will be comparing two samples at a time by comparing their means. The generic hypotheses for each of these comparisons is: H_0: There is no difference between the means of the two samples being compared. H_1: There is a difference between the means of the two samples being compared.	In the Tukey's test, we will be comparing six pairs of mean values. Each of these comparisons is a test of hypotheses. You could write hypotheses for each. For example, H_0: There is no difference between the mean fresh weight for *Lobelia* 'Hannah' grown in compost A compared with compost B, etc. However, common sense suggests that if you do formally record these hypotheses then you use the generic hypotheses only.
2. Have the criteria for using this test been met?	2. Have the criteria for using this test been met? Yes (10.8.2ii).
3. How to work out $T_{calculated}$ Calculate the difference between the means in all possible pairwise combinations. Each of these differences is a $T_{calculated}$ value.	3. How to work out $T_{calculated}$ See Table 10.9.

(continued)

4. How to find $T_{critical}$

i. Examine the ANOVA calculation to find the following:
   n_s: the number of observations in a category or sample.
   a: number of categories or samples.
   MS_{within} is in the ANOVA table. For accuracy, you should use the full value and not a rounded-up value.
   v: degrees of freedom for MS_{within} from the ANOVA calculation table.

ii. See Appendix D, Table D11. Find the q value at $p = 0.05$, in a q table for Tukey's test using a and v.

iii. Calculate $T_{critical}$ where:

$$T_{critical} = q \times \sqrt{\frac{s^2_{within}}{n}}$$

4. How to find $T_{critical}$

i. From Table 10.8.
   $n_s = 10$
   $a = 4$
   $MS_{within} = 1.5566297$
   $v = 36$

ii. To find q for $a = 4$ and $v = 36$, we need to interpolate from the q table:
   $q = 3.814$

iii. $T_{critical} = 3.814 \times \sqrt{\dfrac{1.5566297}{10}}$

   $= 1.50478$ at $p = 0.05$

5. The rule

Compare $T_{critical}$ with each of the differences between the means (i.e. each $T_{calculated}$). If any $T_{calculated}$ is greater than $T_{critical}$, then you may reject the null hypothesis for these two samples.

5. The rule

There are two $T_{calculated}$ values in Table 10.9 that are greater than $T_{critical}$ (1.505) at $p = 0.05$ and these are shown in bold in the table.

6. What does this mean in real terms?

6. What does this mean in real terms?

The general significant difference detected by the ANOVA (Box 10.8) is accounted for by the significant difference ($p = 0.05$) in mean fresh weight of *Lobelia* 'Hannah' growing in compost A compared with compost B and in compost A compared with compost D.
From this information, the student was able to identify which components of the compost appeared to be most important in the weaning of *Lobelia* 'Hannah'.

Further examples relating to the topic including how to use statistical software are included in the Online Resource Centre.

10.8.4 Reporting your findings

In a *post hoc* test such as Tukey's, you are testing many pairs of hypotheses, one for each pair of means. To save you writing each hypothesis out formally at the end, an alternative method for reporting *post hoc* tests is used. Firstly, rank all your mean values from the smallest to the largest. Then underscore those not significantly different from each other, or do not underscore

Table 10.9 The difference between each pair of means from Example 10.5: The fresh weight of *Lobelia* 'Hannah' after 8 weeks' weaning in one of four types of compost (A–D) (each difference is a $T_{calculated}$ value to be used in a Tukey's test)

Compost	A	B	C	D
	$\bar{x} = 4.744$	$\bar{x} = 2.639$	$\bar{x} = 3.649$	$\bar{x} = 2.885$
A $\bar{x} = 4.744$		4.744–2.639 = **2.105**	4.744–3.649 = 1.095	4.744–2.885 = **1.859**
B $\bar{x} = 2.639$			3.649–2.639 = 1.010	2.885–2.639 = 0.246
C $\bar{x} = 3.649$				3.649–2.885 = 0.764
D $\bar{x} = 2.885$				

(by the same line) any two that are significantly different. It requires a bit of thought, but it does work in the end. In our example this would appear as:

B	D	C	A
$\bar{x} = 2.639$	$\bar{x} = 2.885$	$\bar{x} = 3.649$	$\bar{x} = 4.744$

This underscoring is commonly included in the table of results.

An alternative, less cumbersome method is to use superscripts instead of a line. The same superscript is also used for all means that are not significantly different from each other, hence:

B[a]	D[a]	C[ab]	A[b]
$\bar{x} = 2.639$	$\bar{x} = 2.885$	$\bar{x} = 3.649$	$\bar{x} = 4.744$

This alternative approach is particularly useful when you are comparing many means.

10.9 One-way parametric ANOVA with unequal replicates

The ANOVAs are most effective when equal replicates are used. This applies to tests for one or more treatment variables. However, even if experiments are designed with equal replicates occasionally things go wrong and then the analysis has to take this into account. In Example 10.6, we outline an investigation where an undergraduate planned to have equal replicates. However, technical problems meant that this was not achieved.

Example 10.6 Differences in the peroxide levels of popular cooking oils

An increase in peroxide levels in cooking oils is indicative of reduced nutritional value and is associated with an increase in an unpleasant rancid taste. As part of a larger study, an undergraduate assessed the peroxide levels in a number of different oils from samples taken directly from new bottles.

Table 10.10 Peroxide levels (mEq of peroxide per kg sample) in a number of different cooking oils taken as fresh samples from a new bottle. All values $\times 10^{-2}$. Superscript **a** indicates samples not significantly different at $p = 0.05$ using a Tukey–Kramer test (10.10).

There are three replicates in two of the categories but only two replicates in the corn oil category.

There is one treatment variable: 'types of oil' with three nominal categories which have no sampling error associated with them so they are fixed. The dependent variable is peroxide levels, which is measured on an interval scale.

	Peroxide levels (mEq of peroxide per kg sample) in a number of different cooking oils (all values $\times 10^{-2}$)		
	Corn oil[a]	Extra virgin olive oil[a]	Cold pressed rapeseed oil
	14	17	8
	15	13	7
	–	12	8
n_s	2	3	3
$\bar{x}$	14.5	14.0	7.66666
s^2	0.5	7.0	0.33333
$\sum x$	29	42	23

10.9.1 Key trends and experimental design

In Example 10.6 there are two points to note. Firstly the cold pressed rapeseed had much lower levels of peroxide ($\bar{x} = 7.7 \times 10^{-2}$ mEq of peroxide per kg sample) compared to the two other oils. Secondly in this experiment there is considerable variation between the three bottles of extra virgin olive oil ($s^2 = 7.0. \times 10^{-2}$ mEq of peroxide per kg sample^2). The number of replicates here are small and analysis of one of the bottles of corn oil was not successful so there are unequal replicates.

10.9.2 Using this test

i. To use this test you:

1) Need to meet the criteria for using a one-way parametric ANOVA as outlined in 10.7.2ii apart from the requirement for equal replicates.

ii. Does the example meet these criteria?

The data from Example 10.6 meet these criteria in that we wish to examine the effect of one variable (different oils). There are few observations so we will assume that the data are parametric as the observations are measured on an interval scale (Box 5.2). As this is a test that may be used when there are unequal replicates, it does not matter that two of the three treatments have only two rather than three replicates. To test for homogeneity of variances (Box 10.7), $F_{max\ calculated} = 7.0/0.33333 = 21.0$. To find $F_{max\ critical}$, calculate the degrees of freedom for the two variances used in the F ratio. (In this example, these were the variances from the 'extra virgin olive oil' and the 'cold pressed rapeseed oil'.) The degrees of freedom are calculated as before as n_s-1. The degrees of freedom for the cold pressed rapeseed oil are $3-1 = 2$, whilst the degrees of freedom for the extra virgin olive oil are also $3-1 = 2$. Therefore, the degrees of freedom we use to find $F_{max\ critical}$ are 2. With unequal sample sizes, the n_s values may not be the same for each category. For example, if we had used the variance from the corn oil in our F ratio, these degrees of freedom would be $2-1 = 1$. Where this is the case, you use the smallest of the degrees of freedom values. In Appendix D, Table D7, when looking up the critical values, a (number of samples) $= 3$, $v = 2$, therefore $F_{max\ critical} = 87.5$ at $p = 0.05$. As $F_{max\ calculated}$ is less than $F_{max\ critical}$, we may not reject the null hypothesis, the variances are homogeneous, and we may proceed with the one-way ANOVA.

10.9.3 The calculation

There is very little difference in the calculation of a one-way ANOVA for equal replicates (Box 10.8) and unequal replicates (Box 10.10). In this section, we have therefore only highlighted the one step that differs.

BOX 10.10 How to carry out a one-way parametric ANOVA with unequal replicates

This calculation is given in full in the Online Resource Centre. For presentation purposes only all values have been rounded to five decimal places. The calculation is illustrated using data from Example 10.6.

1. **General hypotheses to be tested**

   H_0: There is no difference between the means of the populations from which the samples are taken.

   H_1: There is a difference between the means of the populations from which the samples are taken.

   In our example the hypotheses we are testing are:

   H_0: There is no difference in the mean peroxide levels (mEq of peroxide per kg sample) in different cooking oils.

   H_1: There is a difference in the mean peroxide levels (mEq of peroxide per kg sample) in different cooking oils.

2. **Have the criteria for using this test been met?**

   Yes (10.9.2ii).

3. **How to work out $F_{calculated}$**

   All these steps are the same as those described in Box 10.8 apart from step **3** in '**A. Calculate general terms**'. The only step that is different is as follows:

A. Calculate general terms

3. Add all the observations in a column: $\sum x_s$. Square this value: $\left(\sum x_s\right)^2$. Divide by the number of observations in that column (n_s). Repeat for each column and add these together.

   In our example, the calculation for this step is:

   $$\frac{(29)^2}{2} + \frac{(42)^2}{3} + \frac{(23)^2}{3} = 42.05 + 588.0 + 176.33333$$
   $$= 1184.83333$$

10.10　Tukey–Kramer test following a parametric one-way ANOVA with unequal replicates

When there is a significant outcome from a one-way ANOVA, it is often helpful to carry out Tukey's test to examine in more detail which samples are contributing to the significant difference. However, Tukey's test requires equal replication, which is not present in Example 10.6. To overcome this inconsistent replication within the experiment, a Tukey–Kramer test may be used.

10.10.1　Key trends and experimental design

In Example 10.6 (Table 10.10) it is clear that the cold pressed rapeseed is noticeably different from both the other oils tested and that the corn oil is similar to the extra virgin olive oil. A *post hoc* test would allow us to confirm if this observed trend is statistically significant and so add confidence to the interpretation of these data. However, this test will also have to take into account the unequal replication within this experiment.

10.10.2　Using this test

i.　To use this test you:

1)　Wish to test more specific hypotheses following a significant outcome in a parametric one-way ANOVA.

2)　Do not have equal numbers of observations in each category or sample.

ii.　Does the example meet these criteria?

To illustrate the Tukey–Kramer test, we will use Example 10.6. The criteria for using this test have been met as we have a significant outcome from a parametric one-way ANOVA. (These calculations are shown in full in the Online Resource Centre.) There are unequal numbers of observations in the categories.

10.10.3　The calculation

In a Tukey–Kramer test, each mean value for a sample is compared with every other mean value in the investigation. This difference between the pairs of means is the calculated T value, which you then compare with a critical T value, which you also work out. This is the same approach as Tukey's test. The difference is in the method used for calculating the $T_{critical}$ value when n is not the same for the two samples. We demonstrate this in Box 10.11 with supporting calculations in Table 10.11.

BOX 10.11 How to carry out a Tukey–Kramer test after a significant one-way parametric ANOVA with unequal replicates

GENERAL DETAILS	EXAMPLE 10.6
	This calculation is given in full in the Online Resource Centre. For presentation purposes only all values have been rounded to five decimal places.
1. Hypotheses to be tested You will be comparing two samples at a time by comparing their means. The generic hypotheses for each of these comparisons is: H_0: There is no difference between the means of the two samples being compared. H_1: There is a difference between the means of the two samples being compared.	**1. Hypotheses to be tested** In the Tukey–Kramer test, we will be comparing three pairs of mean values. Each of these comparisons is a test of hypotheses. You could write specific hypotheses for each: for example, H_0: There is no difference between the mean peroxide content (mEq of peroxide per kg sample) of corn oil compared with virgin olive oil. However, common sense suggests that where you are making many comparisons using this test, then it is sensible to report the overall hypotheses only.
2. Have the criteria for using this test been met?	**2. Have the criteria for using this test been met?** Yes (10.10.2ii).
3. How to work out $T_{calculated}$ Calculate the difference between the means in all possible pairwise combinations. Each of these differences is a $T_{calculated}$ value.	**3. How to work out $T_{calculated}$** See Table 10.11.
4. How to find $T_{critical}$ i. Examine the ANOVA calculation to find the following: a: number of samples. MS_{within} from the ANOVA table. (For accuracy you should use the full value and not a rounded-up value.) v: degrees of freedom for the s^2_{within} from the ANOVA calculation table. ii. See Appendix D, Table D11. Find the q value at $p = 0.05$, in a q table for Tukey's test using a and v. iii. Firstly, identify all comparisons between means where the numbers of replicates are the same. Here $T_{critical}$ is calculated as in Tukey's test where: $$T_{critical} = q \times \sqrt{\frac{MS_{within}}{n}}$$	**4. How to find $T_{critical}$** i. In this example, $a = 3$. From the one-way ANOVA, $MS_{within} = 3.033334$ and its degrees of freedom are $v = 5$. ii. When $a = 3$ and $v = 5$, $q = 4.6$ at $p = 0.05$. iii. In our example, this is the comparison between extra virgin olive oil and cold pressed rapeseed oil where $n = 3$ for both samples. $$T_{critical} = 4.6 \times \sqrt{\frac{3.033334}{3}}$$ $$= 4.62549 \text{ at } p = 0.05$$

(continued)

iv. When n is not the same for the two samples, $T_{critical}$ is: $$T_{critical} = q \times \sqrt{\frac{MS_{within}}{2}\left(\frac{1}{n_1}+\frac{1}{n_2}\right)}$$	iv. We have two comparisons where n is not the same for both samples. These are the comparisons between the corn oil ($n=2$) and the rapeseed oil ($n=3$) and the corn oil with the olive oil ($n=3$). Here: $$T_{critical} = 4.6 \times \sqrt{\frac{3.03334}{2}\left(\frac{1}{2}+\frac{1}{3}\right)}$$ $$= 4.6 \times \sqrt{1.51667(0.5+0.33333)}$$ $$= 5.17145$$
5. The rule Compare the correct $T_{critical}$ with each of the differences between the means (i.e. each $T_{calculated}$). If any $T_{calculated}$ is greater than a $T_{critical}$, then you may reject the null hypothesis for this comparison of means.	**5. The rule** There are two $T_{calculated}$ values in Table 10.11 that are greater than their $T_{critical}$ values at $p=0.05$ and these are shown in bold in the table with an asterisk to indicate significance at this level. If we repeat the calculations at $p=0.01$, we find that the hypotheses may be rejected at $0.05 > p > 0.01$.
6. What does this mean in real terms? Standard ways used in reporting results from the Tukey–Kramer test are outlined in 10.8.4.	**6. What does this mean in real terms?** The general significant difference detected by the ANOVA is accounted for by the significant difference ($0.05 > p > 0.01$) in mean peroxide levels (mEq of peroxide per kg sample) in the cold pressed rapeseed oil compared with the other two oils.

Further examples relating to the topic including how to use statistical software are included in the Online Resource Centre.

Table 10.11 The difference between each pair of means from Example 10.6: Peroxide levels (mEq of peroxide per kg sample) in a number of different cooking oils taken as fresh samples from a new bottle. All values $\times 10^{-2}$. (Each difference is a $T_{calculated}$ value to be used in a Tukey–Kramer test.)

	Peroxide levels (mEq of peroxide per kg sample $\times 10^{-2}$)		
Cooking oils	Cold pressed rapeseed oil	Extra virgin olive oil	Corn oil
	$\bar{x} = 7.66667$, $n=3$	$\bar{x} = 14.0$, $n=3$	$\bar{x} = 14.5$, $n=2$
Cold pressed rapeseed oil		14.0–7.66667 = **6.33333***	14.5–7.66667 = **6.83333***
$\bar{x} = 7.66667$			
Extra virgin olive oil			14.5–14.0 = 0.5
$\bar{x} = 14.0$			
Corn oil			
$\bar{x} = 14.5$			

*Significant at $p = 0.05$.

10.10.4 **Reporting your findings**

The method for reporting the result from the Tukey–Kramer test is the same as that outlined in 10.8.4 and the superscript has been added to Table 10.10 to illustrate this.

10.11 ANOVAs for more than one treatment variable

The use of any ANOVA with more than two treatment variables requires you to have an understanding not only of the points considered in 10.6 but also further principles relating both to the design of experiments and the way the ANOVA functions.

10.11.1 **Randomized orthogonal designs**

Data sets for ANOVAs with two or more variables are presented as a table. In the context of experimental designs each cell in a table is called a block. If there are no replicates (e.g. Example 10.9) then there is only one number in each block. With replicates then there is more than one number in a block (e.g. Example 10.7).

To use ANOVAs to test more than one treatment variable you usually use an orthogonal design. In such a design, each category from variable 1 is tested against each category from variable 2. For example if you were developing a technique for extracting DNA from bone material you might wish to compare the amount of DNA (μg) extracted across a range of temperatures and pH values (Table 10.12). This is an orthogonal design with 12 blocks as we wish to examine the effect of two treatment variables (temperature and pH) and each pH treatment is combined with all temperatures.

A further common feature of this design for an ANOVA is that each item is then randomized to these blocks and in our example the bone samples would be allocated at random to a particular extraction method (Table 10.12). As we explained in 2.2.5, this randomization of items within treatments is a way of minimizing the impact of confounding variables and is therefore a powerful tool that can be used when designing an experiment.

Orthogonal
A particular experimental design where every category for one treatment is found in combination with every category for every other treatment.

Table 10.12 An orthogonal experimental design: the effect on DNA extraction of several temperatures is investigated in relation to a number of pH values (each category of one variable is combined with each category of the second variable)

pH	Temperature (°C)			
	5	10	15	20
3	Item 1	Item 7	Item 12	Item 4
7	Item 5	Item 9	Item 2	Item 10
12	Item 11	Item 3	Item 8	Item 6

10.11.2 Confounding variables as a treatment variable

In an investigation into the time spent foraging by two species of lemur at a zoo data were collected over a 3-month period. The treatment variable here is the species of lemur but clearly during the time the data was collected a number of factors including weather, group relationships, seasonal food sources, and visitor numbers are likely to alter. Therefore time of year is a confounding factor that needs to be considered if possible during the evaluation of the data. The approach to take for the purposes of the experiment and analysis is to treat the confounding variable as though it is a treatment variable. For the lemur experiment you would record the dates when the behavior data were collected. These dates become an ordinal fixed treatment variable.

A second common experience in research is to find that laboratory equipment is limited either because there are only a few items of equipment available or because space is at a premium. In a study on the growth of *Arabidopsis thaliana* knockout mutants when infected by a specific pathogen it was not possible to grow all the knockout mutants with their replicates in the same growth cabinet. The plants were therefore assigned at random to different cabinets and the 'cabinets' for the purposes of the analysis are treated as though they are a treatment variable. Of course the hope would be that the cabinets themselves would not significantly affect the results but by recording which plants were in which cabinet and treating this as a nominal treatment variable you can check formally whether this is the case. In Example 10.9 the confounding variable of location in a polytunnel has been addressed in this way.

This approach is a very effective way of dealing with any confounding variables. So in planning any investigation it is important to consider which factors may act as confounding variables and include methods such as this to be able to evaluate their impact. When working with confounding variables in this way you have to take some measurements of the variable and determine if they are random or fixed and if the overall subsequent design is nested (10.16).

10.11.3 More than one confounding variable: a Latin square

When designing experiments, you will seek to identify, control, minimize, and evaluate the effects of confounding variables. For example, you can minimize the effect of confounding variables by the randomization of items to treatments and can evaluate the effects by including them as a treatment variable in your data analysis (10.15). However, there is one specific experimental design called a Latin square that is structured to minimize the effect when there are two known confounding variables that are graded across the area in which you are working. For example, you may need to organize plants in a greenhouse and know that the light levels drop off as you move away from the windows; similarly, you may have a door at one end and know that there is therefore a temperature gradient along the bench caused by a draught from the door. In these circumstances, you can organize your items or blocks in a non-random arrangement. In our example of a Latin square (Fig. 10.3), you can see that there are four categories (A, B, C, and D) for a single variable. Each category is present in only one row and one column. The ANOVA that is used to analyse the results from this design partitions the variation into that due to the confounding variable 1 + variation due to confounding variable 2 + variation due to the treatment variable + sampling error. We have not included a Latin square ANOVA in this book; instead we refer you to Fowler *et al.* (1998).

10.11.4 **Linear confounding variables**

In 10.11.2 and 10.11.3 we have discussed confounding variables that are categorical as they are either nominal or ordinal measures or because they were interval measures divided into categories. In Example 10.9 the polytunnel was divided into lengths by the investigator and these categories have then been formally included in the experimental design and analysis. However, some confounding variables may not be categorical nor is it sensible to treat them as though they are. If you have confounding varaibles that are measurable on an interval scale and if you can confirm that as the confounding variable changes so does the dependent variable in a linear way then you should consider using ANCOVAs. Sadly we are not able to cover all possible statistics tests in one book so you will have to refer elsewhere for explanations of these tests. However, if your data do not meet the criterion of linearity then you can use the blocks approach and ANOVAs instead.

Gradient for non-treatment variable 1

Gradient for non-treatment variable 2

A	D	C	B
C	B	A	D
B	C	D	A
D	A	B	C

Fig. 10.3 A Latin square for two non-treatment variables and one treatment variable with four categories

10.11.5 Repeated measures as a second treatment variable

In some experiments you may use the same item more than once to be tested by more than one treatment in one treatment variable. This generates matched data which is also known as repeated measures. In Example 10.3 this type of experiment resulted in two columns of data—one for a 'before' treatment and one for an 'after' treatment—and these columns of data are then compared to each other. If there are more than two treatments in the treatment variable then a different approach is taken, and for the purposes of the analysis the items themselves are included as a second treatment variable. The analysis then tests whether there is a significant difference between the items themselves as one treatment variable as well as examining the effect of the actual treatment variable (10.11).

10.11.6 Fixed and nested models

We have discussed models in other chapters in relation to other statistical tests (e.g. 8.1 and 9.9). For experiments with two or more treatment variables the terms Model I, Model II, and mixed models in relation to ANOVAs refer to the treatment variables, which can be fixed or random. Table 10.12 is an example of an experiment with two fixed variables. Fixed variables arise when the categories, in this case pH and temperature, do not vary. pH 7 will be pH 7 for all the temperatures tested. Similarly 5°C will be the same for all the pHs tested. Usually categories are fixed like this as they are controlled by the investigator. In Example 10.5 the four composts are a fixed treatment that has been chosen by the investigator and in Example 10.7 both the two varieties of *Cosmos atrosanguineus* and the composts are fixed treatments. Where all the treatment variables are fixed this is a Model I design.

In contrast the categories in some or all of the treatment variables may be randomized. This most often arises when you are collecting samples from a geographical space and introduce a degree of randomization into the sampling method. We see this in Example 10.10, where soil samples have been taken from a number of pits dug in two forests. Clearly, pit 1 in the coniferous forest is not the same as pit 1 in the deciduous forest: the pits as a treatment are randomized within the forests. Treatment variables that are randomized like the 'pits' are known as a Model II design.

However, Example 10.10 has one fixed treatment variable (forests) and one random treatment variable (pits) and as a result the design is not orthogonal but is known as a mixed model where one treatment variable (the main effect) is fixed and the subordinate variable is randomized within the categories of the first variable. A mixed model design should be analysed using a nested ANOVA. There are many versions of mixed models. In this chapter, we have included a mixed model with one fixed (main) variable and one random variable with equal replicates (Table 10.13). For other designs, we refer you to Sokal & Rohlf (2011). Only the nested ANOVA is model dependent, requiring this mixed model design. The other ANOVAs covered in this chapter can be used for a pure Model I (all fixed variables) or a pure Model II (all random variables) design.

 Look at Example 10.11. There are three variables: depths, distances, and bearings from a smelter. Which of these are fixed and which random? What model is this experimental design?

Table 10.13 Mixed model experimental design with one fixed main variable and one random, nested variable, with equal replicates

| Variable 2 | Variable 1 | | | |
| | Category 1 | | Category 2 | |
	Category 1	Category 2	Category 3	Category 4
	BLOCK 1_1	BLOCK 2_1	BLOCK 1_2	BLOCK 2_2
	Observation 1	Observation 1	Observation 1	Observation 1
	Observation 2	Observation 2	Observation 2	Observation 2
	Observation 3, etc.	Observation 3, etc.	Observation 3, etc.	Observation 3, etc.

10.11.7 Interactions

As soon as two variables are being compared in a two-way parametric or non-parametric ANOVA with replicates, then three pairs of hypotheses are tested. Not surprisingly, the first of these three pairs of hypotheses refers to the difference between the samples in relation to the effect of treatment variable 1. In our example on the effectiveness of extracting DNA from bone samples depending on the temperature and pH, the first pair of hypotheses will refer to temperature (Table 10.12). The second pair of hypotheses will refer to the effect of the second variable which in this instance is pH. However, these two-way ANOVAs with replicates will also test for an interaction between these two variables and this is the third pair of hypotheses.

An interaction between treatments can be indicated by a plot of the mean values. In Fig. 10.4, the amount of DNA extracted as the pH increases changes in a similar way for all

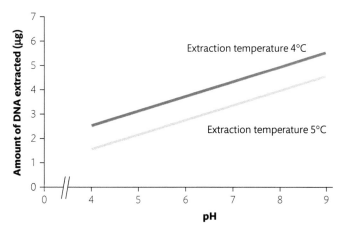

Fig. 10.4 The amount of DNA extracted increases in a similar way as pH increases, for both temperatures, indicating no interaction between pH and temperature (°C)

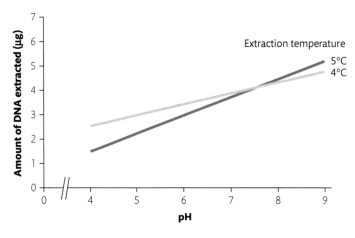

Fig. 10.5 The two temperatures do not have the same effect on the amount of DNA extracted as pH increases, indicating an interaction between temperature and pH

Interaction

The response to a treatment shown by one group of items is reversed in a second sample in such a way that it is clear that two treatment variables are not having the same effect in the two groups.

temperatures, therefore there is no interaction. If there was an interaction, then these lines would not be parallel but would either cross or pull away from each other (Fig 10.5). Plotting the data like this is a quick method of checking whether you are likely to see a significant interaction in your analyses and should be carried out before undertaking either of these ANOVAs.

10.12 Two-way parametric ANOVA with equal replicates

From the points discussed in 10.11 it is clear that you need to think carefully when planning your experiment to ensure you can use an appropriate ANOVA to analyse your data but to also maximize the effectiveness of your analysis in coming to understand whether confounding variables are affecting your experiment. The ANOVA we consider here may therefore be used to examine the effect of two treatment variables or the effect of one treatment variable and one confounding variable.

Example 10.7 The weaning of *Cosmos atrosanguineus* var. 'Pip' and var. 'Christopher' onto one of four composts following propagation by tissue culture

The undergraduate who was working on *Lobelia* 'Hannah' also examined the weaning from tissue culture of two varieties of *Cosmos atrosanguineus*, var. 'Pip' and var. 'Christopher'. As in the *Lobelia* trial (Example 10.6), she transferred plantlets at random into plugs containing one of four different composts (A–D). After 8 weeks, the plantlets were examined and the maximum height of the plants was recorded (Table 10.14).

(continued)

Table 10.14 Final height of two varieties of *Cosmos atrosanguineus* (var. 'Pip' and var. 'Christopher') grown in one of four different composts (A–D) (table includes summary statistics)

> The second treatment variable 'Variety *of C. atrosanguineus*' has two nominal categories which are fixed. There are 10 replicates in each block and the design is orthogonal.

> The first treatment variable 'type of compost' has four nominal categories which are fixed. The dependent variable is height (cm) which is an interval measure.

Variety of *C. atrosanguineus*	Height (cm) of plants in four types of weaning compost							
	A		B		C		D	
C. atrosanguineus var. 'Pip'	20.4	22.0	19.4	22.9	18.7	18.7	20.0	20.1
	20.4	21.6	28.7	25.2	18.5	22.2	25.9	18.1
	24.9	22.7	21.1	27.7	23.2	22.9	17.0	24.1
	19.6	23.5	24.5	26.3	20.6	24.1	22.7	17.5
	19.2	25.4	22.8	18.0	22.3	22.7	25.2	25.2
Summary statistics for var. 'Pip'	$\bar{x} = 21.97$		$\bar{x} = 23.66$		$\bar{x} = 21.39$		$\bar{x} = 21.58$	
	$s^2 = 4.62011$		$s^2 = 12.20266$		$s^2 = 4.39433$		$s^2 = 11.87733$	
C. atrosanguineus var. 'Christopher'	19.5	15.7	8.9	18.1	22.9	10.8	21.6	17.8
	21.9	21.9	8.4	15.8	6.4	17.0	15.2	15.1
	21.9	19.9	15.3	16.2	14.1	21.4	11.0	14.1
	7.9	20.7	13.1	14.8	18.0	15.6	22.9	7.0
	25.5	22.6	13.8	22.6	19.3	17.6	21.0	12.2
Summary statistics for var. 'Christopher'	$\bar{x} = 19.75$		$\bar{x} = 14.7$		$\bar{x} = 16.31$		$\bar{x} = 15.79$	
	$s^2 = 23.65167$		$s^2 = 17.1666$		$s^2 = 24.13656$		$s^2 = 25.80767$	

10.12.1 Key trends and experimental design

The mean values for *C. atrosanguineus* var. 'Pip' indicate that compost B is acting as a better weaning medium ($\bar{x} = 23.66$) whilst the others seem to be very similar. However, there is considerably more variation in performance in the composts B ($s^2 = 12.20$) and D ($s^2 = 11.88$) compared to A and C. For *C. atrosanguineus* var. 'Christopher' the results are quite different. Compost A has given the best results ($\bar{x} = 19.75$) and the variation in performance is much greater across all the treatments. There therefore appears to be a difference in response to this test between the two varieties to these composts. It will be interesting to examine each component (the variety and the composts) and to see if the varieties are responding significantly differently to the composts (an interaction).

10.12.2 **Using this test**

i. To use this test you:

1) Have no *a priori* reason for expecting certain outcomes from your investigation.

2) Have data which are usually presented in tables or summary histograms rather than a scatter plot.

3) Wish to examine the effect of two treatment variables*.

4) Can confirm that the treatment variables* are usually measured on an ordinal and/or nominal scale but could be measured on an interval scale and be arranged into discrete categories.

5) Have a dependent variable which is measured on an interval scale and where the data are parametric.

6) Have the same number of observations (replicates) in each category.

7) Have an experimental design that means that each item is assigned at random to the categories, each item is only measured once and the design is orthogonal.

8) Have samples where the variation is similar (homogeneous).

*If a confounding variable is being tested then it is referred to as a treatment variable for the purposes of this analysis.

ii. Does the example meet these criteria?

The annotations on Table 10.14 confirm that the data from Example 10.7 meet the criteria for using a two-way parametric ANOVA with equal replicates. Each plantlet was allocated at random to the treatments, and the design is orthogonal. You wish to examine the effect of two variables (compost and plant variety), and there are ten plants in each treatment. There are few replicates so it is difficult to confirm that these data are drawn from a normally distributed population. However, height is measured on an interval scale and tends to be assumed to be normally distributed, so we will accept that this criterion is also met. To check whether the variances are similar, an F_{max} test is carried out (Box 10.7). The variances for the data are included in Table 10.15 and $F_{max\ calculated}$ is $25.80767/4.39433 = 5.8729$. At $p = 0.05$, $a = 8$, and $v = 10–1 = 9$, then $F_{max\ critical} = 8.95$ (Appendix D, Table D7). $F_{max\ calculated}$ is less than $F_{max\ critical}$, so we do not reject the null hypothesis; the variances are homogeneous and we may proceed with the ANOVA. If the variances are not homogeneous, you should consider transforming your data (5.10) or using a non-parametric two-way ANOVA (11.5).

10.12.3 **The calculation**

Having satisfied ourselves to the best of our ability that our data meet all the criteria for using this two-way ANOVA, we can then proceed. In Box 10.12, we have organized the calculation of $F_{calculated}$ under a number of subheadings: (A) Calculate general terms; (B) Calculate the sums of squares; and (C) Construct and complete an ANOVA calculation table. At each point, we show how these steps can be applied to Example 10.7. Some of the calculations are included in Tables 10.15 and 10.16.

Table 10.15 Calculation of general terms for two-way parametric ANOVA using data from Example 10.7: The weaning of *Cosmos atrosanguineus* var. 'Pip' and var. 'Christopher' onto one of four composts following propagation by tissue culture

		Height (cm) of plants in four types of weaning compost				Row total
		A	B	C	D	
C. atrosanguineus var. 'Pip'	$\sum x_s$	219.7	236.6	213.9	215.8	$\sum x_R = 886.0$
	$\sum(x^2)$	4868.39	5707.78	4614.87	4763.86	19954.9
	n_s	10	10	10	10	$n_R = 40$
C. atrosanguineus var. 'Christopher'	$\sum x_s$	197.5	147.0	163.1	157.9	$\sum x_R = 665.5$
	$\sum(x^2)$	4113.49	2315.4	2877.39	2725.51	442890.25
	n_s	10	10	10	10	$n_R = 40$
Column totals						Grand totals
	$\sum x_C$	417.2	383.6	377.0	373.7	$\sum x_T = 1551.5$
	$\sum(x^2)$	8981.88	8023.18	7492.26	7489.37	$\sum(x_T^2) = 31986.69$
	n_C	20	20	20	20	N = 80

Table 10.16 ANOVA table for analysis of data from Example 10.7: The weaning of *Cosmos atrosanguineus* var. 'Pip' and var. 'Christopher' onto one of four composts following propagation by tissue culture, showing how some of the terms are calculated

Source of variation	SS	v	MS	$F_{calculated}$	$F_{critical}$	p
Between composts (columns)	59.87137	4–1 = 3	59.87137/3 = 19.95712	19.95712/15.48213 = 1.28904	2.7347	NS
Between varieties (rows)	607.75312	2–1 = 1	607.75312	607.75312/15.48213 = 39.25514	11.7740	0.001
Interaction compost × varieties	114.94938	(2–1)(4–1) = 3	114.94938/3 = 38.31646	38.31646/15.48213 = 2.47488	2.7347	NS
Within-sample variation	1114.713	80–8 = 72	1114.713/72 = 15.48213			
Total		80–1 = 79				

NS, not significant.

BOX 10.12 How to carry out a two-way parametric ANOVA with equal replicates

This calculation is given in full in the Online Resource Centre. For presentation purposes only all values have been rounded to five decimal places. The calculation is illustrated using data from Example 10.7.

1. **General hypotheses to be tested**

   As explained in 10.11 there are three pairs of general hypotheses to be tested.

   H_{0c}: There is no difference between the sample means due to variable 1 (columns).

   H_{1c}: There is a difference between the sample means due to variable 1 (columns).

   H_{0r}: There is no difference between the sample means due to variable 2 (rows).

   H_{1r}: There is a difference between the sample means due to variable 2 (rows).

   H_{0I}: There is no interaction between variable A and variable B in their effects on the sample means.

   H_{1I}: There is an interaction between variable A and variable B in their effects on the sample means.

In our example (Example 10.7), these hypotheses will be:

   H_{0c}: There is no difference between the mean height (cm) of C. atrosanguineus weaned on different composts.

   H_{1c}: There is a difference between the mean height (cm) of C. atrosanguineus weaned on different composts.

   H_{0r}: There is no difference between the mean heights (cm) of C.atrosanguineus var. 'Pip' compared with var. 'Christopher'.

   H_{1r}: There is a difference between the mean heights (cm) of C.atrosanguineus var. 'Pip' compared with var. 'Christopher'.

   H_{0I}: There is no interaction between the weaning composts and C.atrosanguineus varieties in their effects on mean height (cm).

   H_{1I}: There is an interaction between the weaning composts and C.atrosanguineus varieties in their effects on mean height (cm).

2. **Have the criteria for using this test been met?**

Yes, the criteria have been met (10.12.2ii).

3. **How to work out $F_{calculated}$**

A. Calculate general terms

1. First add together every observation in the complete data set (grand total): $\sum x_T$

2. Square each and every observation and add all these squared values together: $\sum(x_T^2)$

3. Add all the observations in a sample (block): $\sum x_s$. This is the sample total. Do this for each sample. Square each sample total: $(\sum x_s)^2$. Add these together: $\sum[(\Sigma x_s)^2]$. Divide this by the number of observations in a sample (n_s).

4. For each column add all the observations together: $\sum x_C$. Square the total: $(\sum x_C)^2$. Add these squared column totals together: $\sum[(\sum x_C)^2]$ and divide this total by the number of observations in the column (n_C).

5. For each row, add all the observations: $\sum x_R$. Square the total: $(\sum x_R)^2$. Add these squared values together and divide this value by the number of observations in the row (n_R).

6. Square the result from step **1** and divide this by N, where N is the total number of observations in all the samples.

For this example (Table 10.15), these calculations are:

1. $\sum x_T = 20.4 + 20.4 + 24.9 + 19.6 + \ldots + 7.0 + 21.0 + 12.2 = 1551.5$

2. $\Sigma(x_T^2) = (20.4)^2 + (20.4)^2 + (24.9)^2 + \ldots (21.0)^2 + (12.2)^2 = 31986.69$

3. There are eight blocks in total and ten observations in each block. Calculate:
   $$\sum[(\Sigma x_s)^2] = (219.7)^2 + (236.6)^2 + (213.9)^2 + (215.8)^2 + (197.5)^2 + (147.0)^2 + (163.1)^2 + (157.9)^2 = 308\,719.77$$
   $$\sum[(\sum x_s)^2]/n_s = \frac{308719.77}{10} = 30871.977$$

4. There are four columns (composts) and 20 observations in each column. Calculate:
   $$\sum[(\sum x_C)^2]/n_C = [(417.2)^2 + (383.6)^2 + (377.0)^2 + (373.7)^2]/20 =$$
   $$= \frac{602985.49}{20} = 30149.2745$$

(continued)

5. There are two rows (var. 'Pip' and var. 'Christopher') and 40 observations in each row. Calculate:

$$\sum\left[\left(\sum x_C\right)^2\right]/n_C = \left[(886.0)^2 + (665.5)^2\right]/40$$
$$= 1227886.25/40$$
$$= 30697.15625$$

6. $\left(\sum x_T\right)^2 / N = (1551.5)^2/80 = 30089.40313$

B. Calculate the sums of squares (SS)

7. SS_{total} = result from step **2** – result from step **6**
8. $SS_{samples}$ = result from step **3** – result from step **6**
9. $SS_{variable\ 1\ columns}$ = result from step **4** – result from step **6**
10. $SS_{variable\ 2\ rows}$ = result from step **5** – result from step **6**
11. $SS_{interaction}$ = result from step **8** – result from step **9** – result from step **10**
12. SS_{within} = result from **7** – result from **8**

For Example 10.7, the calculations are:

7. SS_{total} = 31986.69 – 30089.40313 = 1897.28687
8. $SS_{samples}$ = 30871.977 – 30089.40313 = 782.57387
9. $SS_{variable\ 1\ columns}$ = 30149.2745 – 30089.40313 = 59.87137
10. $SS_{variable\ 2\ rows}$ = 30697.15625 – 30089.40313 = 607.75312
11. $SS_{interaction}$ = 782.57387 – 59.87137 – 607.75312
$$= 114.94938$$
12. SS_{within} = 1897.28687 – 782.57387 = 1114.713

C. Construct and complete an ANOVA calculation table

i. Draw an ANOVA table as illustrated here.

Source of variation	SS	v	MS	F
Variable 1 (columns)	Step **9**			
Variable 2 (rows)	Step **10**			
Interaction	Step **11**			
Within sample variation	Step **12**			
Total	Step **7**			

ii. Transfer the results from the calculations for $SS_{variable\ 1\ columns}$, $SS_{variable\ 2\ rows}$, SS_{within}, and SS_{total}

into the ANOVA table in column 2 (see Table 10.16). (You do not use SS_{sample}.)

iii. Calculate the degrees of freedom (v)
$v_{variable\ 1\ columns}$ = number of columns – 1
$v_{variable\ 2\ rows}$ = number of rows – 1
$v_{interaction}$ = (number of columns – 1) × (number of rows – 1)
v_{within} = N – (number of rows × number of columns)

Enter these results in column 3 (Table 10.16.).

iv. Calculate the mean squares (MS) for the rows, columns, and interaction.

Note: Variances in this context are also known as mean squares (MS).

s^2 = MS = SS/v

For treatment 1 (columns), take the value for SS and divide it by its value for v. Repeat this for treatment 2 (rows), the interaction, and 'within-sample' variation. Put the results from these calculations in column 4 (Table 10.16).

v. Work out the F ratio. As there are three pairs of hypotheses to test, you calculate three F values.

Variable 1 (columns):
$$F_{calculated} = \frac{variable\ 1\ (\mathbf{columns})\ MS}{within\ samples\ MS}$$

Variable 2 (rows):
$$F_{calculated} = \frac{variable\ 2\ (\mathbf{rows})\ MS}{within\ samples\ MS}$$

Interaction: $F_{calculated} = \dfrac{interaction\ MS}{within\ samples\ MS}$

Place the results from these calculations in column 5 (Table 10.16).

4. **To find $F_{critical}$**

See Appendix D, Table D8. Again, as you are testing three pairs of hypotheses, there will be three critical F values to find. Using the F tables for an ANOVA, you may need to interpolate to find the values. The degrees of freedom to use are:

Variable 1 (columns): $v_{variable\ 1\ columns}$ and v_{within}
Variable 2 (rows): $v_{variable\ 2\ rows}$ and v_{within}
Interaction: $v_{interaction}$ and v_{within}
For Example 10.7, the degrees of freedom are:
Variable 1 (columns): 3, 72
Variable 2 (rows): 1, 72

(continued)

Interaction: $v_{interaction}$ and v_{within} 3, 72

The $F_{critical}$ values are included in Table 10.16 for specific p values.

5. **The rule**

If $F_{calculated}$ is greater than or equal to $F_{critical}$, then you may reject the null hypothesis.

In our example, when first considering the columns (composts), $F_{calculated}$ (1.2990) is less than $F_{critical}$ (2.7347) at $p = 0.05$ so we may not reject the null hypothesis. There is no significant difference (NS).

When comparing the two varieties of *C. atrosanguineus*, $F_{calculated}$ (39.255) is greater than $F_{critical}$ (11.774) at $p = 0.001$ so we may reject the null hypothesis.

Finally, when testing for an interaction, $F_{calculated}$ (2.6811) is less than $F_{critical}$ (2.4749) at $p = 0.05$ so

we may not reject the null hypothesis. There is no significant interaction (NS).

6. **What does this mean in real terms?**

There is no significant difference ($F = 1.3, p = 0.05$) between the mean height of *C. atrosanguineus* weaned on different composts and there is no significant interaction ($F = 2.5, p = 0.05$) between the composts and *C. atrosanguineus* varieties in their effects on mean height (cm). However, there is a very highly significant difference ($F = 39.3, p = 0.01$) between the mean heights of *C. atrosanguineus* var. 'Pip' compared with var. 'Christopher'.

Further examples relating to the topic including how to use statistical software are included in the Online Resource Centre.

10.13 Tukey's test following a parametric two-way ANOVA with equal replicates

The two-way ANOVA described in 10.12 has tested three general hypotheses of which one was significant. To test specific hypotheses, you may use a *post hoc* test, such as Tukey's test, which allows particular samples to be compared in pairs and those contributing to the significant difference in the ANOVA can be identified. This additional testing of hypotheses allows you to make much more explicit interpretations of your data and so should always be carried out when you have a significant result in a parametric ANOVA. However, if your interaction term is significant, it is difficult to interpret the results from Tukey's test, as the element of interaction will confound any differences between pairs of means.

10.13.1 Key trends and experimental design

In Example 10.7 there is one variety ('Pip') tested against four composts and a second variety ('Christopher') tested against the same four composts making eight different treatments in total. Comparing these to each other will result in 28 separate comparisons. The ANOVA (10.12) indicated a significant difference between the two varieties and the difference between the means for each treatment in Table 10.17 does reflect this. However, it is not clear if this is true for all the composts tested as the difference in mean height between 'Pip' and 'Christopher' on compost A was only 2.22cm compared to compost B where the difference between the means is 8.96cm.

10.13.2 **Using this test**

i. To use this test you:

1) Wish to test more specific hypotheses following a significant outcome in a parametric two-way ANOVA either for treatment variable 1 and/or for treatment variable 2 but not if there is a significant interaction.

2) Should have equal numbers of observations in each category or sample.

ii. Does the example meet these criteria?

In Example 10.7 there is a very highly significant difference between the mean heights of the varieties but not between the composts nor the interaction (Box 10.12) and there are 10 replicates in each block.

10.13.3 **The calculation**

The method we follow has already been described in Box 10.9. The calculated T values are the differences between the means for two samples. These differences for Example 10.7 are given in Table 10.17. The value for $T_{critical}$ is calculated as described in Box 10.11. For the example we are considering here, $n = 10$, $s^2_{within} = 15.482125$, $a = 8$, $v = 72$, and q when interpolated (6.4.4) from the table of q values at $p = 0.05$ is 4.346. Therefore $T_{critical} = 4.346 \times \sqrt{(15.482125/10)} = 5.4076$.

A significant difference between the two means is indicated where the $T_{calculated}$ value exceeds the $T_{critical}$ value. These significant differences are shown in bold in Table 10.18 and indicate that the difference between the varieties results from the use of compost B and D and to a smaller extent compost C but not to compost A. The commonly used format for reporting these results is outlined in 10.8.4.

Table 10.17 The difference between mean height (cm) ($T_{calculated}$ values) for Tukey's test following a two-way ANOVA on the data from Example 10.7: The weaning of *Cosmos atrosanguineus* var. 'Pip' and var. 'Christopher' onto one of four composts following propagation by tissue culture

		Weaning composts used for *C. atrosanguineus* var. 'Pip'				Weaning composts used for *C. atrosanguineus* var. 'Christopher'			
		A	B	C	D	A	B	C	D
		$\bar{x} = 21.97$	$\bar{x} = 23.66$	$\bar{x} = 21.39$	$\bar{x} = 21.58$	$\bar{x} = 19.75$	$\bar{x} = 14.70$	$\bar{x} = 16.31$	$\bar{x} = 15.79$
Weaning composts used for *C. atrosanguineus* var. 'Pip'	A $\bar{x} = 21.97$		1.69	0.58	0.39	2.22	**7.27**	**5.66**	**6.18**
	B $\bar{x} = 23.66$			2.27	2.08	3.91	**8.96**	**7.35**	**7.87**
	C $\bar{x} = 21.39$				0.19	1.64	**6.69**	5.08	**5.60**
	D $\bar{x} = 21.58$					1.83	**6.88**	5.27	**5.79**
Weaning composts used for *C. atrosanguineus* var. 'Christopher'	A $\bar{x} = 19.75$						5.05	3.44	3.96
	B $\bar{x} = 14.70$							1.61	1.09
	C $\bar{x} = 16.31$								0.52

10.14 Two-way parametric ANOVA with unequal replicates

One of the criteria for using a parametric ANOVA is that replication is equal in all categories. In a one-way ANOVA, an adjustment can be made to overcome this criterion if it is not met (10.9). However, in a more complex design such as a two-way ANOVA, a simple change in the calculation cannot be used.

Example 10.8 Changes in the peroxide levels of cooking oils in relation to storage conditions

The undergraduate extended the small study shown in Example 10.6 by placing aliquots of each cooking oil into small sealed jars and storing the oil for 5 weeks in a number of different commonly used locations. After the 5 weeks, samples were taken from each aliquot and the peroxide levels were assessed.

Table 10.18 The effect of storage conditions on peroxide levels (mEq of peroxide per kg sample) in a number of different cooking oils. All values × 10^{-2}.

> The second treatment variable 'location' has four nominal fixed categories. In most blocks there are three replicates but there is some missing data (–).

> The first treatment variable 'types of oil' has three nominal categories which have no sampling error associated with them so they are fixed. The dependent variable is 'peroxide concentration' measured on a derived interval scale.

	Peroxide levels (mEq of peroxide per kg sample) in a number of different cooking oils (all values × 10^{-2})		
	Corn oil	Extra virgin olive oil	Cold pressed rapeseed oil
Control (before storing)	14.0	17.0	11.0
	16.0	13.0	7.0
	–	12.0	8.0
$\bar{x}$	15.0	14.0	8.66667
s^2	2.0	7.0	4.33333
Fridge	16.0	14.0	14.0
	16.0	13.0	7.0
	14.0	15.0	12.0
$\bar{x}$	15.3333	14.0	11.0
s^2	1.33333	1.0	13.0
Cupboard	18.0	14.0	8.0
	14.0	9.0	10.0
	14.0	8.0	9.0
$\bar{x}$	15.33333	10.33333	9.0
s^2	5.33333	10.33333	1.0
Windowsill (sun)	26.0	30.0	20.0
	24.0	22.0	11.0
	22.0	20.0	–
$\bar{x}$	24.0	24.0	15.5
s^2	4.0	28.0	40.5

10.14.1 Key trends and experimental design

If you examine the mean values in Table 10.18 it is clear that the cold pressed rapeseed has lower peroxide content than the other oils. However, the storage conditions also affect these levels most noticeably when the oil is left on the windowsill the peroxide levels increases. It is therefore important to confirm that these trends are statistically significant yet there is missing data in two of the blocks.

10.14.2 Using this test

i. To use this test you:

Meet all the criteria from 10.12.2 apart from criterion 6. You have unequal replicates or missing data.

ii. Does the example meet these criteria?

The annotation confirms that these data meet the criteria for a two-way parametric ANOVA. There are very few observations so it is difficult to confirm that these are parametric. However, these observations are measured on a derived interval scale so we will assume that criterion 5 is met.

In Example 10.8, results are missing from the corn oil control and the cold pressed rapeseed oil sample stored on the windowsill. One approach we might take to balance out the number of replicates across the whole experiment would be to take values out of the data set at random to correct the imbalance. For example, we might select one of the values from the 'olive oil control' category at random, thus reducing the number of replicates to two and bringing it into line with the number of replicates in the 'corn oil control' category. This could be repeated throughout until each category only contained two replicates. Clearly this would significantly reduce the information in our data set and may introduce a bias not previously present. Alternative approaches are to estimate the missing values using an equation proposed by Shearer (1973) or by including the mean for that category. A discussion of which approach to take and when is given by Zar (2009). No solutions are ideal and should not be considered if many values are missing. For simplicity, we suggest using the mean value as an estimate of the missing value (Table 10.19).

10.14.3 The calculation

The ANOVA then proceeds as described in Box 10.12 until you reach the calculations for the degrees of freedom. The following is an extract from Box 10.12 with the revisions required when 'missing' values have been estimated.

iii. Calculate the degrees of freedom (v) for a two-way ANOVA with unequal replicates

$v_{variable\ 1\ columns} = number\ of\ columns\ -\ 1$

$v_{variable\ 2\ rows} = number\ of\ rows\ -\ 1$

$v_{interaction} = (number\ of\ columns\ -\ 1) \times (number\ of\ rows\ -\ 1)$

$v_{within} = *N - (number\ of\ rows \times number\ of\ columns)$

(*Where only the actual and not the estimated values are counted within this N value: for Example 10.8, two values were estimated so N will be 34 (and not 36).)

Table 10.19 Peroxide levels (mEq of peroxide per kg sample) in a number of different cooking oils and the effect of storage on the peroxide levels. All values × 10⁻². (Missing values have been replaced with the mean where necessary; mean and revised summary values are shown in bold.)

	Peroxide levels (mEq of peroxide per kg sample) in a number of different cooking oils. (All values × 10⁻²)		
	Corn oil	Extra virgin olive oil	Cold pressed rapeseed oil
Control (before storing)	14.0	17.0	11.0
	16.0	13.0	7.0
	$(14+16)/2 = \mathbf{15}$	12.0	8.0
$\bar{x}$	15.0	14.0	8.66667
s^2	**1.0**	7.0	4.33333
Fridge	16.0	14.0	14.0
	16.0	13.0	7.0
	14.0	15.0	12.0
$\bar{x}$	15.3333	14.0	11.0
s^2	1.3333	1.0	13.0
Cupboard	18.0	14.0	8.0
	14.0	9.0	10.0
	14.0	8.0	9.0
$\bar{x}$	15.3333	10.33333	9.0
s^2	5.33333	10.33333	1.0
Windowsill (sun)	26.0	30.0	20.0
	24.0	22.0	11.0
	22.0	20.0	$(20+11)/2 = \mathbf{15.5}$
$\bar{x}$	24.0	24.0	15.5
s^2	4.00	28.00	**20.25**

A common practice if an ANOVA has a significant outcome is to use this and then follow it with a *post hoc* test to obtain a more specific evaluation of the data. If you use this method for estimating missing values in a two-way ANOVA, we do not recommend following the ANOVA with a *post hoc* test as the MS_{within} will be affected by the inclusion of the estimated mean values in the data set.

10.15 Two-way parametric ANOVA with no replicates

There are three occasions when you may find this test relevant to your experiment. Firstly, there may be occasions when you have a similar design to that described in Example 10.7 where you are examining the effect of two treatment variables but without replication, and where you have only taken one observation from any one item. This test, however, can also be used where you know you have a notable confounding variable and you wish to be able to separate the effect of the confounding variable from that of the treatment variable. In this context the confounding variable is considered for statistical purposes to be a second treatment variable. Finally, this test may also be used if you have repeated measures. For example, you may run a number of rats through a number of different mazes and record the time taken. Clearly the results from each animal for all the mazes are likely to be influenced by that animal as well as the maze. To try to identify the effect of the individuals from the mazes (treatment variable), you could use this experimental design and this ANOVA and for the analysis treat the rats and the mazes as treatment variables.

In the first two types of investigation there is independence between all the measures as each item is only included once. This is very different from the repeated measures example where all the observations in row one will be derived from one item. If this was an ANOVA with replication there is a requirement for there to be independence between measures and each item is allocated at random to the blocks but for a two-way ANOVA without replicates this is not the case.

Unlike the other two-way parametric ANOVAs, there will only be two and not three pairs of hypotheses to test. The lack of replication means there are no within category variances to allow us to test for an interaction between the two variables.

 In the Introduction to 10.15 three possible reasons for using this test are given. Which of these reasons applies to Example 10.9?

10.15.1 Key trends and experimental design

In Example 10.9 with only single values for each block it is much harder to get a feel for what is happening in this experiment. However by looking at the row and column totals it does appear that the BB tomato feed is performing better than the other feeds. The location does appear to be adding some background noise but it is not clear if this will be a significant factor.

Example 10.9 The effect of fertilizers on tomato production

A horticulture undergraduate wished to test a number of known tomato feeds to assess the relative effect of each on crop production. However, the only space available was a polytunnel where there was slight micro-environmental variation down the length of the tunnel. To allow some estimate of the effect of the feed as against the effect of environmental variation, the polytunnel was divided into seven lengths (blocks). Within each block, five tomato plants were arranged at random and each was given a particular feed formulation. This was repeated for each block. At the end of the trial, the fruit from each plant was harvested and the total mass recorded (Table 10.20). *(continued)*

Table 10.20 The mass (kg) of tomatoes produced in cultivation in seven locations in a polytunnel and using five different tomato feeds

The second variable is a confounding variable: 'location in polytunnel'. It has seven nominal categories determined by the investigator and so is fixed.

The first treatment variable: 'tomato feeds' has five fixed nominal categories. The dependent variable is mass measured on an interval scale. There are no replicates. Each item (plant) has been assigned at random to both treatments and has only been measured once.

Polytunnel blocks	The mass (kg) of tomatoes produced using different tomato feeds					n_r	$\sum x_r$	$\sum(x_r^2)$
	157a	TomPlus	General	BB	Standard			
1	3.852	3.784	3.963	4.365	3.526	5	19.490	76.34783
2	3.951	3.749	3.852	4.896	3.851	5	20.299	83.30432
3	4.083	3.681	3.795	4.865	3.365	5	19.789	79.61412
4	3.783	3.883	3.762	4.296	3.421	5	19.145	73.70028
5	3.727	3.657	3.844	4.184	3.871	5	19.283	74.53101
6	3.368	3.364	3.347	3.961	3.721	5	17.761	63.39769
7	3.279	3.217	3.236	3.678	3.278	5	16.688	55.84559
n_c	7	7	7	7	7	$N = 35$		
$\sum x_c$	26.043	25.335	25.799	30.245	25.033		$\sum x_T = 132.455$	
$\sum(x^2)_c$	97.41607	92.04034	95.54838	131.87094	89.86511		$\sum(x^2)_T = 506.740853$	

10.15.2 Using this test

i. To use this test you:

1) Have no *a priori* reason for expecting certain outcomes from your investigation.

2) Have data which are usually presented in tables or summary histograms rather than a scatter plot.

3) Either wish to examine the effect of two treatment variables or you wish to examine the effect of one treatment variable and one notable confounding variable or you wish to examine the effect of one treatment variable and have repeated measures.

4) Can confirm that the treatment variables* are usually measured on an ordinal and/or nominal scale but could be measured on an interval scale and arranged into discrete categories.

5) Have a dependent variable which is measured on an interval scale and where the data are parametric.

6) Have no replicates.

7) Have an orthogonal design.

*If a confounding variable is being tested then it is referred to as a treatment variable for the purposes of this analysis. If repeated measures are being tested then the items are also being treated as a variable for the purposes of the analysis.

ii. Does the example meet these criteria?

The data from Example 10.9 meet these criteria as we wish to examine the effect of the feeds (variable) and the different polytunnel blocks (confounding variable) on the production of tomatoes. There are no replicates and the design is orthogonal. There are very few data and no replication within the experiment; therefore, the only indication that these data are parametric is that they are measured on an interval scale (Box 5.2). We will assume that the data are parametric.

10.15.3 The calculation

Having satisfied ourselves that our data meet all the criteria for using this two-way ANOVA without replicates, we can then proceed. In Box 10.13, we have organized the calculation of $F_{calculated}$ under a number of subheadings: (A) Calculate general terms; (B) Calculate the sums of squares; and (C) Construct and complete an ANOVA calculation table. At each point, we show how these steps can be applied to Example 10.9. Some of the calculations are included in Tables 10.20 and 10.21. Where steps have been abbreviated the full calculation is included in the Online Resource Centre.

BOX 10.13 How to carry out a two-way parametric ANOVA with no replicates

This calculation is given in full in the Online Resource Centre. For presentation purposes only all values have been rounded to five decimal places. This calculation is illustrated using the data from Example 10.9.

1. **General hypotheses to be tested**

H_{0c}: There is no difference between the sample means due to variable 1 (columns).

H_{1c}: There is a difference between the sample means due to variable 1 (columns).

H_{0r}: There is no difference between the sample means due to variable 2 (rows).

H_{1r}: There is a difference between the sample means due to variable 2 (rows).

In Example 10.9. these hypotheses will be:

H_{0c}: There is no difference between the mean mass of tomatoes (kg) produced in relation to the specific feed given.

H_{1c}: There is a difference between the mean mass of tomatoes (kg) produced in relation to the specific feed given.

H_{0r}: There is no difference between the mean mass of tomatoes (kg) produced in relation to their location in the polytunnel.

H_{1r}: There is a difference between the mean mass of tomatoes (kg) produced in relation to their location in the polytunnel.

(continued)

2. Have the criteria for using this test been met?

Yes. In this example, the criteria have been met (10.15.2ii).

3. How to work out $F_{calculated}$

A. Calculate general terms

1. First add together every observation in the complete data set (grand total): $\sum x_T$.
2. Square each and every observation and add all these squared values together: $\sum \left(x^2\right)_T$.
3. Add all the observations in a column: $\sum x_c$. Do this for each column. Square each column total: $\left(\sum x_c\right)^2$.

   Add these together: $\sum \left[\left(\sum x_c\right)^2\right]$. Divide this by the number of observations in a column (n_c).
4. For each row add all the observations: $\sum x_r$. Square the total: $\left(\sum x_r\right)^2$. Add these squared values together: $\sum \left[\left(\sum x_r\right)^2\right]$ and divide this value by the number of observations in the row (n_r).
5. Square the result from step **1** and divide this by N, where N is the total number of observations in all the categories.

For this example (Table 10.20), these calculations are:

1. $\sum x_T = 3.852 + 3.784 + \cdots + 3.678 + 3.278 = 132.455$

2. $\sum \left(x^2\right)_T = (3.852)^2 + (3.784)^2 + \cdots + (3.678)^2 + (3.278)^2 = 506.740853$

3. There are five columns and seven observations in each column. The column totals are given in Table 10.20.

   $\sum \left[\left(\sum x_c\right)^2\right] = (26.043)^2 + (25.335)^2 + (25.799)^2 + (30.245)^2 + (25.033)^2 = 3527.09959$

   $\dfrac{\sum \left[\left(\sum x_c\right)^2\right]}{n_c} = \dfrac{3527.09959}{7} = 503.87137$

4. There are seven rows and five observations in each row. The row totals are given in Table 10.20.

   $\sum \left[\left(\sum x_r\right)^2\right] = (19.49)^2 + (20.299)^2 + (19.789)^2 + (19.145)^2 + (19.283)^2 + (17.761)^2 + (16.688)^2 = 2515.8216$

$\dfrac{\sum \left[\left(\sum x_r\right)^2\right]}{n_r} = \dfrac{2515.8216}{5} = 503.16432$

5. $(132.455)^2/35 = 501.266486$

B. Calculate the sums of squares (SS)

6. SS_{total} = result from step **2** – result from step **5**
7. $SS_{variable\ 1\ columns}$ = result from step **3** – result from step **5**
8. $SS_{variable\ 2\ rows}$ = result from step **4** – result from step **5**
9. SS_{error} = result from step **6** – result from step **7** – result from step **8**

For Example 10.9, the calculations are:

6. $SS_{total} = 506.74085 - 501.26649 = 5.47437$
7. $SS_{variable\ 1\ columns} = 503.87137 - 501.26649 = 2.60488$
8. $SS_{variable\ 2\ rows} = 503.16432 - 501.26649 = 1.89783$
9. $SS_{error} = 5.47437 - 2.60488 - 1.89783 = 0.97166$

C. Construct and complete an ANOVA calculation table

i. Draw an ANOVA table as illustrated here.

Source of variation	SS	v	MS	F
Variable 1 (columns)	Step **7**			
Variable 2 (rows)	Step **8**			
Error	Step **9**			
Total	Step **6**			

ii. Transfer the results from the calculations for $SS_{variable\ 1\ columns}$, $SS_{variable\ 2\ rows}$, SS_{error}, and SS_{total} into the ANOVA table in column 2 (see Table 10.21).

iii. Calculate the degrees of freedom (v)

$v_{variable\ 1\ columns} = number\ of\ columns - 1$

$v_{variable\ 2\ rows} = number\ of\ rows - 1$

$v_{error} = (number\ of\ rows - 1) \times (number\ of\ columns - 1)$

$v_{total} = N - 1$

Enter these results in column 3 (Table 10.21).

(continued)

iv. Calculate the mean squares (MS) for the rows, columns and interaction.

Note: Variances in this context are also known as mean squares (MS).

$s^2 = MS = SS/v$

For variable 1 (columns), take the value for SS and divide it by its value for v. Repeat this for variable 2 (rows) and the error SS. Put the results from these calculations in column 4 (Table 10.21).

v. Work out the F ratio. As there are two pairs of hypotheses to test, you calculate two F values.

Variable 1 (columns):

$$F_{calculated} = \frac{\text{variable 1 (columns) MS}}{\text{error MS}}$$

Variable 2 (rows)

$$F_{calculated} = \frac{\text{variable 2 (rows) MS}}{\text{error MS}}$$

Place the results from these calculations in column 4 (Table 10.21).

4. **To find $F_{critical}$**

See Appendix D, Table D8. Again, as you are testing two pairs of hypotheses, there will be two critical F values to find. Using the F tables for an ANOVA, you may need to interpolate to find the values. The degrees of freedom to use are:

Variable 1 (columns): $v_{\text{variable 1 columns}}$ and v_{error}

Variable 2 (rows): $v_{\text{variable 2 rows}}$ and v_{error}

For Example 10.9, these are:

Variable 1 (columns) 4, 24

Variable 2 (rows) 6, 24

The $F_{critical}$ values are included in Table 10.21 for specific p values.

5. **The rule**

If $F_{calculated}$ is greater than or equal to $F_{critical}$ then you may reject the null hypothesis.

In our example, when first considering the columns (tomato feeds), $F_{calculated}$ (16.08) is more than $F_{critical}$ (4.22) at $p = 0.01$ so we may reject the null hypothesis. There is a highly significant difference.

When comparing the locations in the polytunnel where the plants were growing, $F_{calculated}$ (7.81) is greater than $F_{critical}$ (3.67) at $p = 0.01$ so we may reject the null hypothesis.

6. **What does this mean in real terms?**

The confounding variable of 'location in the polytunnel' had a highly significant effect ($F = 7.81$, $p = 0.01$) on the mean mass of tomatoes produced. Despite this there is also a highly significant difference ($F = 16.08$, $p = 0.01$) in the mean mass (kg) of tomatoes produced when given different tomato feeds. We have therefore been able to identify the effect of the tomato feed even though there was a notable confounding variable affecting the experiment.

Further examples relating to the topic including how to use statistical software are included in the Online Resource Centre.

Table 10.21 ANOVA table for analysis of data from Example 10.9: The mass (kg) of tomatoes produced when a plant is treated with a number of different feeds and grown in seven locations (blocks) in a polytunnel

Source of variation	SS	v	MS	$F_{calculated}$	$F_{critical}$	p
Between feeds (columns)	2.60488	5−1 = 4	0.65122	16.08533	4.22	0.01
Between locations (blocks) (rows)	1.89783	7−1 = 6	0.31631	7.81284	3.67	0.01
Error	0.97166	4 × 6 = 24	0.04049			
Total	5.47437	35−1 = 34				

10.16 Two-way nested parametric ANOVA with equal replicates

A nested ANOVA is one where you do not have an orthogonal design. Instead, your second factor is randomized with respect to the first. You can easily tell whether you have a nested design, as no categories within your second treatment variable are quite the same. For example, you may wish to examine the effect of two different growth hormones on the growth of floral meristem explants from *Pharbitis nil* (morning glory). The tissue culture medium was made as one batch, split into two flasks, and different growth hormones were added to each flask. The medium from these flasks was poured into tissue culture jars and left to cool. Four explants were grown in each jar of medium. Within each treatment, the jars were randomized. After 4 weeks, the number of shoots for each explant was recorded. The experimental design is:

Growth medium type 1			Growth medium type 2		
Jar 1	Jar 2	Jar 3	Jar 4	Jar 5	Jar 6
Explant 1	Explant 5	etc.			
Explant 2	etc.				
Explant 3					
Explant 4					

You can tell that this is a nested design as no jar will be exactly the same as any other jar. There will be slight differences in the depth of medium, location of the jar in the growth cabinet, etc. The four explants within each jar are subject to these unique factors and this grouping needs to be taken into account in the analysis. This grouping within the design is recognized by calling the growth media the *group* and the jars the *subgroup*. Clearly, there are many other possible nested designs that you may encounter in your research. There could be more than one main factor or more than one nested factor or unequal replicates. In this book, we illustrate nested ANOVAs with an example with one main variable, one nested variable, and equal replicates. For information on other nested designs, we recommend Sokal & Rohlf (2011).

 In our comparison between shoot production by explants of *Pharbitis nil* on two different tissue culture media, how many groups (*a*) and how many subgroups (*b*) are there? How many items are in each subgroup (n_s) and how many in each group (n_g)? How many observations are there in total (*N*)?

> **Example 10.10** Hydrogen ion concentration in deciduous and coniferous forests
>
> A soil scientist wished to compare the hydrogen ion (H^+) concentration of soils in two different forest types, one coniferous and the other deciduous. Within each forest, three different soil pits were dug at random and from each pit two soil samples were taken for hydrogen ion analysis. The results were recorded as mmol H^+/l per 100g soil (Table 10.22).
>
> Table 10.22 The concentration of hydrogen ions in soil (mmol H^+/l per 100g soil) from three soil pits in two woodlands
>
> > The first treatment variable: 'woodlands' has two fixed nominal categories. As a nested design these are known as groups. The dependent variable is hydrogen ions in soil measured on a derived interval scale.
>
> | Hydrogen ions in soil (mmol H^+/l per 100g soil) in woodlands | | | | | |
> |---|---|---|---|---|---|
> | Deciduous forest | | | Coniferous forest | | |
> | Pit 1 | Pit 2 | Pit 3 | Pit 1 | Pit 2 | Pit 3 |
> | 1.5 | 0.7 | 1.2 | 2.0 | 1.6 | 2.7 |
> | 1.2 | 1.0 | 1.7 | 2.5 | 2.2 | 2.4 |
>
> > The second treatment variable: 'pits' has three nominal categories randomized within the forests. As a nested design these are subgroups within the groups.
>
> > The soil samples from each pit are replicates and are randomized within the soil pits. Each item (soil sample) has only been measured once.

10.16.1 Key trends and experimental design

For the study described in Example 10.10 this experimental design is necessary because of the variable nature of soil. If only one pit was dug in each forest and multiple soil samples taken from this, there would be no assurance that the outcome of an analysis would not depend on the chance of having selected sites with soil that contained particularly high or low values of hydrogen ion concentrations. Hence, soil sampling from a number of sites within each forest is essential to ensure that any differences are due to the different type of forest and not to the location of the site. If no differences, beyond normal soil sample variability, are found between pits within the forest, any differences found can be attributed to the forest type.

Hence, we would expect the difference in soil hydrogen ion concentration due to the forest type to be much greater than the differences due to location within the forests individually. Each of these aspects can be tested in a nested ANOVA.

10.16.2 **Using this test**

i. To use this test you:

1) Have no *a priori* reason for expecting certain outcomes from your investigation.

2) Have data which are usually presented in tables or summary histograms rather than a scatter plot.

3) Wish to examine the effect of two treatment variables, one of which is subordinate or nested within the main factor.

4) Can confirm that the treatment variables are usually measured on an ordinal and/or nominal scale but may be measured on an interval scale and arranged into discrete categories.

5) Have a dependent variable which is measured on an interval scale and where the data are parametric.

6) Have the same number of observations (replicates) in each category.

7) Have assigned each item at random within the subordinate variable.

8) Have samples where the variation is similar (homogeneous).

ii. **Does the example meet these criteria?**

In Example 10.10, it is difficult to tell whether the data are parametric or whether the variances are homogeneous as there are relatively few observations in each subgroup. The scale of measurement is continuous and could be parametric. However, as we are unable to confirm that these criteria are met, we must acknowledge this when reporting the results from the analysis. The other criteria are met in that we wish to examine the effect of two variables (forest and pits) and the pits are nested within the forests, as no one pit will be the same as any other pit. There are two replicates in each sample and these were located at random in the subgroup ('pits').

10.16.3 **The calculation**

Having satisfied ourselves as far as we can that our data meet all the criteria for using this nested ANOVA, we can then proceed. In Box 10.14, we have organized the calculation of $F_{calculated}$ under a number of subheadings: (A) Calculate general terms; (B) Calculate the sums of squares; and (C) Construct and complete an ANOVA calculation table. At each point, we show how these steps can be applied to Example 10.10. Some of the calculations are included in Tables 10.23 and 10.24. Where steps have been abbreviated, the full calculation is included in the Online Resource Centre.

BOX 10.14 How to carry out a nested ANOVA with one main factor and one nested factor with equal replicates

This calculation is given in full in the Online Resource Centre. For presentation purposes only all values have been rounded to five decimal places. This calculation is illustrated using data from Example 10.10.

1. **General hypotheses to be tested**

    H_0: There is no difference between the means of the samples.

    H_1: There is a difference between the means of the samples.

In our example, the two pairs of hypotheses we are testing are:

    H_0: There is no difference in soil hydrogen ion concentrations (mmol H^+/l per 100g soil) due to different forest types.

    H_1: There is a difference in soil hydrogen ion concentrations (mmol H^+/l per 100g soil) due to different forest types.

    H_0: There is no difference in soil hydrogen ion concentrations (mmol H^+/l per 100g soil) among the different soil pits within the forest types.

    H_2: There is a difference in soil hydrogen ion concentrations (mmol H^+/l per 100g soil) among the different soil pits within the forest types.

2. **Have the criteria for using this test been met?**

    Yes. In this example, the criteria have been met (10.16.2ii).

3. **How to work out $F_{calculated}$**

A. Calculate general terms

1. Add together all the observations in all the samples (grand total): $(\sum x_T)$.

2. Square each observation in all the samples and add these together: $\sum (x^2)_T$.

3. Add all the observations in a subgroup: $\sum x_{sg}$. These are the subgroup totals. Do this for all subgroups. Square each subgroup total: $(\sum x_{sg})^2$. Add these together: $\sum (\sum x_{sg})^2$. Divide this total by the number of observations in each subgroup (n_{sg}).

4. Add all the observations in a group: $\sum x_g$. Square these: $(\sum x_g)^2$. Add these together: $\sum (\sum x_g)^2$.

Divide by the number of observations in that group (n_g).

5. Take the result from step **1** and square it: $(\sum x_T)^2$. Divide by N, where N is the total number of observations in all the samples combined.

In our example (Table 10.23), the calculations are:

1. $(\sum x_T) = 1.5 + 1.2 + \ldots\ldots\ldots + 2.7 + 2.4 = 20.7$

2. $\sum (x^2)_T = (1.5)^2 + (1.2)^2 + \ldots + (2.4)^2 = 40.21$

3. $\sum (\sum x_{sg})^2 / n_{sg}$

    $= \dfrac{(2.7)^2 + (1.7)^2 + (2.9)^2 + (4.5)^2 + (3.8)^2 + (5.1)^2}{2}$

    $= 79.29/2 = 39.645$

4. $\sum (\sum x_g)^2 / n_g = \dfrac{(7.3)^2 + (13.4)^2}{6} = 232.82/6 = 38.80833$

5. $(\sum x_T)^2 / N = (20.7)^2 / 12 = 35.7075$

B. Calculate the sums of squares (SS)

6. SS_{total} = result from step **2** – result from step **5**

7. SS_{groups} = result from step **4** – result from step **5**

8. $SS_{subgroups}$ = result from step **3** – result from step **4**

9. SS_{within} = result from step **2** – result from step **3**

In our example, the calculations are:

6. $SS_{total} = 40.21 - 35.7075 = 4.5025$

7. $SS_{groups} = 38.80833 - 35.7075 = 3.10083$

8. $SS_{subgroups} = 39.645 - 38.80833 = 0.83667$

9. $SS_{within} = 40.21 - 39.645 = 0.565$

C. Construct and complete an ANOVA calculation table

i. Draw an ANOVA table as illustrated here.

Source of variation	SS	v	MS	F
Between groups	Step **7**			
Between subgroups within groups	Step **8**			
Within subgroups	Step **9**			
Total variation	Step **6**			

(continued)

ii. Transfer the results from the calculations for SS_{groups}, $SS_{subgroups}$, and SS_{within} into the ANOVA table in column 2 (see Table 10.24).

iii. Calculate degrees of freedom (v).

Let a = the number of groups, b = the number of subgroups, N = total number of observations, and n_{sg} = number of observations in each subgroup.

$$v_{total} = N - 1$$

$$v_{groups} = a - 1$$

$$v_{subgroups} = a(b - 1)$$

$$v_{within} = ab(n_{sg} - 1)$$

Enter these results in column 3.

iv. Calculate the mean squares for the groups, subgroups and within subgroups.

Note: variances in this context are also known as mean squares (MS).

$$s^2 = MS = \frac{SS}{v}$$

In the groups row, take the value for SS and divide it by its value for v. Repeat this for the subgroups row and the within row. Put the results from these calculations in column 4 (Table 10.24).

v. Finally, to calculate the variance ratio to test the first pair of hypotheses, you use the group and subgroup MS, where:

$$F_{calculated} = \frac{group\ MS}{subgroup\ MS}$$

It is usual to place the results from this calculation in column 5 in the first row (Table 10.24).

To test the second pair of hypotheses, you use the subgroups and within subgroups MS. So:

$$F_{calculated} = \frac{subgroup\ MS}{within\ subgroup\ MS}$$

This value is usually placed in the second row in column 5 (Table 10.24).

4. **To find $F_{critical}$**

See Appendix D, Table D8. Identify the critical F value using the F tables for an ANOVA. You may need to interpolate (6.4.4) these values. To find the critical value you need the degrees of freedom relating to the two variances in the F-test.

In our example, to test the first pair of hypotheses, the degrees of freedom are $v_1 = 1$ and $v_2 = 4$.

Therefore, at $p = 0.05$, $F_{critical} = 7.71$, and at $p = 0.01$, $F_{critical} = 21.20$.

To test the second pair of hypotheses, the degrees of freedom are $v_1 = 1$ and $v_2 = 6$.

Therefore, at $p = 0.05$, $F_{critical} = 5.99$.

5. **The rule**

If $F_{calculated}$ is greater than $F_{critical}$, then you may reject the null hypothesis.

In our example, when testing the first pair of hypotheses $F_{calculated}$ (14.82) is greater than $F_{critical}$ (7.71) at $p = 0.05$, but not greater than $F_{critical}$ (21.20) at $p = 0.01$, so you may reject the null hypothesis at $0.01 < p < 0.05$.

For the second pair of hypotheses, $F_{calculated}$ (2.22) is less than $F_{critical}$ (5.99) at $p = 0.05$, so you do not reject the null hypothesis.

6. **What does this mean in real terms?**

There is a significant difference ($F = 14.82$, $0.01 < p < 0.05$) in soil hydrogen ion concentrations due to different forest types. However, there is no difference ($p = 0.05$) in soil hydrogen ion concentrations among the different soil pits within the forest types.

Further examples relating to the topic including how to use statistical software are included in the Online Resource Centre.

Table 10.23 Calculation table for the parametric nested ANOVA on data from Example 10.10: Hydrogen ion concentration in a deciduous and coniferous forest

| | Hydrogen ions in soil (mmol H^+/l per 100g soil) in woodlands | | | | | |
| | Deciduous forest | | | Coniferous forest | | |
	Pit 1	Pit 2	Pit 3	Pit 1	Pit 2	Pit 3
Sample 1	1.5	0.7	1.2	2.0	1.6	2.7
Sample 2	1.2	1.0	1.7	2.5	2.2	2.4
Subgroup totals ($\sum x_{sg}$)	2.7	1.7	2.9	4.5	3.8	5.1
Group totals ($\sum x_g$)		7.3			13.4	
Grand total ($\sum x_T$)				20.7		

Table 10.24 ANOVA table from the analysis of data from Example 10.10: Hydrogen ion concentration in a deciduous and coniferous forest, showing how some of the terms are calculated

Source of variation	SS	v	MS	F
Between forests (groups)	3.10083	2−1 = 1	3.10083	3.10083/0.20917 = 14.82445
Between pits (subgroups)	0.83667	2(3−1) = 4	0.20917	0.20917/0.09417 = 2.22119571
Between replicates in pits (within subgroups)	0.565	$(2 \times 3)(2-1) = 6$	0.09417	
Total variation	4.5025	12−1 = 11		

10.16.4 Tukey's test for nested parametric ANOVAs

It is very difficult to use a *post hoc* test, such as Tukey's, when one variable is nested in the other and therefore any mean values for samples will be means for subgroups within groups. One circumstance where you might use Tukey's test would be, as in our example, when the subgroups are not significantly different but the groups are. Under these circumstances, you could consider all values within a group to be replicates. If you had three or more groups, you could carry out a one-way ANOVA where all subgroup observations are replicates in the groups. The values from this ANOVA could then be used to test specific hypotheses. If you have only two groups, there is no point in carrying out this specific test as it will be clear from the data what the relationship is. So in Example 10.10 it is clear from the ANOVA that the soil hydrogen ion concentrations are significantly greater in the coniferous forest than in the deciduous forest.

10.17 Factorial three-way parametric ANOVA with no replicates

When there are more than two independent factors acting on a sample, the design is called *factorial*. It is possible to have any number of factors and to test for differences arising from these in one large experiment and calculation. When there are many treatment variables the best approach for the analysis is to use the statistical software. However, to use it correctly you need to understand the principles we have been covering in this chapter such as the concepts of fixed and random variables, nested variables, and the independence of observations.

Therefore, we will limit ourselves to one worked example as a demonstration of a three-factor ANOVA (factors A, B, and C) without replicates. Here the effect of the three treatments can be tested in the usual way to determine whether there is a significant difference. We can also determine whether there is an interaction between any or all of the factors. The first three interactions, $A \times B$, $B \times C$, and $A \times C$, are known as first-order interactions and the interaction $A \times B \times C$ is known as a second-order interaction.

There is a difference between a factorial analysis with or without replicates. If there are no replicates and for each combination of treatments there is only one observation, then the interaction term $A \times B \times C$ is used in place of the SS_{within} term in the F ratio. If replicates are present, then an SS_{within} term can be calculated and this is used in the F ratio.

In the same way that a two-way ANOVA can be used to examine the effect of a confounding variable, a three-way ANOVA can also be used. However, for repeated measures you need to examine the design carefully to ensure that the repeated measures only affect one of the variables being examined, and a nested ANOVA may be more suitable.

Example 10.11   Lead levels in soil samples taken at various depths, distances, and bearings from a smelter

Soil samples were taken to discover whether there was a significant difference in the extent of lead contamination due to the proximity of a smelter. Soil samples were taken at various bearings and distances to the smelter and at various depths of soil. Each treatment was combined with every other treatment (Table 10.25).

As there are three variables, the results table would be in three dimensions. As this cannot be represented on paper, a data table is constructed in which the third factor (C) is located within one of the other factors, in this case A. This does not imply that factor C is in any way dependent on factor A (Table 10.26).

Table 10.25 Summary table of the factors ($3 \times 4 \times 4$ factorial ANOVA) in the investigation of lead levels in soil samples

Factor A: Distance (km)	Factor B: Bearing	Factor C: Soil depth (cm)
0.25	South	5
0.5	South-east	10
1.0	South-west	15
2.0		20
		30

(continued)

Table 10.26 Lead concentrations in soil (µg/g) at a number of locations (bearing and distance) and soil depths in the vicinity of a smelter

The first treatment variable (factor A) is 'distance from the smelter'. This is measured as four ordinal points and is fixed. The second treatment variable (factor B) is 'bearing' with three ordinal measures and is also fixed. The third treatment variable (factor C) is 'soil depth' with four fixed ordinal categories.

Distance (km) (factor A)	Soil depth (cm) (factor C)	Lead concentrations in soil (µg/g) at a number of locations		
		Bearing (factor B)		
		South	South-east	South-west
0.25	5	155	96	365
	10	102	73	345
	15	77	23	248
	20	65	12	176
	30	26	8	72
0.5	5	98	54	302
	10	65	24	238
	15	45	14	154
	20	23	12	98
	30	9	10	43
1.0	5	55	32	256
	10	36	24	189
	15	22	14	112
	20	12	8	54
	30	9	5	23
2.0	5	25	18	167
	10	18	14	98
	15	15	13	43
	20	8	6	22
	30	7	4	17

The dependent variable 'Lead concentrations in soil' is measured on a derived interval scale. There are no replicates.

10.17.1 Key trends and experimental design

Not surprisingly there appears in general to be more lead pollution in the top soil and closer to the smelter. There also may be an effect of 'bearing' in that the soils taken from the south-west appear to have noticeably higher levels of lead. Whilst general examination like this can suggest key trends it is not clear if these are significant differences nor is it clear if there are any interactions between the factors. This can be identified using the ANOVA.

10.17.2 Using this test

i. To use this test you:

1) Have no *a priori* reason for expecting certain outcomes from your investigation.
2) Have data which are usually presented in tables or summary histograms rather than a scatter plot.
3) Wish to examine the effect of three treatment variables each with two or more categories.
4) Can confirm that the treatment variables are usually measured on an ordinal and/or nominal scale but may be measured on an interval scale and arranged into discrete categories.
5) Have a dependent variable which is measured on an interval scale and where the data are parametric.
6) Have no replicates.
7) Have an orthogonal design.

ii. Does the example meet these criteria?

The data from this example meet the criteria for using a factorial ANOVA with no replicates as we wish to examine the effect of three variables (soil depth, bearing, and distance from the smelter), each variable has three or four categories, and the design is orthogonal. Each soil sample (the item) was only measured once, so these are not repeated measures. As there are only single observations for each particular set of treatments, it is difficult to tell whether the data are parametric other than that the data are measured on an interval scale (Box 5.2). We shall therefore have to assume that the data are parametric based only on this criterion.

10.17.3 The calculation

In this calculation, we will examine the effects of the treatments (factor A, factor B, and factor C) and the first-order interactions ($A \times B$, $A \times C$, and $B \times C$). The second-order interaction ($A \times B \times C$) is used in place of the variance$_{within}$ term. Therefore, we are testing six pairs of hypotheses and carrying out an F-test for each. The calculation of these variance ratios has been organized into a number of steps: (A) Calculate the general terms; (B) Calculate the sums of squares; and (C) Construct and complete an ANOVA table. In Box 10.15, we illustrate this calculation using the data from Table 10.26.

BOX 10.15 How to carry out a three-way factorial parametric ANOVA without replicates

This calculation is given in full in the Online Resource Centre. For presentation purposes all values have been rounded to five decimal places. This calculation is illustrated using data from Example 10.11.

1. **General hypotheses to be tested**

   As there are six pairs of hypotheses, we have not included general hypotheses but only those from our example.

   H_{0A}: There is no difference between the mean lead concentration in the soil samples ($\mu g/g$) due to distance from the smelter (km) (factor A).

   H_{1A}: There is a difference between the mean lead concentration in the soil samples ($\mu g/g$) due to distance from the smelter (km) (factor A).

   H_{0B}: There is no difference between the mean lead concentration in the soil samples ($\mu g/g$) due to the bearing from the smelter (factor B).

   H_{1B}: There is a difference between the mean lead concentration in the soil samples ($\mu g/g$) due to the bearing from the smelter (factor B).

   H_{0C}: There is no difference between the mean lead concentration in the soil samples ($\mu g/g$) due to the depth at which the soil was sampled (cm) (factor C).

   H_{1C}: There is a difference between the mean lead concentration in the soil samples ($\mu g/g$) due to the depth at which the soil was sampled (cm) (factor C).

   $H_{0(A \times B)}$: There is no interaction between distance (km) (factor A) and bearing (factor B) in their effects on the mean lead concentration in the soil.

   $H_{1(A \times B)}$: There is an interaction between distance (factor A) and bearing (factor B) in their effects on the mean lead concentration ($\mu g/g$) in the soil.

   $H_{0(A \times C)}$: There is no interaction between distance (km) (factor A) and soil sample depth (cm) (factor C) in their effects on the mean lead concentration ($\mu g/g$) in the soil.

   $H_{1(A \times C)}$: There is an interaction between distance (km) (factor A) and soil sample depth

(cm) (factor C) in their effects on the mean lead concentration ($\mu g/g$) in the soil.

   $H_{0(B \times C)}$: There is no interaction between bearing (factor B) and depth of sampling (cm) (factor C) in their effects on the mean lead concentration ($\mu g/g$) in the soil.

   $H_{1(B \times C)}$: There is an interaction between bearing (factor B) and depth of sampling (cm) (factor C) in their effects on the mean lead concentration ($\mu g/g$) in the soil.

2. **Have the criteria for using this test been met?**

   As far as we can tell, these criteria have been met (10.17.2ii).

3. **How to work out $F_{calculated}$**

A. Calculate general terms

1. Calculate the grand total $\left(\sum x_T \right)$, by adding together all the observations in the data set. Note the total number of observations (N).

   $$\sum x_T = 155 + 96 + 365 + ... + 7 + 4 + 17 = 4358.0$$

   $$N = 60$$

2. Square each and every observation and add these together: $\sum \left(x^2 \right)_T$

   $$\sum \left(x^2 \right)_T = (155)^2 + (96)^2 + (365)^2 + \ ... \ + (7)^2 + (4)^2$$
   $$+ (17)^2 = 780172.0$$

3. Summarize the data from the three-way table (Table 10.26) into three two-way tables.

   B × C table

   For the B × C table (Table 10.27), total the four observations for bearing south and 5cm soil depth. This total goes into the first row, second column of the B × C table. Add the four observations for bearing south-east and 5cm soil depth. This total goes into the second row, second column of the A × C table, etc. The number of observations being counted = a = 4.

(continued)

Table 10.27 B × C two-way table (bearing × soil sample depth (cm)) for lead levels in soil samples (μg/g)

Factor B: Bearing from smelter	Lead levels in soil samples (μg/g)					Row total
	Factor C: Depth at which soil samples taken (cm)					
	5	10	15	20	30	
South	333	221	159	108	51	872
South-east	200	135	64	38	27	464
South-west	1090	870	557	350	155	3022
Column total	1623	1226	780	496	233	4358

A × C table

For the A × C two-way table (Table 10.28), add together the three observations for 5cm soil depth and distance 0.25km. This total goes into the first row, second column of the A × C table. Add together the three observations for 5cm soil depth and distance 0.5km. This total goes in the second row, second column, etc. The number of observations being counted = b = 3.

A × B table

For the A × B table (Table 10.29), first add the five values for 0.25km distance and bearing south. The total goes into the first cell (first row first column) of the two way table A × B. Add the five values for distance 0.5km, bearing south. This total goes into the second cell (second row first column) of the A × B table, etc. The number of observations being counted = c = 5.

Table 10.28 A × C two-way table (distance (km) × soil depth (cm)) for lead levels in soil samples (μg/g)

Factor A: Distance from smelter (km)	Lead levels in soil samples (μg/g)					Row total
	Factor C: Depth at which soil samples taken (cm)					
	5	10	15	20	30	
0.25	616	520	348	253	106	1843
0.5	454	327	213	133	62	1189
1.0	343	249	148	74	37	851
2.0	210	130	71	36	28	475
Column total	1623	1226	780	496	233	4358

Table 10.29 A × B two-way table (distance (km) × bearing) for lead levels in soil samples (μg/g)

Factor A: Distance from smelter (km)	Lead levels in soil samples (μg/g)			Row total
	Factor B: Bearing from smelter			
	South	South-east	South-west	
0.25	425	212	1206	1843
0.5	240	114	835	1189
1.0	134	83	634	851
2.0	73	55	347	475
Column total	872	464	3022	4358

(continued)

4. Square each and every row total from the $A \times C$ two-way table (Table 10.28), add these squared values together, and divide by $(b \times c)$:

$$= \frac{(1843)^2 + (1189)^2 + (851)^2 + (475)^2}{3 \times 5}$$

$= 384\,013.0667$

5. Square each and every row total from the $B \times C$ two-way table (Table 10.27), add these squared values together, and divide by $(a \times c)$:

$$= \frac{(872)^2 + (464)^2 + (3022)^2}{4 \times 5} = 505408.2$$

6. Square each and every column total from the $A \times C$ two-way table (Table 10.28), add these squared values together, and divide by $(a \times b)$:

$$= \frac{(1623)^2 + (1226)^2 + (780)^2 + (496)^2 + (233)^2}{4 \times 3}$$

$= 420\,492.5$

7. Square each value in the $A \times B$ two-way table (Table 10.29), add these squared values together, and divide by c:

$$= \frac{(425)^2 + (212)^2 + \cdots\cdots + (55)^2 + (347)^2}{5}$$

$= 600\,678.0$

8. Square each value in the $A \times C$ two-way table (Table 10.28), add these squared values together, and divide by b:

$$= \frac{(616)^2 + (520)^2 + \cdots\cdots + (36)^2 + (28)^2}{5}$$

$= 500\,890.6667$

9. Square each value in the $B \times C$ two-way table (Table 10.27), add these squared values together, and divide by a:

$$= \frac{(333)^2 + (221)^2 + \cdots\cdots + (350)^2 + (155)^2}{4}$$

$= 666\,386.0$

10. Square the grand total from step **1** and divide by the total number of observations (N):

$= (4358)^2 / 60 = 316536.0667$

B. Calculate the sums of squares (SS)

11. SS_A = result from step **4** – result from step **10**

$= 384013.0667 - 316536.0667 = 67477.0$

12. SS_B = result from step **5** – result from step **10**

$= 505408.2 - 316536.0667 = 188872.1333$

13. SS_C = result from step **6** – result from step **10**

$= 420492.5 - 316536.0667 = 103956.4333$

14. SS_{AB} = result from step **7** + result from step **10** – result from step **4** – result from step **5**

$= 600678.0 + 316536.0667 - 384013.0667 - 505408.2$

$= 27792.8$

15. SS_{AC} = result from step **8** + result from step **10** – result from step **4** – result from step **6**

$= 500890.6667 + 316536.0667 - 384013.0667$

$- 420492.5 = 12921.1667$

16. SS_{BC} = result from step **9** + result from step **10** – result from step **5** – result from step **6**

$= 666386.0 + 316536.0667 - 505408.2 - 420492.5$

$= 57021.3667$

17. SS_{ABC} = (result from step **2** + results from steps **4** + **5** + **6**) – (results from steps **7** + **8** + **9** + **10**)

$= (780172.0 + 384013.0667 + 505408.2 +$

$420492.5) - (600678.0 + 500890.6667 + 666386.0$

$+ 316536.0667)$

$= 5595.0333$

C. Construct and complete an ANOVA calculation table

i. Construct and complete an ANOVA table as shown in Table 10.30.

Transfer the results from the calculations for sums of squares (SS) into the table in column 2 (Table 10.30).

ii. Calculate the degrees of freedom (v) where:

$v_A = a - 1$ = number of samples – 1
$v_B = b - 1$ = number of samples – 1
$v_C = c - 1$ = number of samples – 1
$v_{A \times B} = (a - 1)(b - 1)$
$v_{A \times C} = (a - 1)(c - 1)$
$v_{B \times C} = (b - 1)(c - 1)$
$v_{A \times B \times C} = (a - 1)(b - 1)(c - 1)$

(continued)

Enter these results in column 3 (Table 10.30).

iii. Calculate the mean squares (MS)

Note: Variances in this context are also known as mean squares (MS).

$$s^2 = MS = \frac{SS}{v}$$

In the Factor A row, take the value for SS_A and divide it by the value for v_A.

In the Factor B row, take the value for SS_B and divide it by the value for v_B, and so on. Put the results from these calculations in column 4 (Table 10.30).

If the interaction terms are not significant, then you may proceed to test the main effects. In our example, the interaction terms are all significantly different (Table 10.30).

iv. How to work out $F_{calculated}$

As in the F_{max} test, $F_{calculated}$ is a ratio. The F value for each pair of hypotheses to be tested is calculated by the ratio of its mean square over the mean square of $A \times B \times C$ (which represents the 'error' mean square in a factorial analysis with no replicates and assuming that the third factor interaction is zero), e.g.:

$$F_{calculated} = \frac{MS_A}{MS_{ABC}}$$

The results from these calculations are placed in column 5 in the first row and so on (Table 10.30).

4. To find $F_{critical}$

See Appendix D, Table D8. Identify the critical F value using the F tables for an ANOVA. You may need to interpolate (6.4.4) these values, where the degrees of freedom are those relating to the numerator (v_1) and denominator, which is MS_{ABC} (v_2). As you are testing six pairs of hypotheses, you may need to find six values for $F_{critical}$.

5. The rule

The general rule is that if $F_{calculated}$ is greater than or equal to $F_{critical}$, then you may reject the null hypothesis. In this factorial analysis, you should examine the first-order interaction terms first (i.e. $A \times B$, $A \times C$, and $B \times C$). If these are significant, then it is not meaningful to test the main effects of factors A, B, and C as these elements have been shown not to be contributing separately to the overall outcome.

6. Relate your findings back to your investigations

The distance × bearing, distance × soil sample depth, and bearing × soil sample depth are all highly significant interactions ($p = 0.01$). Therefore, the effect of distance from the smelter on the lead concentration in the soil samples depends on the bearing and soil sample depth. In addition, the effect of the soil sample depth on the lead concentration depends on the bearing from the smelter.

Further examples relating to the topic including how to use statistical software are included in the Online Resource Centre.

Table 10.30 ANOVA table for analysis of data from Example 10.11: Lead concentrations in soil (µg/g) at a number of locations (bearing and distance) and soil depths from a smelter

Source of variation	SS	v	MS	$F_{calculated}$	$F_{critical}$ at $p = 0.01$
Factor A: Distance (km)	67477.0	3	22492.3	96.48134	
Factor B: Bearing	188872.1333	2	94436.06	405.08551	
Factor C: Soil sample depth (cm)	103956.43	4	25989.11	111.48083	
Interaction A × B: Distance (km) × bearing	27792.80	6	4632.13	19.86963	3.67
Interaction A × C: Distance (km) × soil sample depth (cm)	12921.1667	12	1076.76	4.61880	3.03
Interaction B × C: Bearing × soil sample depth (cm)	57021.37	8	7127.67	30.5743	3.36
Interaction: A × B × C	5595.03	24	233.13		

10.17.4 Tukey's test and a three-way parametric ANOVA

As we explained in 10.16, any significant interactions between variables will make it very difficult to interpret any *post hoc* comparisons between pairs of means. Therefore, you should restrict your use of Tukey's test to experiments where only the main factors are significant and not the interaction terms. In these cases, you apply Tukey's test as we described in 10.13.

Summary of Chapter 10

- This chapter draws on your understanding of distributions and in particular the normal distribution (5.2.1). The normal distribution is used as the basis for parametric statistics. Data that are parametric (Box 5.2) or have been normalized following transformation (5.10) may be analysed using parametric statistics. Parametric statistics are flexible and powerful.

- The basic elements of parametric statistics are the sum of squares and the variance, which we introduced in Chapter 5 (Box 5.1).

- In this chapter we consider parametric tests for repeated measures or matched data (10.4., 10.11.5, 10.15) and for non-matched data with one treatment variable (10.1, 10.2, 10.3, 10.5) and two and more treatment variables (10.11–10.16).

- The ANOVAs followed by a *post hoc* test such as Tukey's test allow you to test both general and more specific hypotheses. This provides you with considerably more explicit information about the effects of your treatment variables and is a design that should be used whenever possible.

- In Chapter 11, we consider the non-parametric equivalents to the tests covered in this chapter. Therefore, if your data are not parametric (5.8) or cannot be normalized by transforming your data (5.10), you should refer to Chapter 11.

- The Online Resource Centre includes interactive exercises that test your understanding of this chapter with other topics, particularly those considered in Chapters 2 and 8–11.

Answers to chapter questions

A1 There are a number of comments you can make. The first is to accept the outcome from the statistics at face value and say that there really is a difference between Porthcawl and Aberystwyth in the way these periwinkles are evolving. However, at Porthcawl the sample size really is small and is therefore unlikely to be representative of the population. It is possible that sampling error in the small sample from Porthcawl has masked the effect of evolution and a Type II error has occurred. Another possibility could also be that one of the criteria for using this test has not been met. It was difficult to confirm that the population from which the samples were taken at Porthcawl was likely to be normally distributed. There is a clear need for a repeated investigation with much larger samples collected from Aberystwyth and Porthcawl.

A2 All these treatments are under the control of the investigator and fixed. Therefore, this is a Model I design.

A3 In Example 10.9 the location in the polytunnel is a potentially serious confounding variable so it has been incorporated into the design as a variable so that the effect of location on the experiment can be assessed.

A4 $a = 2$ (media type 1 and media type 2).

  $b = 3$ (there are three jars in each group).

  $n_s = 4$ (there are four explants in each jar).

  $n_g = 12$ (there are three jars each with four explants in each group).

  $N = 24$ (there are 24 explants in total).

11 Hypothesis testing:
Do my samples come from the same population? Non-parametric data

In a nutshell

Biological data can be derived from populations that are not normally distributed or it is difficult to confirm from the samples being studied that the distribution is likely to be normal. In these instances, you may wish to test the hypothesis that samples come from the same statistical population using non-parametric tests. In this chapter we have included the tests used most often by our undergraduates and which mirror those covered in the previous chapter for testing the same hypothesis when you have parametric data. There are a number of key principles that need to be understood in relation to the use and interpretation of these tests including how to rank your data, absolute values, matched data, and the interaction between treatment variables. These points are covered when they first become relevant to the statistics tests being considered.

The statistical tests covered in this chapter are:

11.1 Mann–Whitney U test

11.2 Wilcoxon's signed ranks test for matched pairs test

11.3 One-way non-parametric ANOVA (Kruskal–Wallis test)

11.4 *Post hoc* test following a non-parametric one-way ANOVA

11.5 Two-way non-parametric ANOVA

11.6 *Post hoc* test following a two-way non-parametric ANOVA

11.7 Scheirer–Ray–Hare test

This chapter tells you how you may analyse non-parametric data to test the hypothesis that two or more samples come from the same statistical population. The choice of the statistical tests in this chapter is outlined in Chapter 7. Your selection depends on how many variables you are planning to examine, how many categories in each variable, and how many replicates in each category.

For the tests outlined in this chapter it is assumed that you have non-parametric data (5.8 and 5.9). These tests are 'distribution free' and the data therefore do not have to be distributed normally. Non-parametric tests may be used for both parametric and non-parametric data but not the reverse. However, parametric tests (Chapter 10) are more powerful, and more able to confirm a real biological difference when one exists. Therefore if you believe you have parametric data you should make use of parametric statistics. If your data are not parametric, you may be able to transform them (5.10) to 'normalize' the data and so enable you to use a parametric test.

The tests described in this chapter are known as *ranking* tests. Look at the data below. Each sample has been organized into numerical order and you can see that there is a small overlap between the two data sets.

Sample 1	1	3	4	6	7	7	9	10	10	11					
Sample 2									10	11	11	12	13	13	13

If the two data sets are from different statistical populations, then you would expect the overlap to be small. If the two samples are from the same population, then you would expect the overlap to be greater. The ranking tests provide a measure of this overlap and an estimate of the probability that this overlap could occur by chance. If you are not familiar with the process of ranking data, you should refer to Box 5.3 where we give further examples.

In this chapter, we have covered some of the non-parametric equivalents of the parametric tests covered in Chapter 10. These tests cover a range of designs for one or more treatment variables with replication and for repeated measures. Our coverage of these tests is of necessity limited. Therefore as outlined in Chapter 7 you should avoid using a less than perfect statistical test and keep looking for one that is right for your design and data by using all sources available to you. For example, if you have non-parametric data and wish to compare one treatment variable and one confounding variable with no replicates, or one treatment variable and have items as repeated measures, then you should consider the Friedman's test. Similarly, if you have one sample and wish to compare this to a single value and have non-parametric data you should consider the one-sample Wilcoxon's signed rank test.

Seven tests are covered in this chapter. Worked examples are given for each test. If this is the first time you have used these tests, you should cover up these worked examples and use the general information and the data provided to work through these examples. Then check your answers before using the test on your own data. If your answer differs considerably from that given, you should check your calculation by going to the Online Resource Centre. We would expect it to take you about three hours to complete these worked examples and questions. The answers for these exercises are at the end of the chapter. Most of the examples are based on real undergraduate research projects. If these examples are not in your subject area you will find more in the Online Resource Centre.

 The impact of consuming 1.5 units of alcohol was tested on reaction times (seconds) in six men and ten women. What information do you know from this brief description that can help you identify which might be the most suitable statistical test to use? What more do you need to find out before you can make a definite decision about which test to use?

11.1 Mann–Whitney *U* test

We have already introduced you to the idea of ranking observations (Box 5.3) and applied this idea in Chapter 9 when using Spearman's rank test. The Mann–Whitney *U* test uses the rank value of observations and the sum of these ranks for each sample to establish whether the two samples are significantly different. This test is the non-parametric equivalent of the two-sample *t*- (10.2, 10.3) and *z*-tests (10.1) for unmatched data.

We will illustrate the Mann–Whitney *U* test using Example 10.2: The evolution of *Littorina littoralis* at Porthcawl, where we know there is some doubt as to whether the data meet all the criteria for a parametric test (10.2.2ii). In this example, investigators used a systematic random sampling method to collect periwinkles from the mid- and lower shore and measured their shell height (mm). The data from this example are reproduced in Table 11.1. The investigators wished to test the hypothesis that there was a difference between these groups of periwinkles as had been seen in an earlier study carried out at Aberystwyth.

Example 10.2 Height of shells in two putative species of *Littorina* (periwinkles) from the mid- and lower shore at Porthcawl, 2002

Table 11.1 The height of *Littorina* (periwinkle) shells (mm) at two locations on the shore

There is one treatment variable: 'location on the shore', with two categories. The two categories are nominal and neither is a control. The dependent variable is shell height which is measured on an interval scale and the observations can therefore be ranked.

Periwinkles on the lower shore		Periwinkles on the mid-shore	
5.5	4.0	3.3	6.7
8.4	5.0	6.3	5.7
5.0	6.2	6.1	4.2
5.0	5.0	8.0	6.3
5.6	7.7	13.5	6.0
4.8	6.0	5.3	7.2
8.4		6.7	
$n_1 = 13$		$n_2 = 13$	
Median = 5.5		Median = 6.3	
Mode = 5.0		Mode = 6.3; 6.7	
Range = 4.4		Range = 10.5	
Interquartile range = 1.2		Interquartile range = 1.0	

Each periwinkle has only been included once and these are replicates. There are more than five observations in each category. There is one outlier in this sample (13.5mm).

11.1.1 Key trends and experimental design

These data are now being treated as non-parametric so we have added non-parametric descriptive statistics to the table. Clearly the medians are not the same for these two samples; the mid-shore periwinkles appear to be larger than the periwinkles from the lower shore. The outlier which we referred to in Chapter 10 extends the range for the mid-range sample but when the interquartile range is used this spread is seen to be not that different to the lower shore sample. We will include the outlier at this point but would normally expect you to carry out your analysis with the outlier included and then excluded to assess what affect this outlier has.

11.1.2 Using this test

i. To use this test you:

1) Do not have an *a priori* reason for expecting certain outcomes from your investigation.

2) Usually present these data in tables or summary histograms rather than a scatter plot.

3) Wish to examine the effect of one treatment variable with two categories. These categories are usually nominal or ordinal.

4) Have observations (the dependent variable) which are independent and unmatched and measured on an interval or ordinal scale. These observations do not need to be drawn from a normally distributed population.

5) Have data that can be ranked (5.1).

6) Have two samples that both have a similar-shaped distribution. For example, if one distribution is skewed to the left and the other to the right, then you should not use this test. If this does arise you could try transforming the data.

7) Do not need equal sample sizes.

8) Should not use this test if one sample has only one observation or if both samples have fewer than five observations each.

9) Can use the Mann–Whitney U test when n for each sample is less than 20. If n is greater than 20 you can use a modified z-test but only if there are few tied ranks. We have included a brief outline of this z-test in the text below Box 11.1. When $n > 20$ and there are many tied ranks you should either refer to other texts which explain how to modify the calculation to allow for the tied ranks (e.g. Zar, 2009) or use statistical software which should make this modification for you.

ii. Does the example meet these criteria?

This data set meets all the criteria for using a Mann–Whitney U test as there is one variable (periwinkles) and two samples (mid- and lower shore periwinkles). The data are unmatched as each periwinkle was measured only once. The data can be ranked and both samples have more than five observations. There are a few tied ranks. For example in the sample from the lower shore there are four observations of 5.0mm which all therefore have the same rank (Table 11.2). However, as both n values are less than 20 this does not matter. There is a very slight difference in skew but this is negligible (Figs 11.1 and 11.2).

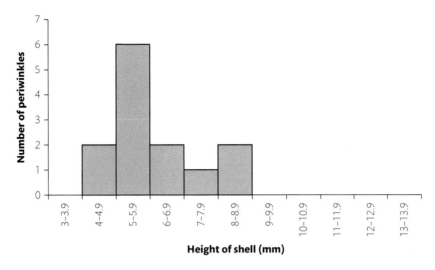

Fig. 11.1 The distribution of heights of *Littorina* (periwinkle) shells (mm) on the lower shore of Porthcawl in 2002

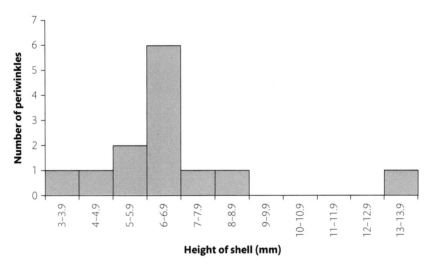

Fig. 11.2 The distribution of heights of *Littorina* (periwinkle) shells (mm) on the mid-shore of Porthcawl in 2002

11.1.3 The calculation

We show you the general process and a specific example of a Mann–Whitney *U* test in Box 11.1. In addition, you will need to refer to the calculation table (Table 11.2). If your sample sizes are large (*n* > 20) then *U* approximates to a normal distribution and an estimate of *z* may be calculated (see text after Box 11.1).

Table 11.2 Calculation of ranks for the height of shells in two groups of periwinkles from the mid- and lower shore at Porthcawl, 2002

The height of *Littorina* (periwinkle) shells (mm) at two locations on the shore			
Periwinkles from the lower shore		Periwinkles from the mid-shore	
Observations in numerical order	Rank	Observations in numerical order	Rank
		3.3	1.0
4.0	2.0		
		4.2	3.0
4.8	4.0		
5.0	6.5		
5.0	6.5		
5.0	6.5		
5.0	6.5		
		5.3	9.0
5.5	10.0		
5.6	11.0		
		5.7	12.0
		6.0	13.5
6.0	13.5		
		6.1	15.0
6.2	16.0		
		6.3	17.5
		6.3	17.5
		6.7	19.5
		6.7	19.5
		7.2	21.0
7.7	22.0		
		8.0	23.0
8.4	24.5		
8.4	24.5		
		13.5	26.0
	$R_1 = 153.5$		$R_2 = 197.5$

BOX 11.1 How to carry out a Mann–Whitney *U* test

GENERAL DETAILS	EXAMPLE 10.2
	This calculation is given in full in the Online Resource Centre. For presentation purposes only all values have been rounded to five decimal places.
1. Hypotheses to be tested H_0: There is no difference between the median values in sample 1 and sample 2. H_1: There is a difference between the median values in sample 1 and sample 2.	**1. Hypotheses to be tested** H_0: There is no difference between the median shell heights (mm) of the periwinkles from the mid- and lower shore at Porthcawl. H_1: There is a difference between the median shell heights (mm) of periwinkles from the mid- and lower shore at Porthcawl.
2. Have the criteria for using this test been met?	**2. Have the criteria for using this test been met?** Yes (11.1.2ii).
3. To work out $U_{calculated}$ i. Decide which of the samples is sample 1 and which sample 2. ii. Combine all observations (sample 1 and sample 2) and arrange in numerical order. iii. Assign ranks to these values. Where there are tied values, take the middle (average) rank for these values (Box 5.3). iv. Add up the ranks $(\sum r_1 = R_1)$ for sample 1. Repeat for sample 2 $(\sum r_2 = R_2)$. v. Record n_1, the number of observations in sample 1, and n_2, the number of observations in sample 2. vi. You now calculate two *U* terms, U_1 and U_2. Watch the formulae closely. A common error is to use R_1 when calculating U_1. $$U_1 = n_1 n_2 + \frac{n_2(n_2+1)}{2} - R_2$$ $$U_2 = n_1 n_2 + \frac{n_1(n_1+1)}{2} - R_1$$ vii. Examine the two *U* values. The smaller *U* value is $U_{calculated}$. viii. You can check your maths at this point since $U_1 + U_2 = n_1 \times n_2$.	**3. To work out $U_{calculated}$** i. Let the lower shore periwinkles be sample 1 and the mid-shore periwinkles be sample 2. ii.–iv. See Table 11.2. $R_1 = 153.5$ and $R_2 = 197.5$ v. In this example: $n_1 = 13, n_2 = 13$ vi. Taking each calculation in stages: $n_1 n_2 = 13 \times 13 = 169$ $n_2(n_2+1) = 13(13+1) = 13 \times 14 = 182$ Therefore: $$U_1 = 169 + \frac{182}{2} - 197.5 = 62.5$$ $$U_2 = 169 + \frac{182}{2} - 153.5 = 106.5$$ vii. Therefore $U_{calculated} = 62.5$ viii. $(62.5 + 106.5) = (13 \times 13)$ correct.

(Continued)

4. To find $U_{critical}$ when $n \leq 20$	**4. To find $U_{critical}$ when $n \leq 20$**
See Appendix D, Table D12. To find $U_{critical}$ using the table of critical values, you need to know n_1 and n_2. Where the n_1 row intersects the n_2 column for $p = 0.05$, this is the critical value. This is a two-tailed test.	$n_1 = n_2 = 13$ and at $p = 0.05$, $U_{critical} = 45.0$
5. The rule	**5. The rule**
If the calculated value of U is less than the critical value of U, then you may reject the null hypothesis (H_0).	In our example, $U_{calculated}$ (62.5) is greater than $U_{critical}$ (45.0) and therefore you do not reject the null hypothesis.
6. What does this mean in real terms?	**6. What does this mean in real terms?**
	There is no significant difference ($U = 62.5$, $p = 0.05$) between the median shell height (mm) of the periwinkles from the mid- and lower shore at Porthcawl. The results from this investigation do not support the notion that sympatric speciation is occurring. This concurs with the analysis of the data using a t-test (Boxes 10.3 and 10.4).

Further examples relating to the topic including how to use statistical software are included in the Online Resource Centre.

As the sample sizes increase the U tends to be normally distributed so you may use a modified z-test to test the significance of the U value. There are a number of methods that have been proposed and we report the one most frequently encountered but slightly different versions may be used in statistical software. This z-test can be further modified when you have tied ranks. The ranks and U are calculated as previously.

For a large sample ($n > 20$) and no tied ranks

$$z_{calculated} = \left| \frac{U - mean\ of\ U}{\sqrt{variance\ of\ U}} \right|.$$

Where the mean of $U = \dfrac{n_1 n_2}{2}$ and the variance of $U = \dfrac{n_1 n_2 (n_1 + n_2 + 1)}{12}$

$z_{critical}$ can be found in Appendix D, Table D5 for a p value such as $p = 0.05$.

This modified z-test does not work well if there are tied ranks and a further modification is then needed.

For a large sample ($n > 20$) and tied ranks

A correction factor must first be calculated followed by a revised variance of U.

$$CF = \sum (t^3 - t\ )$$

We will use the observations in the table below to explain how to calculate this correction factor. Firstly you would pool all the observations in your experiment from all treatments and place these in numerical order. The ranks are then assigned as usual as this example shows.

Observations	1	2	2	2	3	4	5	5	6
Rank	1	3	3	3	5	6	7.5	7.5	9
Observations	7	7	7	8	9	11	13	13	14
Rank	11	11	11	13	14	15	16.5	16.5	18
Observations	16	16	18	19	21	22	22	24	26
Rank	19.5	19.5	21	22	23	24.5	24.5	26	27

Six numbers (2, 5, 7, 13, 16, and 22) have tied ranks. There are three number 2s so t for this value is 3. There are two number 5s so t for number 5 is 2.

When $t = 3$, $t^3 - t = (3^3 - 3) = 27 - 3 = 24$

When $t = 2$, $t^3 - t = (2^3 - 2) = 8 - 2 = 6$

The t values are calculated for each group of ties and then added together so that the correction factor $CF = \sum (t^3 - t) = 24 + 6 + 24 + 6 + 6 + 6 = 72$.

The revised variance of U is $\left(\dfrac{n_1 n_2}{N^2 - N} \right) \left(\dfrac{N^3 - N - CF}{12} \right)$.

When $N = n_1 + n_2$

The z calculation is now completed as previously.

11.2 Wilcoxon's signed ranks test for matched pairs test

Occasionally you may design an investigation in which you observe an item before and after a particular event or treatment. This type of data is called matched or repeated measures and has to be handled in a different way to that described in 11.1, because the 'before' measure is not independent of the 'after' measure. In Example 10.2, none of the periwinkles was measured more than once. However, in Example 11.1, each item is observed twice (before and after the treatment). In Wilcoxon's signed ranks test, this lack of independence is recognized and allowed for. For this test the differences in the ranks for each pair are calculated. If the total positive difference is similar to the total negative difference then this indicates that these two samples are from the same statistical population. This test is the non-parametric equivalent of the t- and z-tests for matched data (10.4).

Example 11.1 Enjoyment of consuming chocolate at two times in the day

A small randomized study was carried out in 2015 where undergraduates were asked to eat a particular chocolate bar at 7 a.m. and at 6 p.m. On each occasion, the students were asked to rate their enjoyment on a scale of 1 (low) to 10 (high). The first three columns of Table 11.3 show the results.

Table 11.3 Enjoyment rating for consumption of chocolate bars at two times in the day with calculations for the Wilcoxon's signed ranks test

There are seven students and these are the items and the replicates. Each item has been measured twice for the one treatment variable so these data are not independent but are repeated measures.

There is one treatment variable: 'time of day', and two categories. The two categories are nominal and neither is a control. The dependent variable is the rating of the enjoyment of eating chocolate which is measured on an ordinal scale and can be ranked.

The 'enjoyment' rating of eating chocolate at two different points in one day

Student	Enjoyment rating at 7 a.m.	Enjoyment rating at 6 p.m.	Difference (d)	Ranks for absolute d values	Signed rank for d
1	2.0	10.0	−8.0	6	−6
2	0.0	6.0	−6.0	2	−2
3	1.0	7.5	−6.5	3	−3
4	8.0	1.0	7.0	4	4
5	1.0	1.0	0.0		
6	0.0	7.5	−7.5	5	−5
7	1.5	7.0	−5.5	1	−1
8	1.0	9.5	−8.5	7	−7
9	0.0	9.0	−9.0	8	−8
Median	1.0	7.5	7.25		
Range	8	9			
Interquartile range	1.5	3.0			

11.2.1 Key trends and experimental design

From Table 11.3 we can see that most people prefer to eat chocolate in the evening, one person preferred the chocolate in the morning, and one person did not like chocolate. The two medians reflect this difference in preference (median = 1.0 enjoyment score in the morning and median = 7.0 enjoyment score in the evening). However, the range and interquartile range demonstrate that the students' responses in the evening were more variable than in the morning. This design is very limited and the number of replicates is surprisingly small given that the study involved eating chocolate.

11.2.2 Using this test

i. To use this test you:

1) Do not have an *a priori* reason for expecting certain outcomes from your investigation.

2) Usually present these data in tables or summary histograms rather than a scatter plot.

3) Wish to examine the effect of one treatment variable with two categories. These categories are usually nominal or ordinal.

4) Have two observations (the dependent variable) for each item measured for the one treatment variable. These data are therefore matched. Each pair must be independent of any other pair.

5) Can confirm that the differences between the observations for each item are continuous and rankable. This usually means that the dependent variable must be measured on an interval scale though observations measured on some ordinal scales will also meet this criterion. (If your design does not meet this criterion you should consider the Cochran *q* test.) The observations do not need to be drawn from a normally distributed population.

6) Exclude the zero differences between the two observations for each item and then confirm that the remaining differences have a symmetrical distribution. (If the distribution is skewed try a logarithmic transformation of the original observations. The differences between the transformed observations are then used in the calculation.)

7) Have at least six pairs of observations and fewer than 20 pairs* where the difference between the observations is greater than zero. If you have only six pairs of these observations all differences will have to be either positive or negative for a significant difference to be detected.

8) Have no tied ranks.*

*If $n > 20$ and/or you have tied ranks there are modified z-tests that can be calculated. We have included these modifications in the text below Box 11.2.

ii. Does the example meet these criteria?

The data from Example 11.1 meet the criteria for using a Wilcoxon's matched pairs test in that there is one variable (time of day) and two samples (7 a.m. and 6 p.m.). Two measurements are made for each student. These two measurements are not independent of each other and the data are 'matched'. The differences can be ranked and there are no tied ranks. There are eight pairs of observations where the difference is greater than zero. The distribution of the differences, although platykurtic, is symmetrical.

11.2.3 The calculation

In this test, the calculation is dependent on the difference between each pair of observations and the sum of these differences. The general principles and a worked example are shown in Box 11.2. Some steps in the calculation are included in Table 11.3, columns 4–6.

BOX 11.2 How to carry out Wilcoxon's matched pairs test

GENERAL DETAILS	EXAMPLE 11.1
	This calculation is given in full in the Online Resource Centre. All values have been rounded to five decimal places.
1. Hypotheses to be tested H_0: There is no difference between the median of sample 1 and the median of sample 2. H_1: There is a difference between the median of sample 1 and the median of sample 2.	**1. Hypotheses to be tested** H_0: In a small group of undergraduates the median 'enjoyment scores' of the two times of day are equal. H_1: In a small group of undergraduates the median 'enjoyment scores' of the two times of day are not equal.
2. Have the criteria for using this test been met?	**2. Have the criteria for using this test been met?** Yes (11.2.2ii).
3. To work out $T_{calculated}$ i. Decide which is sample 1 and which is sample 2.	**3. To work out $T_{calculated}$** i. Let the enjoyment scores recorded at 7 a.m. be sample 1 and the enjoyment scores recorded at 6 p.m. be sample 2.
ii. Calculate the difference (d) between each pair of observations. Where the observation for sample 2 is larger than the observation for sample 1, the difference will be negative.	ii. See Table 11.3, column 4.
iii. Ignoring any d values where $d = 0$, rank the absolute d values. When working with absolute values you ignore the negative signs and treat all values as positive. If you are not familiar with ranking values, see Box 5.3.	iii. See Table 11.3, column 5. There is one $d = 0$ which we ignore. $d = |-5.5|$ is the lowest value and has the rank 1, etc.
iv. Assign to each rank the positive or negative signs from the d value.	iv. See Table 11.3, column 6.
v. Sum all the ranks that are negative in sign. The absolute values are used.	v. The sum of negative ranks (T^-) $= 6+2+3+5+1+7+8 = 32$
vi. Sum all the ranks that are positive in sign.	vi. The sum of positive ranks (T^+) $= 4$
vii. Examine the two values. For a two-tailed test the smaller value is the calculated value of T.	vii. $T_{calculated} = 4$

(Continued)

4. To find $T_{critical}$	4. To find $T_{critical}$
See Appendix D, Table D13. This is a two-tailed test. Use a T table of critical values where N is the number of pairs of observations used to provide ranks (i.e. not those where $d = 0$).	For this example, one of the differences is zero, therefore $N = 8$ and $T_{critical} = 3$ at $p = 0.05$.
5. The rule	5. The rule
When $T_{calculated}$ is equal to or less than $T_{critical}$, then you may reject the null hypothesis (H_0).	In this example, $T_{calculated}$ (4) is greater than $T_{critical}$ (3) and therefore you do not reject the null hypothesis.
6. What does this mean in real terms?	6. What does this mean in real terms?
	There is no significant difference ($T = 4$, $p = 0.05$) between the median chocolate enjoyment scores at 7 a.m. compared with 6 p.m. in a small group of undergraduates.

Further examples relating to the topic including how to use statistical software are included in the Online Resource Centre.

When $n > 20$ this distribution moves towards the normal so a modified z-test may be used. There are a number of methods given to achieve this. We report the one most frequently encountered. This z-test can be further modified when you have tied ranks. Any differences that equal zero are not included in the calculation and the value of n is adjusted to the number of pairs that are included. The ranks and T are calculated as previously.

For a large sample ($n > 20$) and no tied ranks

$$z_{calculated} = \left| \frac{T - mean_T}{\sqrt{variance_T}} \right|$$

Where:

$$mean_T = \frac{n(n+1)}{4}$$

$$variance = \frac{n(n+1)(2n+1)}{24}$$

$z_{critical}$ is found in Appendix D, Table D13.

For a large sample ($n > 20$) and tied ranks

The variance is adjusted using a correction factor (CF). So that

$$CF = -0.5\left(\sum (t_j(t_j - 1)(t_j + 1))\right)$$

All the observations are pooled so that the ranks can be assigned to all the values. The observations are placed in numerical order and ranks assigned as this example shows, where six numbers (2, 5, 7, 13, 16, and 22) have tied ranks.

Observations	1	2	2	2	3	4	5	5	6
Rank	1	3	3	3	5	6	7.5	7.5	9
Observations	7	7	7	8	9	11	13	13	14
Rank	11	11	11	13	14	15	16.5	16.5	18
Observations	16	16	18	19	21	22	22	24	26
Rank	19.5	19.5	21	22	23	24.5	24.5	26	27

There are three number 2s so t for this value is 3. There are two number 5s so t for number 5 is 2. When $t = 3, (t\,(t-1)(t+1)) = 3(3-1)(3+1) = 3 \times 2 \times 4 = 24$
When $t = 2, (t\,(t-1)(t+1)) = 2(2-1)(2+1) = 2 \times 1 \times 3 = 6$
Complete these calculations for each tied group of values. The correction factor for this example would be:

$$CF = 0.5\left(\sum (t(t-1)(t+1))\right) = 0.5\left(\sum (24+6+24+6+6+6)\right) = 0.5 \times 72 = 36$$

This would then be subtracted from the variance and the z calculation completed as previously.

11.3 One-way non-parametric ANOVA (Kruskal–Wallis test)

The acronym ANOVA stands for analysis of variance. Whilst parametric ANOVAs are more powerful and flexible, it is common in science to find yourself dealing with data that do not fit the criteria for a parametric ANOVA and cannot be normalized by transforming them. Under these circumstances, non-parametric tests may be used. In these tests, you will be comparing the distributions of the samples by using ranks and sums of ranks. The one-way non-parametric ANOVA was first described by Kruskal and Wallis (1952) and is usually called the Kruskal–Wallis test.

The test statistic for a Kruskal–Wallis test is often called either the K or H statistic. The calculated value of K or H is determined using the following equation:

$$K_{calculated} = \left(\Sigma\left(\frac{R^2}{n_s}\right) \times \frac{12}{N(N+1)}\right) - 3(N+1)$$

R^2 is found by summing the rank values. When there is little variation in the data, there will be many values with the same rank. If n is relatively large, the result can then be a negative value for K or H. In these circumstances, there is insufficient variation in the data compared with the sample size to detect any difference between categories (samples).

The critical test statistic can be found in two tables: H (Appendix D, Table D14) and chi-squared (Appendix D, Table D1). H is the exact test statistic and a specific critical value can be found in relation to a (the number of categories or samples) and n (the number of observations in a category). Where there are equal replicates, n will be the same for all categories, but where there are unequal replicates, n will differ between categories and the H value will depend on all the n values. By contrast, the chi-squared critical value is found in relation to

p and the degrees of freedom. The latter is not dependent on sample size. With relatively large sample sizes, the critical values of H approximate to the critical values of chi-squared. We have included a table of critical H values (Appendix D, Table D14) for a design with equal replicates up to 25. Therefore, if n is less than 25 and there are equal numbers of observations in each category, we recommend using the H table. However, if you have unequal numbers of observations in each category or n is greater than 25, we suggest using the chi-squared distribution to find the critical value. An alternative approach in large samples is to test for normality and, if all criteria are met, use a parametric ANOVA. The Kruskal–Wallis test is the non-parametric equivalent of parametric one-way ANOVAs for equal (10.7) and unequal (10.9) replicates.

11.3.1 Key trends and experimental design

In Example 11.2, the medians allow us to order these areas in terms of their density of daisies from highest to lowest as Lawn > Quadrangle > Rugby pitch > Cricket pitch. The modes suggest that the distribution of densities is more platykurtic from the Cricket pitch and the Quadrangle. This is also evident by the variation around the median indicated by the range and interquartile range. In contrast the distribution of density of daisies on the Rugby pitch is more mesokurtic with one clear central point and a moderate spread of observations around this, whilst the density of daisies on the Lawn has a more leptokurtic distribution with one central point and less variation around this. There are only eight replicates in each location and there are no controls.

Example 11.2 The density of *Bellis perennis* (daisy plants) at four different locations on the University of Worcester campus, 2002

Students in their first year at the University of Worcester randomly sampled four locations on campus using a 0.5m² quadrat. The number of *Bellis perennis* in each of eight quadrats in the four areas was recorded (Table 11.4). Is there a significant difference between these four areas?

(Continued)

Table 11.4 The number of *Bellis perennis* growing at four locations on the campus of the University of Worcester

> There is one treatment variable: 'location' and four nominal categories. None of the categories is a control. The dependent variable is the number of daisies, which is an ordinal measure that can be ranked.

	Number of *B. perennis* at four locations on the University of Worcester campus			
	Cricket pitch	Lawn	Quadrangle	Rugby pitch
	8	15	10	15
	9	13	16	10
	9	15	18	15
	12	18	13	12
	4	11	12	8
	5	12	16	5
	5	13	8	10
	7	13	12	10
Median	7.5	13.0	12.5	10.0
Mode	5;9	13	12;16	10
Range	8	6	10	10
Interquartile range	4	2.25	4.5	3.25

> Each quadrat is an item. There are eight items and these are replicates. Each item has only been measured once and therefore the observations are independent of each other.

11.3.2 Using this test

i. To use this test you:

1) Have no *a priori* reason for expecting certain outcomes from your investigation.

2) Have data which are usually presented in tables or summary histograms rather than a scatter plot.

3) Wish to examine the effect of one treatment variable with more than two categories. The data table therefore has three or more columns of values.

4) Have categories for the treatment variable which are usually derived from an ordinal and/or nominal scale but could be measured on an interval scale and arranged into discrete categories.

5) Have a dependent variable which is measured on an interval or ordinal scale and observations can be ranked.

6) Do not need an underlying population distribution that is normal. This test may be used for parametric data but this is not recommended (10.6).

7) Do not need equal sample sizes.

8) Have an experimental design that means that each item is independent from all other items and has therefore only been measured once.

9) Must, if there are only three samples, have more than five observations per sample.

10) Can confirm that all groups have a similar-shaped distribution. If not you should consider using Welch's ANOVA.

11) May not use this test if you have little variation in a relatively large sample size (see at start of 11.3). Where some tied ranks are present, then a correction should be calculated (Box 11.3).

ii. Does the example meet these criteria?

 Do the data from Example 11.2 meet the criteria for a non-parametric one-way ANOVA (Kruskal–Wallis test)?

11.3.3 The calculation

The method for testing general hypotheses is outlined in Box 11.3 and the calculation table (Table 11.5). Statistical software can return slightly different outcomes depending on whether and how tied ranks are corrected for.

BOX 11.3 How to carry out a one-way non-parametric ANOVA (Kruskal–Wallis test)

GENERAL DETAILS	EXAMPLE 11.2
	This calculation is given in full in the Online Resource Centre. For presentation purposes only all values have been rounded to five decimal places.
1. Hypotheses to be tested H_0: There is no difference between the medians of the samples. H_1: There is a difference between the medians of the samples.	1. Hypotheses to be tested H_0: There is no difference between the median density of *Bellis perennis* at the four locations (cricket pitch, lawn, rugby pitch, and quadrangle). H_1: There is a difference in the median density of *Bellis perennis* at the four locations (cricket pitch, lawn, rugby pitch, and quadrangle).
2. Have the criteria for using this test been met?	2. Have the criteria for using this test been met? Yes (A2, at end of chapter).

(Continued)

3. To work out $K_{calculated}$

i. Combine all the observations and arrange them in numerical order. Assign the appropriate rank to each observation.

ii. Sum the ranks for each sample: $\sum r = R_s$

iii. Record n_s, the number of observations in each sample.

iv. Square each R value (R^2). Divide each R^2 by its n_s value: R^2/n_s

v. Add all the R^2/n_s values together: $\sum(R^2/n_s)$

vi. Add all the n_s values together: $\sum n_s = N$

vii. The test statistic K is calculated as:

$$K_{calculated} = \left(\sum\left(\frac{R^2}{n_s}\right) \times \frac{12}{N(N+1)} \right) - 3(N+1)$$

(If you have a negative K value at this point, see explanation in 11.3.)

3. To work out $K_{calculated}$

i–iv. See Table 11.5.

v. $\sum\dfrac{R^2}{n_s} = 392 + 4278.125 + 3549.0313 + 1755.2813$

$\qquad = 9974.4376$

vi. $\sum n_s = N = 8 + 8 + 8 + 8 = 32$

vii. $K_{calculated} = \left(9974.4376 \times \dfrac{12}{32(32+1)} \right) - 3(32+1)$

$\qquad = 14.345882$

---

4. When there are tied ranks

If there are no tied ranks in your data, you may proceed to the next step (5). When there are tied ranks, a correction may be applied.

i. First record the numbers of each tied rank value (t). Then calculate $T^{kw} = \sum(t^3 - t)$. Each of these calculations is straightforward but tedious so we have calculated a number of these for you:

t	2	3	4	5
$(t^3 - t)$	6	24	60	120
t	6	7	8	9
$(t^3 - t)$	210	336	504	720

ii. The correction factor is

$$C = 1 - \left(\frac{T^{kw}}{(N^3 - N)} \right)$$

where N is the total number of observations.

iii. The corrected K value (K_C) is the ratio $K_C = K/C$ When a corrected K value has been calculated, this is then the K value used in the hypothesis testing.

4. When there are tied ranks

i. For our example the numbers of each tied rank (t) are:

Density value with tied rank	The rank for this value	How many observations with this rank (t)
5	3	3
8	7	3
9	9.5	2
10	12.5	4
12	18	5
13	22.5	4
15	26.5	4
16	29.5	2
18	31.5	2

Then calculated $T^{kw} = \sum(t^3 - t)$.
$T^{kw} = (3^3 - 3) + (3^3 - 3) + (2^3 - 2) + (4^3 - 4) + (5^3 - 5)$
$+ (4^3 - 4) + (4^3 - 4) + (2^3 - 2) + (2^3 - 2)$
$= 24 + 24 + 6 + 60 + 120 + 60 + 60 + 6 + 6 = 366$

ii. The correction factor for this example is:

$$C = 1 - \left(\frac{366}{(32^3 - 32)} \right) = 1 - \left(\frac{366}{32768 - 32} \right)$$

$$= 1 - \left(\frac{366}{32736} \right) = 1 - 0.01118 = 0.98882$$

iii. $K_C = 14.34588 / 0.98882 = 14.50809$

5. To find $K_{critical}$

$K_{critical}$ may be found in one of two tables depending on sample size and whether there are equal numbers of observations in each category.

i. If n is less than 25 and there are equal numbers of observations in each category, use Appendix D, Table D14.

Locate $H_{critical}$ for a, the number of categories (samples), and n, the number of observations in each category. This is a two-tailed test.

ii. If you have unequal numbers of observations in each category or n is greater than 25, use Appendix D, Table D1. Using a table of critical values for chi-squared and $p = 0.05$. The degrees of freedom (v) = number of categories (samples) −1. This is a two-tailed test.

5. To find $K_{critical}$

i. In our example, there are four categories (cricket pitch, lawn, rugby pitch, and quadrangle), so $a = 4$. There are eight observations in each category so we may use the H table of critical values.
$H_{critical} = 7.534$ at $p = 0.05$

ii. If we wished to use the chi-squared table to find our critical value, then $v = 4 − 1 = 3$ and chi-squared at $p = 0.05 = 7.81$ Clearly there is little difference between these two values, although the chi-squared value tends to be higher and therefore more conservative when used in the hypothesis testing (i.e. some significant differences will not be detected when using the chi-squared value compared with the H value).

6. The rule

If the calculated value of K is greater than the critical value of H or chi-squared, then you may reject the null hypothesis.

6. The rule

$K_{calculated}$ (14.508) is greater than $H_{critical}$ (7.534) and we may therefore reject the null hypothesis. In fact, at $p = 0.01$, $H_{critical} = 10.42$. Therefore, we can reject the null hypothesis at $p = 0.01$.

7. What does this mean in real terms?

7. What does this mean in real terms?

There is a highly significant difference ($K = 14.508$, $p < 0.01$) between the median densities of *Bellis perennis* at the four locations (cricket pitch, lawn, rugby pitch, and quadrangle).

Further examples relating to the topic including how to use statistical software are included in the Online Resource Centre.

11.4 *Post hoc* test following a non-parametric one-way ANOVA

The one-way non-parametric ANOVA tests the general hypothesis 'There is no difference between the categories'. If this is significant, it is useful to test more specific hypotheses. There are several ways to do this. In this section, we examine a *post hoc* test that allows the comparison of two of the categories at a time. This test finds the mean of the rank values and compares pairs of means using a Q statistic.

The Kruskal–Wallis test followed by a *post hoc* test like this is the non-parametric equivalent of the parametric one-way ANOVA and Tukey's test. There are alternative approaches to making more specific tests of hypotheses following a Kruskal–Wallis test. One proposed by Meddis (1984) (see also Barnard *et al.*, 2001) is included in the Online Resource Centre.

Table 11.5 Calculation for one-way non-parametric ANOVA (Kruskal–Wallis test) for Example 11.2: The density of *Bellis perennis* growing at four sites on the campus of the University of Worcester

Number of *B. perennis* at four locations on the University of Worcester campus							
Cricket pitch		Lawn		Quadrangle		Rugby pitch	
Number	Rank	Number	Rank	Number	Rank	Number	Rank
8	7.0	15	26.5	10	12.5	15	26.5
9	9.5	13	22.5	16	29.5	10	12.5
9	9.5	15	26.5	18	31.5	15	26.5
12	18.0	18	31.5	13	22.5	12	18.0
4	1.0	11	15.0	12	18.0	8	7.0
5	3.0	12	18.0	16	29.5	5	3.0
5	3.0	13	22.5	8	7.0	10	12.5
7	5.0	13	22.5	12	18.0	10	12.5
R	56.0		185.0		168.5		118.5
n	8		8		8		8
R^2	3136.00		34225.00		28392.25		14042.25
R^2/n	$R^2/n = \frac{3136}{8}$ $= 392.00$		4278.125		3549.0313		1755.2813
Means of ranks: R/n	7.00		23.12		21.0625		14.8125

11.4.1 Key trends and experimental design

To illustrate this *post hoc* test, we will use Example 11.2. The medians indicate that the greatest difference lies between the lawn (median = 13.0) and the cricket pitch (median = 7.5) (Table 11.4). A *post hoc* test will enable us to extend this interpretation to see which, if any, of the locations are statistically distinctive.

11.4.2 Using this test

i. To use this test you:

1) Meet the criteria for using a Kruskal–Wallis test and have a significant outcome from this test.

ii. Does the example meet these criteria?

To illustrate this *post hoc* test, we will use Example 11.2 and the Kruskal–Wallis test in Box 11.3. We know from 11.3.2ii that these data meet the criteria for a Kruskal–Wallis test, and the outcome from the analysis in Box 11.3 confirms a significant difference between these locations in the density of *Bellis perennis* ($K = 14.508$, $p < 0.01$).

11.4.3 **The calculation**

The method for making these *post hoc* comparisons after a Kruskal–Wallis test is outlined in Box 11.4 and the calculation table (Table 11.6).

BOX 11.4 **How to carry out a *post hoc* test after a Kruskal–Wallis test**

GENERAL DETAILS	EXAMPLE 11.2
	This calculation is given in full in the Online Resource Centre. For presentation purposes only all values have been rounded to five decimal places.
1. Hypotheses to be tested	**1. Hypotheses to be tested**
H_0: There is no difference between the means of the ranks of the samples. H_1: There is a difference between the means of the ranks of the samples.	As in Tukey's test (Chapter 10), you could write specific hypotheses for each comparison of means for example: H_0: There is no difference between the means of the ranks for the density of *Bellis perennis* on the rugby pitch compared with the quadrangle. H_1: There is a difference between the means of the ranks for the density of *Bellis perennis* on the rugby pitch compared with the quadrangle. However, common sense suggests that where you are making many comparisons using this test, then it is sensible to report the overall hypotheses only.
2. Have the criteria for using this test been met?	**2. Have the criteria for using this test been met?** Yes (11.3.2ii and 11.4.2ii).
3. To work out $Q_{calculated}$ i. Calculate the difference between each pairwise combination of means using the mean of the ranks.	**3. To work out $Q_{calculated}$** i. The mean of ranks for the cricket pitch is 7.00 (Table 11.5). The mean of ranks for the lawn is 23.12. The difference between these two values is 16.12. In total, six comparisons between means can be made. Table 11.6 gives all the differences between these pairs of means.
ii. For each pair of means, calculate the standard error of the difference of the means (SE). a. If there are few tied ranks, the calculation is: $$SE = \sqrt{\left(\left(\frac{N(N+1)}{12} \right) \times \left(\frac{1}{n_a} + \frac{1}{n_b} \right) \right)}$$ where N is the total number of observations, n_a is the number of observations in the first of the two categories being compared, and n_b is the number of observations in the second of the two categories being compared. If n is the same for all categories (samples), then the SE will be the same for all pairwise combinations and only needs to be calculated once.	ii. In our example there were tied ranks and from Box 11.3 $T^{kw} = 366$, $N = 8 + 8 + 8 + 8 = 32$; all n values for this example are the same as there are eight observations in each category so we only need to calculate the standard error of the difference once and the value will be the same for all subsequent comparisons of means. Therefore, using the adjusted equation $$SE = \sqrt{\left[\left(\left(\frac{32(32+1)}{12} \right) - \left(\frac{366}{12(32-1)} \right) \right) \times \left[\left(\frac{1}{8} \right) + \left(\frac{1}{8} \right) \right] \right]}$$ $$= \sqrt{\left[(88 - 0.98387) \times (0.125 + 0.125) \right]}$$ $$= \sqrt{\left[87.01613 \times 0.25 \right]} = \sqrt{21.75403}$$ $$= 4.66412$$

(Continued)

b. If there are tied ranks, then the calculation for the standard error is modified.

$$SE = \sqrt{\left[\left(\left(\frac{N(N+1)}{12}\right)-\left(\frac{T^{kw}}{12(N-1)}\right)\right)\times\left[\left(\frac{1}{n_a}\right)+\left(\frac{1}{n_b}\right)\right]\right]}$$

where N is the total number of observations, n_a is the number of observations in the first of the two categories being compared, n_b is the number of observations in the second of the two categories being compared, and T^{kw} is the value calculated during the Kruskal-Wallis test (Box 11.3), reflecting the numbers of tied values in the data.

Again, if n is the same for all categories (samples), then this adjusted SE will be the same for all pairwise combinations.

iii. Divide the difference in the means by its SE. This is the $Q_{calculated}$ value for the comparison between that particular pair of means.

iii. For the comparison between the means of the ranks from the cricket pitch and the lawn:

$$Q_{calculated} = 16.12 / 4.66412 = 3.45617$$

The $Q_{calculated}$ values for the other comparison are given in Table 11.6.

---

4. To find $Q_{critical}$

See Appendix D, Table D15. $Q_{critical}$ may be found using a, the number of categories or samples and p. This critical value for Q will be the same for each comparison of means.

4. To find $Q_{critical}$

In our example, there are four categories (cricket pitch, lawn, rugby pitch, and quadrangle), so $a = 4$. Therefore:

$Q_{critical} = 2.639$ at $p = 0.05$

---

5. The rule

If the calculated value of Q is greater than the critical value of Q, then you may reject the null hypothesis.

5. The rule

When comparing the means of ranks for the cricket pitch and the lawn. $Q_{calculated}$ (3.456) is greater than $Q_{critical}$ (2.639) and we may therefore reject the null hypothesis. In fact, at $p = 0.01$, $Q_{critical} = 3.144$. Therefore, we can reject the null hypothesis at this higher level of significance. At $p = 0.001$, $Q_{critical} = 3.765$, so the null hypothesis cannot be rejected at this level of significance.

All pairwise combinations have been tested against the $Q_{critical}$ value. The results are shown in Table 11.6.

---

6. What does this mean in real terms?

6. What does this mean in real terms?

There is a highly significant difference ($Q = 10.868$, $0.01 > p > 0.001$) between the means of the ranks of the densities of *Bellis perennis* on the cricket pitch and the lawn and the cricket pitch and the quadrangle ($0.05 > p > 0.01$). This suggests that the cricket pitch with very few daisies (median = 7.5 daisies) is the most distinctive location in this study.

Further examples relating to the topic including how to use statistical software are included in the Online Resource Centre.

Table 11.6 Calculation of differences between the means of ranking values ($\bar{x}_{ranks}$) and Q values for samples from Example 11.2: The density of *Bellis perennis* at four locations at the University of Worcester

Location		Differences between the means of ranking values ($\bar{x}_{ranks}$) at four locations at the University of Worcester			
		Rugby pitch $\bar{x}_{ranks} = 14.8125$	Quadrangle $\bar{x}_{ranks} = 21.0625$	Lawn $\bar{x}_{ranks} = 23.12$	Cricket pitch $\bar{x}_{ranks} = 7.00$
Rugby pitch $\bar{x}_{ranks} = 14.8125$	Difference between means		6.25	8.3075	7.8125
	Q		1.34002	1.78115	1.67502
Quadrangle $\bar{x}_{ranks} = 21.0625$	Difference between means			2.0575	14.0625
	Q			2.0575/4.66412 = 0.41134	14.0625/4.66412 = **3.01504***
Lawn $\bar{x}_{ranks} = 23.12$	Difference between means				16.12
	Q				16.12/4.66412 = **3.45617**
Cricket pitch $\bar{x}_{ranks} = 7.00$	Difference between means				
	Q				

*$p \leq 0.05$; **$p \leq 0.01$.

The outcome from a multiple comparisons test allows you to be more specific in your interpretation of the significant results found from a Kruskal–Wallis test. In our example, it would lead us to examine the microhabitat particularly of the cricket pitch to enable us to identify factors that may explain the observations. For *post hoc* tests the methods described in 10.8.4 and 12.1.7 can be used for reporting these results. Here, we have placed the significant values in Table 11.6 in bold and indicated the level of significance using the asterisk system.

11.5 Two-way non-parametric ANOVA

There are a number of important points that relate to two-way ANOVAs and you should read 10.11 before using this test as most of the points apply here. The test we cover in this section is an extension of the one-way ANOVA (Meddis, 1984; Barnard *et al.*, 2001), which allows you to examine the effect of two variables and their interaction. All these tests are carried out within one calculation so you reduce the likelihood of generating a Type I error.

This test uses the same equation for calculating K as that used in the one-way non-parametric ANOVA. Therefore, it is also possible to obtain a negative $K_{calculated}$ value when N is large compared with the variation in the data. If this occurs, you should consider using the Scheirer–Ray–Hare test (11.7).

Example 11.3 The effect of fertilizer treatment on the density of *Bellis perennis* (daisy) at two locations on the University of Worcester campus

At the University of Worcester, the lawn and cricket pitch were divided and half of each plot was treated with fertilizer. Students extended their original study (Example 11.2) and, using stratified random sampling with quadrats, recorded the numbers of *Bellis perennis* in these four locations (cricket pitch with fertilizer, cricket pitch without fertilizer, lawn with fertilizer, and lawn without fertilizer) (Table 11.7). The data are organized into blocks and these will be referred to as blocks A–D.

Table 11.7 The density of *Bellis perennis* (daisies) in two locations, with or without fertilizer treatments at the University of Worcester

The second treatment variable: 'fertilizer treatment' has two nominal categories. The 'no fertilizer' treatment can be considered to be a control. This is an orthogonal design with four blocks (A–D).

This is the first treatment variable: 'location' which has two nominal categories. None of the categories is a control. The dependent variable is the number of daisies which is an ordinal measure that can be ranked.

Fertilizer treatment	The density of *B. perennis* in each quadrat in two locations	
	Lawn	Cricket pitch
Fertilizer	8	15
	9 (A)	13 (B)
	9	15
	12	18
Median	9	15
Range	4	5
No fertilizer	4	11
	5	12
	5 (C)	13 (D)
	7	13
Median	5	12
Range	3	2

Each quadrat is an item. There are four items in each block, these are replicates. Each item has only been measured once and therefore the observations are independent of each other.

11.5.1 Key trends and experimental design

The four treatments can be arranged in descending order of density as:
'Cricket pitch + fertilizer' (median = 15 daisies) > 'Cricket pitch + no fertilizer' (median = 13 daisies) > Lawn + fertilizer (median = 9 daisies) > 'Lawn + no fertilizer' (median = 5 daisies).

This suggests that the density of *B. perennis* is dependent on the location but it is less clear if the fertilizer treatment is important. The range indicates that there is more variation in the results from the fertilizer treatment compared to the control.

11.5.2 **Using this test**

i. To use this test you:

1) Have no *a priori* reason for expecting certain outcomes from your investigation.

2) Have data which are usually presented in tables or summary histograms rather than a scatter plot.

3) Wish to examine the effect of two treatment variables.*

4) Have treatment variables* which are usually measured on an ordinal and/or nominal scale but could be measured on an interval scale and arranged into discrete categories.

5) Have a dependent variable that may be measured on an interval or ordinal scale and the observations can be ranked.

6) Do not need an underlying population distribution that is normal. This test may be used for parametric data but this is not recommended (10.6).

7) Can confirm that both treatment variables have two or more categories. The data table therefore has two or more columns and two or more rows; the design is orthogonal.

8) Have more than one observation in each block but each block has the same number of observations in it.

9) Have an experimental design that means that each item is only measured once—there are no matched values or repeated measures.

10) Can confirm that all groups have a similar-shaped distribution. If not you should consider using Welch's ANOVA.

11) May not use this test if you have little variation in a relatively large sample size. (Many tied values in your data may result in a negative $K_{calculated}$ value.) Where some tied values are present, a correction may be applied.

*If a confounding variable is being included then it is referred to as a treatment variable for the purposes of this analysis (10.11).

ii. **Does the example meet these criteria?**

The annotations on Table 11.7 confirm Example 11.3 meets the criteria for using this test. This is an orthogonal design as every treatment is compared against every other treatment. There are four equal replicates in each block. Each quadrat was only measured once and there are no repeated measures in this design. There are tied values and the calculation is therefore adjusted as shown in Boxes 11.5 A–C.

11.5.3 **The calculation**

As in the two-way parametric ANOVA for an orthogonal design, three pairs of hypotheses are tested: the differences between the medians due to treatment 1, the differences between the medians due to treatment 2, and the interaction between the two treatment variables.

The idea of an interaction is explained in Chapter 10 (10.11). The following set of boxes and calculation tables is arranged to take you through testing one pair of hypotheses at a time:

a. Is there a difference between the columns (i.e. locations)? Go to Box 11.5(A) and Table 11.8.

b. Is there a difference between the rows (i.e. fertilizer treatment)? Go to Box 11.5(B) and Table 11.9.

c. Is there an interaction between variable 1 and variable 2, i.e. is there an interaction between the locations and the effect of the fertilizer? Go to Box 11.5(C) and Table 11.10.

BOX 11.5A How to carry out a two-way non-parametric ANOVA test: columns

GENERAL DETAILS	EXAMPLE 11.3
	This calculation is given in full in the Online Resource Centre. For presentation purposes all values have been rounded to five decimal places.
1. Hypotheses to be tested H_0: There is no difference between the median values for samples given treatment 1 (columns). H_1: There is a difference between the median values for samples given treatment 1 (columns).	**1. Hypotheses to be tested** H_0: There is no difference between the median density of *Bellis perennis* on the lawn and on the cricket pitch. H_1: There is a difference between the median density of *Bellis perennis* on the lawn and on the cricket pitch.
2. Have the criteria for using this test been met?	**2. Have the criteria for using this test been met?** Yes (11.5.2.ii).
3. To work out $K_{calculated}$ i. Combine all the observations in all the blocks and arrange them in numerical order. Assign the appropriate rank to each observation. ii. Sum the ranks for all values in column 2 (R_{c1}) and then all values for column 3. Repeat for each column of data (R_{c2}) etc. iii. Sum the number of observations for column 1 of treatment 1 (n_{c1}) and column 2 of treatment 1 (n_{c2}) etc. (Remember to test general hypotheses all n values should be the same.) iv. Using the column totals. For column 2, square the sum of ranks (R^2_{c1}) and divide by the number of observations in that column (R^2_{c1} / n_{c1}). In a similar way, calculate R^2_{c2} / n_{c2} for column 2. Continue for each column of data. Add together all these R^2/n values: $\Sigma(R^2/n)$ Add all the n_c values together (N).	**3. To work out $K_{calculated}$** i–iii. See Table 11.8. There are two columns, 'cricket pitch' and 'lawn', and eight observations in each column. iv. $$\frac{R^2_{c1}}{n_{cl}} = 175.78125$$ $$\frac{R^2_{c2}}{n_{c2}} = 1212.7813$$ $\Sigma(R^2 / n) = 175.78125 + 1212.7813 = 1388.5626$ $N = 8 + 8 = 16$

(Continued)

v. The test statistic K is calculated as:

$K_{\text{calculated columns}}$

$$= \left(\sum \left(\frac{R^2}{n_s} \right) \times \frac{12}{N(N+1)} \right) - 3(N+1)$$

v.

$K_{\text{calculated columns}}$

$$= \left[\left(1388.5626 \times \left(\frac{12}{16(16+1)} \right) \right) - 3(16+1) \right]$$

$$= 10.26012$$

4. Where there are tied ranks

If there are no tied ranks in your data, you may proceed to the next step (5). When there are tied ranks, a correction may be applied.

i. First record the number of each tied rank (t). Then calculate
$T^{kw} = \sum(t^3 - t)$.
Each of these calculations is straightforward but tedious so we have calculated a number of these for you:

t	2	3	4	5
$(t^3 - t)$	6	24	60	120
t	6	7	8	9
$(t^3 - t)$	210	336	504	720

4. Where there are tied ranks

i. For our example, the numbers of each tied rank (t) are:

Density value with tied rank	The rank for this value	How many observations with this rank (t)
5	2.5	2
9	6.5	2
12	9.5	2
13	12.0	3
15	14.5	2

Then calculate $T^{kw} = \sum(t^3 - t)$.

$$T^{kw} = (2^3 - 2) + (2^3 - 2) + (2^3 - 2)$$
$$+ (3^3 - 3) + (2^3 - 2)$$
$$= 6 + 6 + 6 + 24 + 6 = 48$$

ii. The correction factor is

$$C = 1 - \left(\frac{T^{kw}}{(N^3 - N)} \right)$$

where N is the total number of observations.

ii. The correction factor for this example is

$$C = 1 - \left(\frac{48}{(16^3 - 16)} \right) = 1 - \left(\frac{48}{4096 - 16} \right)$$

$$= 1 - \frac{48}{4080} = 1 - 0.01177 = 0.98824$$

iii. The corrected K value (K_C) is the ratio $K_{C\text{ columns}}$
$= K_{\text{columns}}/C$ (When a corrected K value has been calculated, this is then the K value used in the hypothesis testing.)

iii. $K_{C\text{ columns}} = 10.26012 / 0.98824 = 10.38226$

5. To find K_{critical}

See Appendix D, Table D1. Using a table of critical values for $\bar{x}^2$ and $p = 0.05$, the degrees of freedom (v) = number of columns − 1.

5. To find K_{critical}

As $v = 2 - 1 = 1$
$K_{\text{critical columns}} = 3.84$ at $p = 0.05$

6. The rule

If the calculated value of K is greater than the critical value of K, then you may reject the null hypothesis.

6. The rule

$K_{\text{calculated columns}}$ (10.38) is greater than K_{critical} (3.84). You may therefore reject the null hypothesis (H_0). In fact, at $p = 0.01$, $K_{\text{critical}} = 6.64$ and at $p = 0.001$, $K_{\text{critical}} = 10.83$. Therefore, we can reject the null hypothesis at $p = 0.01$, but not at $p = 0.001$.

7. What does this mean in real terms?

Now go on to Box 11.5(B).

7. What does this mean in real terms?

There is a highly significant difference ($K = 10.38$, $0.01 > p > 0.001$) in the median density of *Bellis perennis* on the lawn and on the cricket pitch.

Table 11.8 Calculation table for two-way non-parametric ANOVA test (columns) for Example 11.3: The density of *Bellis perennis* per quadrat in two locations (cricket pitch and lawn) at the University of Worcester

Fertilizer treatment	The density of *B. perennis* in each quadrat in two locations and their associated rank values			
	Cricket pitch		Lawn	
	Number	Rank	Number	Rank
Fertilizer	8	5.0	15	14.5
	9	6.5	13	12.0
	9	6.5	15	14.5
	12	9.5	18	16.0
No fertilizer	4	1.0	11	8.0
	5	2.5	12	9.5
	5	2.5	13	12.0
	7	4.0	13	12.0
Totals for the columns:				
n		8		8
R		37.5		98.5
R^2		1406.25		9702.25
R_c^2		175.78125		1212.7813
n_c				

BOX 11.5B How to carry out a two-way non-parametric ANOVA test: rows

GENERAL DETAILS	EXAMPLE 11.3
	This calculation is given in full in the Online Resource Centre. For presentation purposes only all values have been rounded to five decimal places.
1. Hypotheses to be tested H_0: There is no difference in the median values for samples in response to treatment 2 (rows). H_1: There is a difference in the median values for samples in response to treatment 2 (rows).	1. Hypotheses to be tested H_0: There is no difference in the median density of *Bellis perennis* between the areas treated with fertilizer and those not treated. H_1: There is a difference in the median density of *Bellis perennis* between the areas treated with fertilizer and those not treated.
2. Have the criteria for using this test been met?	2. Have the criteria for using this test been met? Yes (11.5.2ii).

(Continued)

3. To work out $K_{calculated}$

i. Sum the ranks for each row (R_{r1}) and (R_{r2}), etc.
ii. Sum the number of observations for a row (n_r).

iii. From the row totals.
   a. For row 1, square the sum of ranks (R^2_{r1}) and divide by the number of observations in that first row: $\left(\dfrac{R^2_{r1}}{n_{r1}}\right)$ In a similar way, calculate $\dfrac{R^2_{r2}}{n_{r2}}$ for row 2.
      Continue for each row of data.
   b. Add together all these R^2/n values: $\Sigma(R^2/n)$

   c. Add all the n_r values together (N).
iv. The test statistic K is calculated as before.

$$K_{calculated\ rows} = \left(\Sigma\left(\frac{R^2}{n}\right) \times \frac{12}{N(N+1)}\right) - 3(N+1)$$

3. To work out $K_{calculated}$

i, ii. See Table 11.9.
   In our example, there are two rows, 'fertilizer' and 'no fertilizer', each with eight observations in the row.

iii.
   a. $\dfrac{R^2_{r1}}{n_{r1}} = 892.53125$
      $\dfrac{R^2_{r2}}{n_{r2}} = 331.53125$

   b. $\Sigma\dfrac{R^2}{n} = 892.53125 + 331.53125$
      $= 1224.0625$
   c. $\Sigma n_r = N = 8 + 8 = 16$
iv. $K_{calculated\ rows}$
   $= \left(1224.0625 \times \dfrac{12}{16(16+1)}\right) - 3(16+1)$
   $= 3.00260$

4. Where there are tied ranks

A correction may be used where there are tied ranks as outlined in Box 11.5(A). The correction factor (C) will be the same and $K_{C\ row} = K_{rows} / C$

4. Where there are tied ranks

From Box 11.5(A), $C = 0.98824$.
$K_{C\ rows} = 3.00260 / 0.98824 = 3.03833$

5. To find $K_{critical}$

See Appendix D, Table D1. Using a table of critical values for χ^2 and $p = 0.05$, the degrees of freedom (v) = number of rows − 1

5. To find $K_{critical}$

As $v = 2 - 1 = 1$, $K_{critical\ rows} = 3.84$ at $p = 0.05$.

6. The rule

If the calculated value of K is greater than the critical value of K, then you may reject the null hypothesis (H_0).

6. The rule

$K_{calculated\ rows}$ (3.04) is less than $K_{critical}$ (3.84), so you may not reject the null hypothesis (H_0).

7. What does this mean in real terms?

7. What does this mean in real terms?

There is no significant difference ($K = 3.04$, $p = 0.05$) in the median density of *Bellis perennis* in areas treated by fertilizer and those not treated.

Now go on to Box 11.5(C).

Table 11.9 Calculation table for two-way non-parametric ANOVA test (rows) for Example 11.3: The number of *Bellis perennis* per quadrat with or without fertilizer treatment at the University of Worcester

Fertilizer treatment	The density of *B. perennis* in each quadrat with and without a fertilizer treatment and their associated rank values					
	Cricket pitch		Lawn		Total for rows	
	Number	Rank	Number	Rank		
Fertilizer	8	5.0	15	14.5	n	8
	9	6.5	13	12.0	R	84.5
	9	6.5	15	14.5	R^2	7140.25
	12	9.5	18	16.0	$\dfrac{R_{r1}^2}{n_{r1}} = 892.53125$	
No fertilizer	4	1.0	11	8.0	n	8
	5	2.5	12	9.5	R	51.5
	5	2.5	13	12.0	R^2	2652.25
	7	4.0	13	12.0	$\dfrac{R_{r2}^2}{n_{r2}} = 331.53125$	

Table 11.10 Calculation table for a two-way non-parametric ANOVA test (interaction) for Example 11.3: The number of *Bellis perennis* per quadrat in two locations with or without fertilizer treatment at the University of Worcester

Fertilizer treatment	The density of *B. perennis* in each quadrat at two locations (Cricket pitch and Lawn) with and without a fertilizer treatment and their associated rank values			
	Cricket pitch		Lawn	
	Number	Rank	Number	Rank
	BLOCK A		BLOCK B	
With fertilizer	8	5.0	15	14.5
	9	6.5	13	12.0
	9	6.5	15	14.5
	12	9.5	18	16.0
	n	4	n	4
	R	27.5	R	57.0
	R^2	756.25	R^2	3249.00
	R^2/n	189.0625	R^2/n	812.2500
	BLOCK C		BLOCK D	
No fertilizer	4	1.0	11	8.0
	5	2.5	12	9.5
	5	2.5	13	12.0
	7	4.0	13	12.0
	n	4	n	4
	R	10.0	R	41.5
	R^2	100.00	R^2	1722.25
	R^2/n	25.0000	R^2/n	430.5625

BOX 11.5C How to carry out a two-way non-parametric ANOVA test: interaction

GENERAL DETAILS	EXAMPLE 11.3
	This calculation is given in full in the Online Resource Centre. For presentation purposes only all values have been rounded to five decimal places.
1. Hypotheses to be tested H_0: There is no difference between the median values of all samples due to an interaction between treatment variables 1 and 2. H_1: There is a difference between the median values of all samples due to an interaction between treatment variables 1 and 2.	**1. Hypotheses to be tested** H_0: There is no difference between the median density of *Bellis perennis* due to an interaction between location and fertilizer treatment. H_1: There is a difference between the median density of *Bellis perennis* due to an interaction between location and fertilizer treatment.
2. Have the criteria for using this test been met?	**2. Have the criteria for using this test been met?** Yes (11.5.2ii.).
3. To work out $K_{calculated}$ i. For each block, calculate R, R^2, n, and R^2 / n. ii. Add all the R^2 / n values together: $\Sigma\left(R^2 / n\right)$ iii. Add all the n values from the blocks: $\Sigma n = N$ iv. $K_{calculated\ total}$ is calculated as before. $$K_{calculated\ total} = \left(\Sigma\left(\frac{R^2}{n_s}\right) \times \frac{12}{N(N+1)}\right) - 3(N+1)$$ v. Where there are tied ranks, $K_{calculated\ total}$ may be corrected using the same correction term calculated in Box 11.5(A) and where $K_{C\ total} = K_{total} / C$ vi. You now have three values for K: $K_{columns}$ (Box 11.5A), K_{rows} (Box 11.5B), and K_{total} (step iv. or v. above). Using these values, calculate $K_{interaction}$: $K_{interaction} = K_{total} - K_{rows} - K_{columns}$	**3. To work out $K_{calculated}$** i. See Table 11.10. There are four blocks A–D, each with four observations. ii. $\Sigma\left(R^2 / n\right) = 189.0625 + 821.25 + 25.00$ $+ 430.5625 = 1456.875$ iii. $N = 4 + 4 + 4 + 4 = 16$ iv. $K_{calculated\ total} = \left(1456.875 \times \dfrac{12}{16(16+1)}\right) - 3(16+1)$ $= 13.27380$ v. From Box 11.5(A) $C = 0.98824$ So $K_{C\ total} = K_{total} / C = 13.27380 / 0.98824 = 13.43176$ vi. $K_{columns} = 10.38226$ $K_{rows} = 3.03833$ $K_{total} = 13.43176$ $K_{interaction} = 13.43176 - 3.03833 - 10.38226$ $= 0.01117$
4. To find $K_{critical}$ The degrees of freedom (v) are v for K_{total} = number of blocks − 1 v for K_{rows} = number of rows − 1 v for $K_{columns}$ = number of columns − 1 v for $K_{interaction} = v_{total} - v_{rows} - v_{columns}$ See Appendix D, Table D1. Use the v for $K_{interaction}$ and $p = 0.05$ to locate the critical value of K in a chi-squared table.	**4. To find $K_{critical}$** v for $K_{total} = 4 - 1 = 3$ v for $K_{fertilizer} = 2 - 1 = 1$ v for $K_{location} = 2 - 1 = 1$ v for $K_{interaction} = 3 - 1 - 1 = 1$ Therefore, the critical value for $K_{interaction}$ at $p = 0.05$ and $v = 1$ is 3.84.

(Continued)

5. The rule	5. The rule
If the calculated value of K is greater than the critical value of K, then you may reject the null hypothesis (H_0).	$K_{calculated\ interaction}$ (0.01) is less than $K_{critical}$ (3.84), so you may not reject the null hypothesis (H_0).
6. What does this mean in real terms?	6. What does this mean in real terms?
	There is no significant interaction ($K = 0.011$, $p = 0.05$) between the fertilizer treatment and location in the median density of *Bellis perennis* at the University of Worcester.

Further examples relating to the topic including how to use statistical software are included in the Online Resource Centre.

Our analysis of these data shows us that only the location is a significant factor in the density of daisies in this experiment. Neither the fertilizer treatment nor the interaction between fertilizer and location had a significant effect on the median density of daisies. However, this has only been a test of general hypotheses. To be able to draw more explicit conclusions about the trends in the data, we should consider using a *post hoc* test (11.6). This becomes particularly relevant in experiments with more than two categories for each treatment variable.

11.6 *Post hoc* test following a two-way non-parametric ANOVA

If a two-way non-parametric ANOVA has been carried out and there is a significant difference in at least one of the outcomes from Boxes 11.5(A–C), then you may wish to carry out a *post hoc* test to allow you to identify more specifically where the differences lie in your data. The *post hoc* test we outlined in 11.4 can be extended to cover the comparisons for a two-way non-parametric ANOVA. In this case, there are more pairwise comparisons that can be made as each block counts as one category to be compared with every other different category. This is the non-parametric equivalent to the two-way parametric ANOVA followed by Tukey's test.

In Example 11.3 (Boxes 11.5A–C), we can see that in fact only the location had a significant effect on the density of *Bellis perennis*. There are only two locations in this comparison, and we can see by looking at the data that the density of *Bellis perennis* is greater on the lawn than on the cricket pitch. There is therefore little reason for carrying out a formal *post hoc* test on

---

Example 11.4   Retrieval of pollen grains from different fabrics

In an investigation into forensic techniques, an undergraduate examined the effectiveness and reliability of pollen retrieval from a range of difference fabrics using the pollen from two species (Table 11.11). A two-way non-parametric ANOVA was carried out. This calculation is shown in the Online Resource Centre. The outcome of the analysis was that there was no significant difference as a result of the fabrics, nor was there a significant interaction between the two variables (fabrics and species). However, there was a highly significant difference between the species in the pollen retrieved from the fabric ($K = 23.27$, $p = 0.001$).

Table 11.11 The number of pollen grains retrieved from four different fabrics (denim, fleece, viscose, and polycotton) using pollen from two species (*Phleum pratense* and *Ambrosia artemisiifolia*).

> The second treatment variable: 'species of pollen' has two nominal categories. Neither is a control. This is an orthogonal design with eight blocks.

> This is the first treatment variable: 'fabrics' which has four nominal categories. None of the categories is a control. The dependent variable is the number of pollen grains which is an ordinal measure that can be ranked.

Species used as pollen sources	The number of pollen grains retrieved from four different fabrics			
	Denim	Fleece	Viscose	Polycotton
A. artemisiifolia	8998	6670	6551	4858
	3019	8012	7557	12,099
	3839	12,108	8311	4754
	3354	8097	3613	4555
Median	3596.5	8054.5	7054	4806
Range	5159	5438	4698	7544
P. pratense	2886	2316	1002	628
	402	1804	712	1104
	529	1018	392	509
	554	1513	506	698
Median	541.5	1658.5	609	663
Range	2584	1298	610	595

> Each piece of fabric that has been tested is an item. There are four items in each block, these are replicates. Each item has only been measured once and therefore the observations are independent of each other.

these data. We have therefore included Example 11.4 to illustrate the use of a *post hoc* test after a two-way non-parametric ANOVA.

11.6.1 Key trends and experimental design

It is clear that there is considerable variation in the retrieval of pollen grains from the fabrics. This is evident for all fabric types but in the test using *P. pratense* on denim there appears to be an outlier in one of the replicates. Although this is being included in the analysis at this point this does need to be considered. The medians do indicate that it is easier to retrieve *A. artemisiifolia* compared to *P. pratense* and this was confirmed by the significant two-way ANOVA. However, as there are four fabrics it is of interest to evaluate the data further and see if all fabrics contribute to this difference.

11.6.2 **Using this test**

i. To use this test you:

1) Meet the criteria for using a Kruskal–Wallis and have a significant outcome from this test.

ii. Does the example meet these criteria?

The data in this example meet the criteria for using a two-way non-parametric ANOVA in that we wish to test the effect of two variables (fabric and species). The design is orthogonal (every combination of one variable has been compared with the other variable). The data are non-parametric (numbers of pollen grains) and can be ranked. There are four observations

BOX 11.6 **How to carry out a *post hoc* test after a two-way non-parametric ANOVA**

GENERAL DETAILS	EXAMPLE 11.4
	This calculation is given in full in the Online Resource Centre. For presentation purposes only all values have been rounded to five decimal places.
1. Hypotheses to be tested H_0: There is no difference between the means of the ranks of the samples. H_1: There is a difference between the means of the ranks of the samples.	1. Hypotheses to be tested As in Tukey's test and in the *post hoc* test in 11.4, you could write specific hypotheses for each comparison of means, for example: H_0: There is no difference between the mean of the ranks for the number of pollen grains retrieved from *P. pratense* pollen on polycotton compared with *P. pratense* pollen on viscose. H_1: There is a difference between the mean of the ranks for the number of pollen grains retrieved from *P. pratense* pollen on polycotton compared with *P. pratense* pollen on viscose. However, common sense suggests that where you are making many comparisons using this test, then it is sensible to report the overall hypotheses only.
2. Have the criteria for using this test been met?	2. Have the criteria for using this test been met? Yes (11.6.2ii).
3. To work out $Q_{calculated}$ i. Combine all the observations across all blocks and assign ranks. For each block calculate the mean of the ranks. (This step is usually completed as part of the two-way non-parametric ANOVA.)	3. To work out $Q_{calculated}$ i. For example the mean of ranks for the *P. pratense* pollen on polycotton is 7.5. The mean of ranks for *P. pratense* pollen on viscose is 5.25 (Table 11.12).

(Continued)

ii. Calculate the difference between each pairwise combination of means using the mean of the ranks (R/n).

iii. For each pair of means, calculate the standard error of the difference of the means (SE).

   a. If there are few tied ranks the calculation is:

$$SE = \sqrt{\left[\left(\frac{N(N+1)}{12}\right) \times \left(\frac{1}{n_a} + \frac{1}{n_b}\right)\right]}$$

   where N is the total number of observations, n_a is the number of observations in the first of the two categories being compared, and n_b is the number of observations in the second of the two categories being compared. If n is the same for all categories (samples), then this value for SE will be the same for all pairwise combinations.

   b. If there are tied ranks then the calculation for the standard error is modified:

$$SE = \sqrt{\left[\left(\left(\frac{N(N+1)}{12}\right) - \left(\frac{T^{kw}}{12(N-1)}\right)\right) \times \left[\left(\frac{1}{n_a}\right) + \left(\frac{1}{n_b}\right)\right]\right]}$$

   where N is the total number of observations, n_a is the number of observations in the first of the two categories being compared, n_b is the number of observations in the second of the two categories being compared, and T^{kw} is the value calculated during the two-way non-parametric ANOVA (11.5) reflecting the numbers of tied values in the data. If n is the same for all categories (samples), then this value of SE will be the same for all pairwise combinations.

iv. Divide the difference in the means by its SE. This is the $Q_{calculated}$ value for the comparison between that particular pair of means.

ii. The difference for this pair is therefore 2.25 (Table 11.13).

In total, 28 comparisons between means can be made. Table 11.13 gives all the differences between these pairs of means.

iii. In our example, there are no tied ranks so we may complete the calculation in step iii (a). Where $N = 32$ and $n_{P.\ pratense\ polycotton} = 4$ and $n_{P.\ pratense\ viscose} = 4$
(All n values in this example are 4 as there are four observations in each category.)

$$SE = \sqrt{\left[\left(\frac{32(32+1)}{12}\right) \times \left(\frac{1}{4} + \frac{1}{4}\right)\right]}$$
$$= \sqrt{(88.0 \times 0.5)} = \sqrt{44.0}$$
$$= 6.63325$$

iv. For the comparison between the means of the ranks from the *P. pratense* polycotton and *P. pratense* viscose, $Q_{calculated} = 2.25 / 6.63325 = 0.33920$.
The $Q_{calculated}$ values for the other comparison are given in Table 11.13.

4. To find $Q_{critical}$

See Appendix D, Table D15. $Q_{critical}$ may be found using a, the number of categories or samples, and p. This critical value for Q will be the same for each comparison of means.

4. To find $Q_{critical}$

In our example there are eight categories (*P. pratense* polycotton, *P. pratense* viscose … *A. artemisiifolia* denim), $a = 8$.
$Q_{critical} = 3.124$ at $p = 0.05$.

5. The rule

If the calculated value of Q is greater than the critical value of Q, then you may reject the null hypothesis.

5. The rule

When comparing the means of ranks for the *P. pratense* pollen on polycotton and *P. pratense* pollen on viscose, $Q_{calculated}$ (0.339) is less than $Q_{critical}$ (3.124) so we do not reject the null hypothesis.

(Continued)

	If we examine the means of ranks for *A. artemisiifolia* fleece and *P. pratense* denim, then $Q_{calculated}$ (3.128) is just greater than $Q_{critical}$ (3.124) at $p = 0.05$ so we may reject the null hypothesis. There are only two significant outcomes from these pairwise comparisons (Table 11.13).
6. What does this mean in real terms?	6. What does this mean in real terms? There is a significant difference ($Q = 3.128$, $p = 0.05$) in recovery of pollen between *A. artemisiifolia* pollen on fleece and *P. pratense* pollen on denim, and between *A. artemisiifolia* pollen on fleece and *P. pratense* pollen on viscose ($Q = 3.354$, $p = 0.05$).

Further examples relating to the topic including how to use statistical software are included in the Online Resource Centre.

Table 11.12 Calculation of ranks and means of ranks for a *post hoc* test following a two-way non-parametric ANOVA on the number of pollen grains retrieved from four different fabrics (denim, fleece, viscose, and polycotton) using pollen from two species (*Phleum pratense* and *Ambrosia artemisiifolia*)

Species used as pollen sources	The number of pollen grains retrieved from four different fabrics							
	Denim		Fleece		Viscose		Polycotton	
	Number of pollen grains	Rank	Number of pollen grains	Rank	Number of pollen grains	Rank	Number of pollen grains	Rank
A. artemisiifolia	8998	30	6670	25	6551	24	4858	23
	3019	17	8012	27	7557	26	12,099	31
	3839	20	12,108	32	8311	29	4754	22
	3354	18	8097	28	3613	19	4555	21
$\Sigma r = R$		85		112		98		97
n		4		4		4		4
Mean of ranks (R/n)		85/4 = 21.25		28.00		24.50		24.25
P. pratense	2886	16	2316	15	1002	10	628	7
	402	2	1804	14	712	9	1104	11
	529	5	1018	12	392	1	509	4
	554	6	1513	13	506	3	698	8
$\Sigma r = R$		29		54		23		30
n		4		4		4		4
Mean of ranks (R/n)		7.25		13.5		5.75		7.5

Table 11.13 The difference between the means of ranks ($\bar{x}_{ranks}$) and the Q values for a *post hoc* test between the number of pollen grains retrieved from four fabrics (denim, fleece, viscose, and polycotton) using pollen from two species (*Phleum pratense* and *Ambrosia artemisiifolia*)

		Denim A. artemisiifolia $\bar{x}_{ranks} = 21.25$	Fleece A. artemisiifolia $\bar{x}_{ranks} = 28.00$	Viscose A. artemisiifolia $\bar{x}_{ranks} = 24.50$	Polycotton A. artemisiifolia $\bar{x}_{ranks} = 24.25$	Denim P. pratense $\bar{x}_{ranks} = 7.25$	Fleece P. pratense $\bar{x}_{ranks} = 13.5$	Viscose P. pratense $\bar{x}_{ranks} = 5.75$	Polycotton P. pratense $\bar{x}_{ranks} = 7.5$
Denim A. artemisiifolia $\bar{x}_{ranks} = 21.25$	Difference in means		28.0 – 21.25 = 6.75	24.5 – 21.25 = 3.25	24.25 – 21.25 = 3.0	21.25 – 7.25 = 14.0	21.25 – 13.5 = 7.75	21.25 – 5.75 = 15.5	21.25 – 7.5 = 13.75
	Q		6.75/6.63325 = 1.01760	3.25/6.63325 = 0.48996	3.0/6.63325 = 0.45227	14.0/6.63325 = 2.11058	7.75/6.63325 = 1.16836	15.5/6.63325 = 2.33671	13.75/6.63325 = 2.07289
Fleece A. artemisiifolia $\bar{x}_{ranks} = 28.00$	Difference in means			28.0 – 24.5 = 3.5	28.0 – 24.25 = 3.75	28.0 – 7.25 = 20.75	28.0 – 13.5 = 14.5	28.0 – 5.75 = 22.25	28.0 – 7.5 = 20.5
	Q			3.5/6.63325 = 0.52764	3.75/6.63325 = 0.56533	20.75/6.63325 = 3.12818*	14.5/6.63325 = 2.18596	22.25/6.63325 = 3.35431*	20.5/6.63325 = 3.09049
Viscose A. artemisiifolia $\bar{x}_{ranks} = 24.50$	Difference in mean				24.5 – 24.25 = 0.25	24.5 – 7.25 = 17.25	24.5 – 13.5 = 11.0	24.5 – 5.75 = 18.75	24.5 – 7.5 = 17.0
	Q				0.25/6.63325 = 0.03769	17.25/6.63325 = 2.60076	11.0/6.63325 = 1.65831	18.75/6.63325 = 2.82667	17.0/6.63325 = 2.56285
Polycotton A. artemisiifolia $\bar{x}_{ranks} = 24.25$	Difference in means					24.25 – 7.25 = 17.0	24.25 – 13.5 = 10.75	24.25 – 5.75 = 18.5	24.25 – 7.5 = 16.75
	Q					17.0/6.63325 = 2.56285	10.75/6.63325 = 1.62062	18.5/6.63325 = 2.78898	16.75/6.63325 = 2.52516
Denim P. pratense $\bar{x}_{ranks} = 7.25$	Difference in means						13.5 – 7.25 = 6.25	7.25 – 5.75 = 1.5	7.5 – 7.25 = 0.25
	Q						6.25/6.63325 = 0.94222	1.5/6.63325 = 0.22613	0.25/6.63325 = 0.03769

Table 11.13 *(Continued)*

		Denim	Fleece	Viscose	Polycotton	Denim	Fleece	Viscose	Polycotton
		A. artemisiifolia $\bar{x}_{ranks} = 21.25$	*A. artemisiifolia* $\bar{x}_{ranks} = 28.00$	*A. artemisiifolia* $\bar{x}_{ranks} = 24.50$	*A. artemisiifolia* $\bar{x}_{ranks} = 24.25$	*P. pratense* $\bar{x}_{ranks} = 7.25$	*P. pratense* $\bar{x}_{ranks} = 13.5$	*P. pratense* $\bar{x}_{ranks} = 5.75$	*P. pratense* $\bar{x}_{ranks} = 7.5$
Fleece *P. pratense* $\bar{x}_{ranks} = 13.50$	Difference in means							13.50 − 5.75 = 7.75	13.50 − 7.5 = 6.00
	Q							7.75/6.63325 = 1.16836	6.00/6.63325 = 0.94749
Viscose *P. pratense* $\bar{x}_{ranks} = 5.75$	Difference in means								7.5 − 5.75 = 1.75
	Q								1.75/6.63325 = 0.26382
Polycotton *P. pratense* $\bar{x}_{ranks} = 7.5$	Difference in means								
	Q								

*$p \leq 0.05$

362

in each category. There are no tied values so we do not expect to obtain a negative K value. There is a highly significant difference between the species in the pollen retrieved from the fabric ($K = 23.27, p = 0.001$).

11.6.3 The calculation

This *post hoc* test is outlined in Box 11.6 and the calculation tables 11.12 and 11.13.

As in our one-way *post hoc* test, the outcome allows our interpretation of the data to be more specific. In this example, although there was a significant difference between species in the retrieval of pollen, this was mainly due to a difference between the fleece material for *A. artemisiifolia* and the smoother materials (denim and viscose) for *P. pratense*. This information can now be used to develop further more specific experiments to examine the factors that are resulting in this difference in pollen retrieval.

There are a number of methods that can be used for reporting results from statistical tests. We review these in 10.8.4 and 12.1.7. Here, we have placed the significant values in Table 11.13 in bold and indicated the level of significance using the asterisk system.

11.7 Scheirer–Ray–Hare test

The ANOVAs we have described have a number of criteria for use. One of these relates to the numbers of tied ranks, especially in large sample sizes. Under these conditions, the one- and two-way ANOVAs may produce a negative test statistic, which indicates insufficient variation within the data for these tests to function adequately as tests of hypotheses.

An alternative method, the Scheirer–Ray–Hare test, may then be helpful. This is similar to a parametric ANOVA in that it tests general hypotheses only, but uses rank values rather than the actual observations (Scheirer *et al.*, 1976).

11.7.1 Key trends and experimental design

To illustrate this test we will use Example 11.3. The key trends are considered in 11.5.1.

11.7.2 Using this test

i. To use this test you:

1) Have no *a priori* reason for expecting certain outcomes from your investigation.
2) Have data which are usually presented in tables or summary histograms rather than a scatter plot.
3) Wish to examine the effect of two treatment variables.*
4) Have treatment variables* that are usually measured on an ordinal and/or nominal scale but could be measured on an interval scale and arranged into discrete categories.
5) Have a dependent variable that is measured on an interval or ordinal scale and the observations can be ranked.

6) Do not need an underlying population distribution that is normal. This test may be used for parametric data but this is not recommended (10.6).

7) Can confirm that both treatment variables have two or more categories. The data table therefore has two or more columns and two or more rows; the design is orthogonal.

8) Have more than one observation in each block but each block has the same number of observations in it.

9) Have an experimental design that means that each item is only measured once—there are no matched values or repeated measures.

10) Can confirm that all groups have a similar-shaped distribution.

*If a confounding variable is being included then it is referred to as a treatment variable for the purposes of this analysis (10.11).

ii. Does the example meet these criteria?

To illustrate this test, we will use the data from Example 11.3. The criteria here are almost identical to those for the Kruskal–Wallis test apart from the issues relating to tied ranks so we know this example meets these criteria (11.5.2ii).

11.7.3 The calculation

The ranks have already been assigned and it is these that are used in the analysis (Table 11.14). The three pairs of hypotheses being tested are those given in Boxes 11.5(A–C).

A two-way parametric ANOVA as described in Box 10.12 is carried out on these data, generating the values shown in the ANOVA table (Table 11.15). For more computation details, see the Online Resource Centre. At this point, the analysis diverges from that described for a two-way parametric ANOVA in Chapter 10. Firstly, a mean square (MS) is calculated by adding all the sum of squares (SS) values together and dividing the total by the total degrees of freedom (Table 11.15). An H value is calculated instead of an F value as SS/MS_{total}. The critical values for

Table 11.14 Ranking values assigned to Example 11.3: The density of *Bellis perennis* from two locations ('Cricket pitch' and 'Lawn') treated or not treated with fertilizer, for the Scheirer–Ray–Hare test

Fertilizer treatment	The density of *B. perennis* in each quadrat with and without a fertilizer treatment and their associated rank values	
	Cricket pitch	Lawn
Fertilizer	5.0	14.5
	6.5	12.0
	6.5	14.5
	9.5	16.0
No fertilizer	1.0	8.0
	2.5	9.5
	2.5	12.0
	4.0	12.0

Table 11.15 ANOVA table for Scheirer–Ray–Hare analysis of data from Example 11.3: The density of *Bellis perennis* (daisy) on the University of Worcester campus, 2003

Source of variation	SS	v	MS	$H_{calculated}$	$H_{critical}$ for a given p value
Variable 1 (location)	232.5625	1		232.5625/22.4 = 10.382	$p = 0.01$ $H_{critical} = 6.64$
Variable 2 (fertilizer)	68.0625	1		68.0625/22.4 = 3.0385	$p = 0.05$ $H_{critical} = 3.84$
Interaction	0.25	1		0.25/22.4 = 0.01116	$p = 0.05$ $H_{critical} = 3.84$
Within samples	35.125	12			
Total	336.0	15	336/15 = 22.4		

H are found in a chi-squared table (Appendix D, Table D1) for the degrees of freedom associated with the sum of squares (SS) (Table 11.15). The rule is the same as that given for the parametric two-way ANOVA (Box 10.12) in that if $H_{calculated}$ is greater than $H_{critical}$, then you may reject the null hypothesis. In our example, it can be seen that there is a significant difference ($H = 10.38$, $p = 0.01$) between the median density of *Bellis perennis* in the two locations, but there is no significant difference for the fertilizer treatment or the interaction. The outcome is therefore the same as that for the non-parametric ANOVA described in Boxes 11.5(A–C).

 We are designing an experiment with one treatment variable and three categories, one of which is a control. Which test should we consider using and can we make a comparison that will allow us to analyse the difference between the control and other treatments?

Summary of Chapter 11

- We consider several non-parametric tests in this chapter. The Mann–Whitney U test for unmatched data (11.1) and Wilcoxon's matched pairs test (11.2) for matched data may both be used to analyse data from experiments with one treatment variable, with two categories. The one-way non-parametric ANOVA (Kruskal–Wallis test, 11.3) is particularly appropriate for a design with one treatment variable but three or more categories, and the two-way non-parametric ANOVA (11.5) and Scheirer–Ray–Hare tests (11.7) may be appropriate for designs with two or more variables.

- The non-parametric ANOVAs followed by a *post hoc* test allow you to test both general and specific hypotheses (11.4 and 11.6). This provides you with considerably more explicit information about the effects of your treatment variables and is a design that should be used whenever possible.

- In general, these non-parametric tests can be applied to more restricted experimental designs compared with the parametric equivalents considered in Chapter 10, especially when an experiment sets out to investigate the effect of two or more variables.

- These non-parametric tests are distribution-free and therefore the criteria for their use are often easier to meet than for the parametric tests (Chapter 10). Most of the tests are ranking tests: central steps are the determination of ranks and sum of ranks as outlined in Chapter 5 (Box 5.3).

- In Chapter 10, we consider the parametric equivalents to the tests covered in this chapter. If your data are parametric (Box 5.2) or can be normalized by transforming them (5.10), you should refer to Chapter 10.

- The Online Resource Centre includes interactive exercises that test your understanding of this chapter with other topics, particularly those considered in Chapters 6–11.

Answers to chapter questions

A1 You wish to test the effect of one variable (consumption of alcohol). The data are matched, with observations on a total of 16 volunteers before and after consumption of alcohol. The observations (time taken) are measured on a continuous scale and the data may therefore be parametric. Gender is a possible confounding variable. To decide which test to use and whether the experiment needs to be modified, you need to decide whether the data are parametric. You cannot confirm this until you have carried out the investigation or a pilot experiment. If the data are parametric, you may consider a t-test for matched pairs. If the data are not parametric, then you may consider using Wilcoxon's signed ranks test for matched pairs test. It would be worth analysing the male and female results separately to obtain some information about this potential confounding variable.

A2 Yes. The data from Example 11.2 are suitable for analysis by a non-parametric ANOVA as the dependent variable is 'number of plants', an ordinal measure that can be ranked. There is one treatment variable (location on campus), four categories (cricket pitch, lawn, quadrangle, and rugby pitch), and eight replicates (observations) in each category. Each item has only been measured once. When plotted these distributions look similar in terms of spread and having two peaks. Box 11.3 explains how to adjust the calculation for the tied ranks.

A3 With this design, if the data are non-parametric, we should consider using a non-parametric one-way ANOVA followed by a post hoc test to test both general and specific hypotheses (11.3 and 11.4). This will allow us to ask specific questions about the control and other treatments. If the data are parametric, we should consider a parametric one-way ANOVA, which, if significant, could be followed by Tukey's test (e.g. 10.7 and 10.8).

Reporting your results

12 Reporting your research

→ In a nutshell

There is little point in carrying out research unless you communicate the findings to other researchers. In this chapter, we take you through the most common approaches taken to achieve this: writing a report, presenting a poster, and giving a presentation. These all have similar elements: Title, Introduction, Methods, Results, and Discussion. We explain what each section is for, approaches to take when writing each section, common errors, and issues specific to each form of communication.

Although the aim of a scientist's professional life is to make discoveries it is then critical to pass on what you have found so that your ideas and findings can be tested, appreciated, and further developed by other scientists. The most common ways to do this are to give a verbal presentation at a conference, to write a report or paper, or to present a poster. In this chapter we consider all three with an emphasis on written forms of communication as this underpins all these formats.

12.1 Writing a research paper or report

Generally, publications of practical research in science journals adhere to a specific format. This is the structure most undergraduates are also asked to follow when preparing a report. In this section, we include a guide to the general format common in scientific papers and reports. The advice outlined in this chapter should be used in conjunction with local guidelines, either those relating to the journal you wish to publish your article in or your course or department requirements. Additional helpful comments on writing scientific reports are included in texts such as Johnson & Scott (2014).

When learning how to write a scientific report or improving your skills, it is best if you have a journal article to look at. Try to have one from your subject area to hand, either as hardcopy or online, as we will use it to illustrate our points. We are conscious that there is some variation in the approach taken in different countries around the world to scientific writing. What is described here could therefore be said to be the European approach to writing a scientific report. Where we are aware of differences in practice we have flagged this up in the text.

If you work through this chapter and the questions, it should take about 4 hours. Unlike previous chapters, only some of the answers to the questions are at the end of the chapter. The remaining answers are integrated into the main body of text. All our examples are based on real undergraduate research projects.

12.1.1 General format

In general, there are nine sections to a full journal article: Title, Abstract, Keywords, Acknowledgements, Introduction, Methods, Results, Discussion, and References. Of these, the abstract, keywords, and acknowledgements are invariably not required for most undergraduate practical reports. When writing an honours project report or thesis, you may be required to include an abstract, and most people choose to include acknowledgements. In addition, in unpublished reports such as honours projects and theses, and for some journals, you may include an appendix for additional material such as the names of suppliers of chemicals, details about how stock solutions were made up, raw data, etc.

 Look at the journal article you have selected. What sections are there? How do they differ from the list given? What information do these extra sections contain that explains their use?

Your whole report should usually be written in the passive, past tense. The passive tense means that you are not referring to anyone in particular so you do not use 'I', 'you', 'we', etc.

The 'past tense' means that you write about what was done and not what is being done or will be done. Have a look at the introduction and methods in the article you are examining and note the tense that is used.

 Rephrase the following sentences so that they are in the passive, past tense.

1. I added 5mg of sodium chloride to water and stirred until it was dissolved.
2. In my reading, I found that ice is less dense than water.

Your report must be referenced correctly throughout. This is very important. Failure to do so is called plagiarism and usually carries severe penalties. We discuss plagiarism under 'cheating' (12.1.6) and referencing systems are discussed in 12.1.13.

In larger studies, you may have carried out more than one investigation. Some small elements may be redundant to the main thrust of your report. If this is the case, then leave them out or include this detail in an appendix. If your research extends over several months or years, you may tend to write about your work in the historical order in which it was carried out. This chronology may not be relevant to the actual outcome from the research and can be detrimental to the structure of your report or paper. Reports should not normally therefore be solely based on the order the work was carried out.

Striving to be succinct in your writing is an important skill which may be undermined by having to work to specified word limits especially where these are several thousand words. If you are conscious of writing just for the sake of meeting a word limit then it is advisable to discuss this with your supervisor or tutor. It may indicate that you are not including the correct material in your report such as sufficient background information or that the discussion has insufficient depth.

12.1.2 Title

This is the first thing a reader looks at and it should capture their attention. After all your hard work, you want people to read about your investigation. The title should be concise and should relate to the aim and conclusion. If you are publishing in a journal, your title will be part of the information made available for electronic searches so you need to consider who you want to read your article and include the terms they are likely to use in an Internet search. This can be easily achieved by looking at the research papers you have read. What drew your attention to these papers?

Common errors

Too long It is not necessary to write 'An investigation into . . .'. Editors of journals like to save ink and will not wish to include unnecessary words. Your title will be more forceful if you go right to the subject under investigation. In undergraduate reports, the same is true.

Too short You are not writing tabloid headlines. Your title does need to be complete and informative, and to use scientifically appropriate language.

 Read the title of the journal article you have. Does it do everything we recommend? If not, how could it be improved?

12.1.3 Abstract

The purpose of an abstract is to provide the reader with a clear concise overview of the main points of your study. Readers use the abstract as a way of determining whether to read the whole report or skip on to another research report or paper.

An abstract is defined in the *Concise Oxford Dictionary* as a 'brief statement of content'. Most journal publishers do not want the 'brief statement' to be an extract copied from part of the report, so you have to write an abstract, not cut and paste it. This is also usually the case for undergraduate work where you are asked to include an abstract.

The format for an abstract varies among publications. In some journals, the key findings are given as a list; in others, the abstract is written as prose with sentences organized into paragraphs. Whichever format is required, you need to include enough background information so that your aim, which you also include, can be understood. This is followed by a summary of your data and your conclusion. There needs to be enough detail in the abstract for your conclusion to be understood. You should not include any new information in the abstract that is not elsewhere in your report. The abstract focuses on your study and it is therefore not usual to include reference citations. You would normally, however, include *p* values or similar detail from the analysis of your data to support your points.

Common errors

The most common error in undergraduate reports where an abstract is required is insufficient detail. This can include forgetting to include the aim or the conclusion. Inevitably errors in the main report will be transferred into the abstract. The most noticeable of these tends to be an incorrect conclusion, but can include an incorrect evaluation or a poorly written aim. To avoid this try to polish the rest of your report before you write the abstract.

 Do not look at the abstract of the journal article you are examining. Instead, read the paper and write your own abstract. How does it compare with the author's version? What are the differences? Which is the better version and why?

Keyword

A word or short phrase used by potential readers when searching for relevant research publications. A list of keywords may be requested by publishers.

12.1.4 Keywords

Keywords are terms or phrases which are used as search terms and therefore need to be chosen with care to ensure that your work is not overlooked and is read by the relevant research community. When submitting papers to some journals for publication you may be asked to provide a short list of keywords which is included just below the abstract. If asked to provide keywords there will be guidelines concerning them in the journal's guide to authors.

Keywords are also found in the main body of a report, as electronic searching allows the full text of an article to be searched for the words determined by the reader. Therefore when you are writing for publication it is important that you give some thought to which words are most relevant to the audience who you wish to read your work and make sure these are included.

When considering keywords you can use single very specific words or linked words. For example, linked words might be 'resource protection' or 'agri-environment schemes'. To identify the most suitable keywords consider the words and phrases you have found most helpful in locating literature relevant to your study.

 Q5 Does the paper you are reading have a list of keywords? How effective are these at indicating the content of the paper? If there are no keywords which terms would be useful?

12.1.5 Acknowledgements

Many people feel that they would like to include acknowledgements at the completion of a significant piece of work. Those who may be acknowledged fall into three categories: the funding body, those who provided professional assistance, and those who provided personal support. It is usually a requirement as well as a courtesy to include an acknowledgement to any organization or person who has provided you with funding or support in kind (e.g. seed supplies, an item of equipment) that has enabled you to carry out the research you are now reporting. You may also have received professional assistance, such as being given permission to carry out your work in a particular location, help with technical aspects of the research, or help in the identification of species. Again, it is important that the input of these people is acknowledged. In journal articles, it would be unusual to find personal acknowledgements, although in honours project reports and theses, it is common practice to thank friends and/or family for their support. This is personal to you, and most universities have regulations that allow the inclusion of these acknowledgements. It is not appropriate to make these acknowledgements too personal in either a positive or negative way, nor should your acknowledgements include tenuous supporters such as drinking friends and distant cousins!

12.1.6 Introduction

An introduction is the part of a report that sets the scene and leads up to a statement of your aim. The introduction has three elements: the background, rationale, and the aim. The background and rationale are integrated to form the main body of the introduction. In this section we discuss each element of an introduction and then the approach to take when preparing an introduction.

i. The background

Your introduction should cover enough background information for a reader to understand the aim. If you are not sure how to achieve this, first look at your aim and note down critical words. Your introduction should include details about all these words.

ii. The rationale

Woven into the points you are making about the background to the study needs to be a clear explanation of why you are undertaking this specific study. This is the rationale. Part of your personal rationale is usually that you want to investigate this topic as no one else knows the answer to the question you are posing. However, this is generally taken for granted so the main part of the rationale you need to explain is the biological reason as to why you are posing this particular question.

iii. The aim

Writing an introduction is like telling a joke: you must not give the punchline first or no one will understand it. The punchline in an introduction is the aim. It will be uppermost in your mind so it is tempting to place it first, but if you do this, your reader, when reading the aim, will not understand the reasons for the investigation or fully understand the aim itself. Writing the aim first is a format often used in schools, but you should now use the format current in the scientific literature instead. The aim therefore is something you lead up to and should be in the last paragraph at the end of the introduction. How to write an aim and how it differs from objectives are discussed in 1.1.1. How the aim can be developed during your planning process is considered in Chapter 2.

iv. A literature review

The length of the background within a report is clearly dependent on the overall size of the report. In both short and longer reports you will be asked to include a review of the literature. There are a number of approaches that can be taken with a literature review and a poor approach can be very evident especially in a longer report where the literature review may feature as a subsection or be separate to an introduction. The role of the literature review is to establish the background to, and rationale for, the investigation, as research rarely functions in complete isolation and nearly always develops from previous work in the same or similar areas.

There are two approaches that are commonly taken when reading in preparation for writing an introduction. Which approach do you take?

a. You plan the introduction and then find a few articles on each aspect and use these papers to write your introduction.

b. You find all the literature you can using very broad search terms and save all the titles. You skim read through the abstracts or papers and from this overall comprehensive body of literature gain an overview of the current research in the area. You sort the articles in terms of their relevance and usefulness. You use these as the basis for your introduction.

Literature review
To identify all items of literature in the area you are considering followed by a review and selection process which identifies the research that best reflects the current thinking in the area.

The second approach is what is meant by the term 'literature review' and the end result is very different from the first approach. In the first approach you box yourself in right at the outset and you cannot tell if the papers you have read truly reflect current thinking and are mainstream in this scientific field. The second approach starts out by taking the overview and only then do you determine which the most relevant studies are. If, as an undergraduate, you take the first approach when preparing a report for an assessment you are unlikely to achieve more than an average grade in your work as the limitations in your knowledge and understanding will be evident.

v. A critical reader

A competent scientist has the skills of a critical reader, never taking at face value the conclusions presented in a research paper but confirming and checking that these conclusions have a sound basis. We examined these skills in relation to assessing the quality of an experimental design in Chapters 1 and 2 and in relation to the data analysis in Chapter 7. As a student, to demonstrate that you have this skill is important. In longer works such as an honours project report or thesis you can convey alternative sides to a debate or concerns about the quality of research in your report. Your introduction may therefore include key studies that are not very good which you include as part of your discussion of the background literature, and their weaknesses can form part of your rationale. This shows that you have the capacity for independent critical thought and is an important life skill.

Critical reader
Making a balanced assessment of the quality of the experimental design, analysis, and reported evaluation for each item of published work and reflecting this evaluation in your own work.

vi. Writing your introduction

Having completed your literature review process and considered the strengths and weaknesses of the relevant studies you are now ready to write your introduction.

 Have a look at the format of the introduction in your journal article. How many paragraphs are there? How does this compare with the discussion? Where is the aim? Are there any figures or tables? What information is contained in brackets throughout the introduction?

This quick look at your journal article should have drawn your attention to several format features of an introduction: (i) it is written as prose with fully referenced sentences and paragraphs; (ii) there are not usually any figures or tables; and (iii) the aim is at the end. Other format features to note about introductions are the lack of quotations and length of the introduction. We consider these points in the section on 'common errors'.

Be guided by the words you have identified in your aim and prepare a plan that takes the reader step by step through the background to the research. The background needs to incorporate the information that clearly establishes the rationale and leads directly to a statement of the aim of the research. On completion of the first draft of your introduction you should then go back over it checking that the keywords and phrases are included.

Common errors

The introduction reads like an essay An introduction is not an essay. It performs a very specific and different function. Even if you are asked to include an extensive literature review, it should normally be written so that each paragraph takes you one step closer to being able to state your aim. One feature that tends to be associated with 'essay' introductions is their length. If in doubt, count how many paragraphs you have in the introduction compared with the discussion. If the introduction is considerably longer than the discussion, then you have probably made this error.

The introduction lacks detail This is a feature that has become more common in recent years with the advent of electronic literature searching facilities. Many publishers make abstracts from journal articles available online at no cost and there is a temptation to rely on these as the prime source of information when reading about a topic. Abstracts do not provide enough detail for anyone to be able to appreciate fully the value or rigour of the research or the detailed arguments presented within the paper. A reliance on abstracts will lead to a lack of critical detail in your report, leading to errors in your understanding and a weak introduction (and discussion).

The introduction is aimless It is surprising how often students forget to include their aim in their introduction. This leaves the reader in utmost confusion, usually for the remainder of their time reading the report. Given report formats, it is extremely difficult to fully comprehend what the aim of the research is unless you explicitly state it.

The introduction is not focused on the actual piece of research that has been carried out This usually occurs because you are using sources of information that are looking at a slightly different topic and you can get swept along. Sometimes you may find a lot of information on a small specific part of your study and so you tend to note down everything you have found about that topic at the expense of areas where information is harder to find. You can avoid this by planning your introduction based on your literature review before you start writing. For example, one of our undergraduates investigated the opinions of students in relation to over-the-counter genetic testing. She might have written an introduction all about genetic testing and never touched on 'over-the-counter' aspects or how people develop their opinions. All of these topics should be covered in the introduction, but as information is easiest to find on genetic tests, it would not be surprising if the student was led astray.

You use quotations In Europe in the biological sciences, we are rarely interested in how someone else has phrased a sentence; we are interested in the facts within the sentence. Therefore, unlike reports written for subjects such as English, you will not usually see quotations being used. When you are reading around the subject, you should extract the facts as bullet points only. You then use these succinct notes to write your own introduction. This makes for a much more readable piece of work. It also avoids inadvertent plagiarism (see Cheating (plagiarism), p.377). The exception to the use of quotes tends to be if you are defining a word, and in published articles not even then.

In some countries the use of quotations is more widespread. Therefore if you are studying in a country other than the one you trained in you should discuss the local approaches to the use of quotations.

 Look at your journal article. How many quotations are included in the introduction?

Predictions It appears to be becoming increasingly common for students to declare, in their introduction, what they are expecting to find from their investigation. (We use the term 'expectation' here as it might be used in general spoken English. Elsewhere, we use the term in a statistical way, e.g. Chapters 8 and 9.) Making predictions like this is a really bad idea for two reasons. Firstly, you should approach any investigation with an enquiring and open mind. If you start out by predicting the outcome, than you are likely to bias your design, your execution of the investigation, and your evaluation of the data. This can lead to a flawed piece of research and/or completely incorrect conclusions. Secondly, if you include your prediction in your report, your readers will know that you have not come to this work with an enquiring and open mind and be prejudiced against your work as it is more likely to be biased.

More about the 'punchline too early' phenomenon Another error we come across is where detailed elements, usually relating to the methods, appear in the introduction. In your introduction, you need to keep yourself aloof from what you actually did. The nearest you get to referring to what you did is the aim. The introduction is for background information and establishing the rationale only.

Cheating (plagiarism) The computer age has brought easier access to sources of information through electronic searching and the 'cut-and-paste' facility. It has also tempted some undergraduates to cheat, either by downloading whole reports or essays, or by cutting and pasting paragraphs from articles. Assessments, including reports of investigations, are supposed to be your own work, which means that they should be your own ideas and your own words. There is no excuse for the words not being your own. However, if the ideas are not your own, you should indicate the source by referencing them. Failure to do this is called plagiarism. Not only will you not learn anything by cheating but you will also be subject to your institution's or the journal's penalties. Scientists make their living by the facts they discover and the communication of these facts. This is why you must always acknowledge the sources of your information and this is why there are severe penalties if you do not. We explain how to reference correctly in 12.1.13.

If you are still tempted, please bear in mind that if you can find these articles so can others and it is surprisingly easy to spot the use of essay bank work and content that has been copied from another source.

 In 1.1.1 we explained that the aim and title are very similar. Look at the aim and the title in the paper you are reading. What are the important words that have been used? Go through the introduction and see how these different aspects have then been covered. Are there any gaps or any additional material not directly linked to the aim or title? If there are, could the introduction, aim, or title be revised to improve the relationship between these three aspects of the report?

12.1.7 Method

The method outlines the procedure(s) you have undertaken during your investigation. Methods can vary in their content and you should check your local regulations or guide to authors to confirm the content. A method section can include a site description, the method, and details about the statistical analyses used.

i. Site

This section is included in the method when the location is an important part of the research and its evaluation. You may either have carried out your research at a particular site or have sampled from a particular place, and there is a possibility that the site has given your sample or study unique characteristics. The site description is written in such a way that your site can be located and the features of the site that impinge on your investigation need to be reported.

For example, an investigation was carried out into the percentage cover of five plant species growing on and/or off anthills in grassland. In this study, the site is important because the surrounding habitat and its management will affect the flora found in the locality. Therefore, you need to include a description of the area and its geology, typical plant species, and details about the management, e.g. grazing, presence of rabbits, cutting regimen, spraying, etc.

The location of your site may be given in terms of Ordnance Survey (OS) coordinates or latitude and longitude. You therefore need to indicate which coordinate system you are using. The latitude and longitude will be identified from the equipment such as a GPS recorder or equivalent application (app) on other hardware. These coordinates are usually reported to four decimal places. For example, the location of the University of Worcester would be recorded as 52.1971, –2.2426. Ordnance Survey (OS) coordinates are found using an OS map of the area. The first set of numbers comes from the x axis (west to east) and the second set of numbers refers to the y axis (north to south). The complete coordinates consist of two letters and six to ten numbers, e.g. the University of Worcester is at SO 83 508 55 452. If using a printed map you will find the letters are printed on the map itself and are large and pale blue, e.g. Worcester is in the 'SO' compartment. The numbers are derived from the numbers printed at the sides of the map. Your accuracy will depend on whether you are using a printed or electronic map, the electronic maps being the better of the two.

ii. Method

The methods section describes the processes you carried out to investigate your aim. There needs to be enough detail included so that someone else can repeat your work. There is some variation in how much detail is required. For example, in undergraduate work you do not usually have to indicate the name of the company that supplied your chemicals, but you may explain how you made up stock solutions that were then used to make other solutions. In journal articles, the reverse tends to be the case. For example, some substances (e.g. compost) vary from one supplier to another and in these instances details about the supplier and the full trade name of the substance should be given.

As this information about materials is an essential part of a method, this section may be titled Materials and Methods. However, the materials detail is usually integrated into the methods and is not given as a separate section in the methods.

 Look at the methods section in the article you have been examining and imagine yourself repeating the experiment. Is there enough detail included for you to do this? If not, what more is needed?

The methods, as you will have noticed from examining the journal article (Q9), are written as paragraphs of prose usually in the passive, past tense. Often in school you are required to write a method as a list of steps, often with a separate section for materials. This format is no longer appropriate.

Some investigations are time sensitive. Most commonly, this is where you have collected samples from the field or carried out work outside the laboratory. If the outcome from your method is likely to be influenced by the seasons, you should include the date in your method. Some investigations are influenced by the time of day, for example, the behaviour of animals. In this instance, you need to report the time at which you made your observations.

iii. Statistical analyses

In some methods sections where experiments have been carried out, you will also see a brief outline of the statistical tests used to test the hypotheses. This needs to be no more than a brief statement of the test used and a reference indicating the resource used to help you carry out this test. For example, in Chapter 2 we discussed how the effect of wind speed on seed dispersal might be investigated and we explained which test would be appropriate. When reporting this in a methods section, you would state that the distance travelled by the seed in relation to the wind speed was 'analysed using a Kruskal–Wallis test (Holmes *et al.*, 2016)'.

Common errors

Too little or too much information Methods most often suffer from too little or too much information. Like Goldilocks, you need to get it 'just right'. To check whether you have included too little information, give your method to a friend and ask them to tell you how the investigation was carried out. This will quickly identify the gaps. Too much information can arise when you are using the 'Materials and Methods' format from school or you include unnecessary statements, e.g. 'The data were collected and summarized'. If you have developed a method during the research process and so in each use there are slight differences, you do not have to report the whole method each time. Reporting the full method once and indicating the subsequent revisions will provide sufficient detail for your work to be understood.

It is difficult to find the correct balance when writing a method, but you can learn to do this by checking with your lecturers/supervisor and learning from the feedback you have received about earlier reports.

Symbols   As most reports are now word-processed, errors relating to the use of symbols have become common. Most versions of Microsoft Word have the following facilities that enable you to write symbols correctly:

- *Subscripts* (e.g. H_0) and *superscripts* (e.g. m^2) are generally located under 'right click' Font
- *Symbols* can usually be found either in Insert–Symbol (e.g. μ, σ, Σ) or as part of Equation Editor at Insert–Object–Microsoft Equation (e.g. Σ, $\bar{X}$).

When using symbols, be consistent in their use in terms of the font you choose and other text characteristics such as italics.

Cheating   If you are provided with a protocol for a method, this does not necessarily mean that it is written in the format appropriate for a report or that you necessarily have permission to report it verbatim in your report. You may also have made some revisions. Methods that you have been given need to be considered with as much care when you are preparing a report as a method you have developed yourself. Failure to acknowledge the source constitutes plagiarism; failure to follow the appropriate format or provide accurate details indicates learner incompetence.

 Read through the method in your article. Does it follow our guidelines? If not, in what way does it differ? Why do you think this is? Check your department's or the journal's guide to authors. What do they say about a method? How does it compare with the guidance given in this chapter?

12.1.8 Results: text

The whole point of a results section is that you draw to your reader's attention the main trends you have found in your data. You achieve this by first organizing your data in tables (12.1.9) or figures (12.1.10) and/or using summary statistics (5.3–5.7). By taking an overview of your data, you are more likely to be able to identify the apparent trends in your data. Secondly, if you have hypotheses to test, carry out the statistical analysis and draw inferences from this. Finally, look at the raw data to see whether any specific further points can be made. This information is what needs to be communicated in your results section. It is not easy to write a results section: it requires practice. Like the introduction, it helps if you plan it out first.

 Have a look at the results section in your article. How many paragraphs of text are there? How many figures and how many tables?

A scientific report is written as words. This seems to be stating the obvious but it is amazing how many people forget to write any words in their results section. Instead, they include a series of tables and/or figures and leave the reader to work out what is going on for themselves. To write a results section, look at the information you have about the trends in your data and prepare a list of all the points, ordered in a coherent way. If you have a number of objectives, you will often find it useful to arrange the results under subheadings relating explicitly to each objective. When you have a suitable plan, you then write your results and only then should you consider whether to use any figures or tables. The role of figures and/or tables in a results section is a supporting one only. They allow you to include detail and to illustrate your trends. To review good practice in relation to figures and tables, see 12.1.9 and 12.1.10.

Common errors

'Therefore' It can often feel very awkward when writing a results section to limit yourself to highlighting the trends seen in the data without referring to possible explanations for these results. If you find yourself including 'because' or 'therefore', this is probably what you are

doing. The discussion section is the place for all these explanations. The reason there is a need to have separate sections is that explanations for the trends in your data are rarely straight-forward and conclusive. If you embarked on lengthy discussions in the middle of presenting your data, the reader will lose any overall sense of your findings. It is better to keep all discussion separate from the presentation of your results.

Repetition   Many students suffer from overenthusiasm for tables and figures. The tendency, which again appears to come from early training in school, is to include a table of your data and then represent this with a figure. Such repetition shows that you have not thought through how best to support the points in your text. It is very rare indeed that a table and figure are needed to illustrate the same data. The exception to this would be a desire to demonstrate the overall trend using a figure but a need then to identify details in the data, which is best done using a table.

12.1.9 Results: tables

A long string of numbers does not help you to understand your raw data or help you to explain it to someone else. Therefore, it is common practice to organize the raw data in a summary table. There are a number of formats for summary tables depending on the type of data you have. For some investigations you will collect data that fall into two or more discrete categories (e.g. Table 9.3).

If you have collected data measured in a continuous scale, you may select your own categories, within certain constraints. These divisions are often artificial in that they have no scientific significance or meaning. The categories have to demonstrate clearly and fairly the trend you wish to illustrate, but not overlap, and should normally all be the same size (e.g. Table 12.4). You then record the number of observations that fall within each category.

As you evaluate your data, you may also construct a calculation table; we have seen many examples of these in Chapters 7–11 (e.g. Table 10.9). In scientific papers or reports, you may include a summary table and occasionally a calculation table. Which table will depend on the points you have made in your results section. You do not normally, however, include the calculation itself.

When writing for a particular journal or preparing other reports, there may be local requirements that you need to follow when producing a table. For example, it is usual when reporting summary statistics for parametric data to include the mean, the standard error of the mean (5.7), and n. In doing so, you have provided sufficient detail for a reader if they wish to be able to calculate confidence intervals for your data.

A table should always:

- *Have a title* This should indicate clearly what the data are. The title may be followed by additional details about the contents of the table, the outcome of the analysis, or a further comment about the method so that the table can be taken out of context and still be understood. This additional detail is usually placed in italics to distinguish it from the title. In most scientific journals, the title for a table is placed above the table. The additional details may be placed either following the title or below the table.

- *Be numbered* Numbering tables makes it much easier for you to explain which table you are referring to. Table numbers should be sequential.

- *Be labelled (rows and columns)* Most tables are organized into rows and columns. It is important to make it clear what information each row and each column contains.

- *Indicate the units of measurement* If a row or column includes data, you should indicate the units of measurement, e.g. mm, °C, etc., in the column and/or row headings. Units should also be included in the table title (if they apply to the entire table) or in column or row headings. Having indicated the units in the title or column and row heading, you do not then include units in the central body of the table.

- *Be simple* The table is being used to illustrate the points you are making in the text. If a table is very complicated your reader may not be able to grasp the points you are making. In this instance you should consider presenting the information in several different tables.

- *Distinguish between zero and missing data* Sometimes investigations do not go according to plan and you have 'missing data'. You may then need to distinguish between a zero measurement and missing data (e.g. Table 12.1 in Example 12.1).

Example 12.1   The response of tobacco explants to auxin

In an undergraduate investigation into the effect of auxin on the growth of tobacco in tissue culture, the relative increase in diameter (mm) of leaf explants was recorded after 2 weeks. Some of the samples, e.g. explant 1, auxin 1, showed no increase after 2 weeks. Two of the samples became contaminated and died. These missing data are indicated by a dash (–) in Table 12.1.

Table 12.1 The relative increase in diameter of explants of tobacco after treatment with auxins

Explant	Relative increase in diameter (mm) after 2 weeks growth in tissue culture	
	Auxin 1	Auxin 2
1	0	1
2	2	0
3	–	1
4	1	–

- *Use appropriate classes in a frequency table* Where data are measured on a continuous scale, 'classes' may be used to enable you to present a summary table. Two common errors relating to the use of frequency tables are choosing the wrong number of size classes and choosing size classes that overlap.

In Example 2.3, we described some research where the natural variation in growth of *Secale cereale* (rye) seedlings was examined and the researcher recorded heights (mm) after 3 weeks for 15 seedlings. In her work book, the researcher recorded the following:

12.0, 12.0, 11.5, 18.0, 14.0, 11.0, 14.5, 11.5,

10.0, 10.0, 19.5, 19.0, 21.0, 15.5, 14.5

These are continuous data and are therefore best summarized in a frequency table with size classes of your choosing. We illustrate three possible sets of size class for these data so that you can see which class sizes best illustrate the trend in the data (Tables 12.2–12.4).

It is also a common failing to choose size classes that overlap. If you were planning a frequency table for the rye data (Example 2.3), you may have chosen classes 10.5–11.0mm, 11.0–11.5mm, etc. With these classes there is an overlap. If you had a value of 11.0mm, which class would you place it in? The classes in Table 12.4 do not overlap and are therefore correct.

Table 12.2 Frequency distribution of leaf length (mm) in *Secale cereale* (rye) seedlings after 3 weeks' growth

	Leaf length (mm)											
	10.0–10.4	10.5–10.9	11.0–11.4	11.5–11.9	12.0–12.4	12.5–12.9	13.0–13.4	13.5–13.9	14.0–14.4	14.5–14.9	15.0–15.4	15.5–15.9
Number of individuals	2	0	1	2	2	0	0	0	1	2	0	1
	16.0–16.4	16.5–16.9	17.0–17.4	17.5–17.9	18.0–18.4	18.5–18.9	19.0–19.4	19.5–19.9	20.0–20.4	20.5–20.9	21.0–21.4	
Number of individuals	0	0	0	0	1	0	1	1	0	0	1	

Too many size classes. It is not possible to see any overall trend in this data.

Table 12.3 Frequency distribution of leaf length (mm) in *Secale cereale* (rye) seedlings after 3 weeks' growth

	Leaf length (mm)	
	10.0–19.9	20.0–29.9
Number of individuals	12	1

Too few size classes. This summary does not allow you to see any detail.

Table 12.4 Frequency distribution of leaf length (mm) in *Secale cereale* (rye) seedlings after 3 weeks' growth

	Leaf length (mm)					
	10.0–11.9	12.0–13.9	14.0–15.9	16.0–17.9	18.0–19.9	20.0–21.9
Number of individuals	5	2	4	0	3	2

Just about right. This summary table allows you to see some detail (compared with Table 12.3) whilst allowing the trend to be evident (which it was not in Table 12.2).

● *Rounding up* Rounding up is where the value is abbreviated. When carrying out a calculation you should use the complete value known to you including all decimal places to avoid errors in the calculation. (In our chapters of choosing and using statistical tests, all calculations were carried out using the full values but for the purposes of reporting the values in this book they were rounded up to five decimal places.)

However, when reporting numerical values, it is common and sensible practice to round these values up. You can round up values either to a given number of significant figures or to a given number of decimal places. As a general rule of thumb, if you are reporting a mean or variance for data, you round the values up to the same number of decimal places or significant figures used in your observations. When reporting calculated statistical values, these would be rounded up to the same number of decimal places as the critical values.

Q12 Examine the tables in your paper. Do they follow all these guidelines? Find a table you have prepared. Are there any improvements you could make?

12.1.10 Results: figures

If you believe that a figure best illustrates the points you are making in your results section, then you first need to decide which is the correct figure to use and then check that you draw the figure in accordance with the standard conventions relating to information, axes, representation of variation, and lines.

i. Type of figure

The five most common types of figures for data are a scatter plot, pie chart, bar chart, line graph, and histogram. For a review of more figure types, see Hawkins (2009). The type of figure you use depends on the point you are making, as some styles will illustrate the points you are interested in more clearly than others.

Pie chart
A circle is divided up to reflect the relative proportions (or percentages) of observations in each category or sample.

Bar chart
A figure where each category from a nominal or ordinal scale is represented as a discrete (separate) bar on the figure.

> ### Example 12.2   The behaviour of captive orang-utans
>
> An undergraduate examined the behaviour of orang-utans at a zoo in the UK. She categorized the various behaviours, e.g. grooming, playing, solitary, eating, and showing aggression, and recorded the time spent on each.

The data for Example 12.2 can be presented in a number of ways. A summary for any one of the orang-utans' behaviour can be presented as a pie chart (Fig. 12.1). A pie chart is a circle which is divided up to reflect the relative proportions (or percentages) of observations in each category or sample. A bar chart, however, would allow more data to be presented. In a

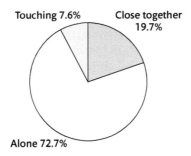

Fig. 12.1 Relative number of occasions (%) that particular behaviours were shown by one female orang-utan during a 4-hour observation period

bar chart each category from a nominal or ordinal scale is represented as a discrete (separate) bar on the figure. In Example 12.2 using this format it is possible to illustrate the differences between three orang-utans in their behaviours (Fig. 12.2).

Both bar charts and histograms are drawn as bars on a figure. It is therefore tempting to think that they are the same, but bar charts have a gap between the bars on the *x* axis whereas histograms do not. The gaps indicate, on a bar chart, that the data are measured on a discrete scale (e.g. blood groups). If the data are nominal then the order of bars on the *x* axis of a bar chart is imposed by the reporter and does not indicate any meaningful order. A histogram with no gaps shows that the data represented on the *x* axis are measured on a continuous scale (e.g. height in mm); the order of the bars is derived from the data and is meaningful. For example, Fig. 12.2 which illustrates the data from Example 12.2 has a gap between each

Histogram

A figure where each category of data is represented as a bar and each bar abuts the next, indicating that the scale of measurement on the *x* axis is continuous not discrete

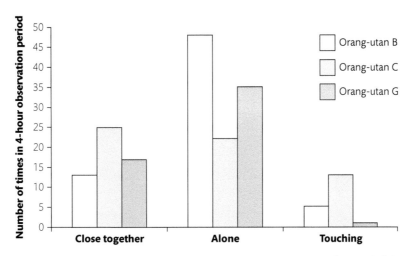

Fig. 12.2 Number of times particular behaviours were shown by three orang-utans (B, C, and G) during a 4-hour observation period

'behaviour' to indicate that this is not a continuous scale. The height of 87 male students was recorded (5.2.1) and presented in a figure (see Fig. 5.1). The scale of measurement is centimetres which is a continuous interval scale. In this representation of the data, there are no gaps between the bars so that the continuous nature of this scale is shown diagrammatically. Figure 5.1 is therefore a histogram.

Example 12.3 The effect of petrol on the growth of *Lolium perenne*

As part of a study of a polluted site, an undergraduate investigated the effect of petrol on the relative increase in leaf length of a grass (*Lolium perenne*) over a 6-week period. Three concentrations were examined (Fig. 12.3).

Line graph

Each observation or mean from a group of observations for a given *x* value is plotted as a point on the figure and may be joined together with a trend line.

Scatter plot

When observations for two continuous variables are recorded for each item and marked as a point on the figure.

In Example 12.3, the effect of only one variable (concentration of petrol) is being examined. Concentration (ml/g soil) is measured on a continuous scale. However, only a few (three) discrete points along this scale have been examined. The variable (concentration of petrol) was determined by the investigator. These results are best shown as a line graph where each observation or mean from a group of observations for a given *x* value is plotted as a point on the figure (Fig. 12.3). These points may be joined together with a trend line as the intervals between the points are meaningful.

The final type of figure we consider is a scatter plot. When you have two treatment variables with continuous scales the nature of the data on both axes lends itself to graphical rather than tabular representation and is a format most often used when data have been analysed using either a regression and/or correlation, for example Fig. 9.1.

Choosing the correct format for a figure therefore depends on the type of data and numbers of variables you have. Table 12.5 gives a guide to choosing the correct figure.

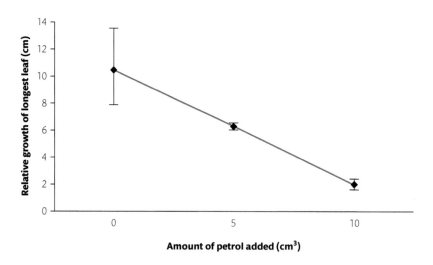

Fig. 12.3 Relative growth of longest leaf (cm) of *Lolium perenne* exposed to petrol in the soil

Table 12.5 Guide to the correct figures to use

Type of data	Pie chart	Bar chart	Histogram	Line graph	Scatter plot
One treatment variable with nominal categories. For a pie chart this determines the number of categories. If using a bar chart this would be plotted on the x axis. A dependent variable with ordinal or interval measures. For a pie chart this determines the relative sizes of each category. For a bar chart these values are plotted on the y axis.	✓	✓			
One treatment variable with ordinal categories. For a pie chart this determines the number of categories. If using a bar chart this would be plotted on the x axis. A dependent variable with ordinal or interval measures. For a pie chart this determines the relative sizes of each category. For a bar chart these values are plotted on the y axis.	✓	✓			
One treatment variable with an interval scale that is divided into categories by the investigator and is plotted on the x axis. A dependent variable with ordinal or interval measures plotted on the y axis.			✓		
One treatment variable with a continuous scale. The treatment variable is usually under the control of the investigator and few points are examined. These are plotted on the x axis. The dependent variable is a continuous measure plotted on the y axis.		✓		✓	
One treatment variable with a continuous scale. The treatment variable is not usually under the control of the investigator and many points have been recorded which are plotted on the x axis. The dependent variable is a continuous scale plotted on the y axis.				✓	
Two treatment variables where one or both are measured on an interval scale. If there is an independent variable this is plotted on the x axis.					✓
Any other designs especially where there are variables with categories are usually best presented in a table.					

Q13 Represent the data relating to the growth of *Secale cereale* (rye) seedlings (Table 12.4) by an appropriate figure. You may wish to repeat this exercise using a computer package. How does the output compare?

ii. Information

More errors in figures are appearing with the use of computer software, often, although not always, through user inexperience. If you are going to use a software package, it is worth roughing out the figure beforehand to be sure that you are quite clear in your own mind what is required. As with tables, you are likely to have local regulations about how you should present figures in your reports, papers for publication, etc. The main aim is always to make figures explicit and appropriate for the data. Therefore, a figure should:

- *Include a title* Most journals require figure titles to contain enough detail so that they can be taken out of context and still be understood. You do not need to include such things as 'A figure to show …'. It is clear that this is a figure, so you do not need to include this in your title. In most scientific reports the standard practice is to place the title below the figure. You may also add brief additional information about the figure and outcomes from any statistical analysis.

- *Be numbered* This enables you to explain clearly which figure you are referring to at any one time. The numbering should be sequential. For some studies, e.g. microbiology, molecular biology, and ecology, you may have photographs that illustrate your results. These used to be called plates but more recently have tended to be included in with the figures and are numbered accordingly.

- *Have clearly labelled axes, stating the units in use* If you fail to label your axes, then your figure will convey no useful information at all.

- *Provide a key if necessary* You may be using your figure to allow easy comparison between several samples or treatments. Figure 12.2 shows the behaviours of several orang-utans. Each column is shaded in a different way and a key is given so that you can identify which individual is which. Where shading or colour is used to identify particular samples or treatments, this coding should be consistent for all figures. Thus, if sample 1 is shaded blue in the first figure, it should be shaded blue in all subsequent figures.

iii. Axes

For some data it is possible to identify a dependent and an independent variable. These terms are used by different authors in different ways. When we use these terms, we mean that the independent variable is taken without sampling error, is usually under the investigator's control, and that there is probably an association between the dependent and independent variables but this may not be causal. In Chapter 9, we considered two examples: Example 9.5: Heavy metal contamination of soil under electricity pylons, and Example 9.7: Lower arm (cm) and lower leg (cm) length of a small cohort of female undergraduates. In both these investigations, two variables were being considered. In Example 9.5, soil samples were taken at regular points away from the electricity pylon. This treatment variable (distance from pylon) is under the investigator's control and is the independent variable. The level of lead contamination is not under the investigator's control and is said to be the dependent variable. In Example 9.7, however, neither parameter (arm length or leg length) is being manipulated within the experiment and neither could sensibly be said to be dependent on, or independent of, the other.

Where you have a dependent and an independent variable, the dependent variable is plotted on the *y* axis and the independent variable is plotted on the *x* axis. Where there is no clear case for the variables being classed as dependent and independent then either variable may be plotted on the *x* or *y* axis.

The scale selected for the axes is very important in enabling the figure to fulfil its role of illustrating a particular point. Care should be taken in selecting a scale that will allow the critical trend to be seen. However, if you are then presenting a number of related figures where the data cannot be presented as one combined figure you may need to compromise on the scale as the reader will wish to compare one figure to the next and to do this the scales must be the same.

 Cows in calf were fed known amounts of supplementary corn. The calves were weighed when born. The investigator wanted to test whether there was any association between the amount of corn the cows were fed and the calves' birth weight. Is there a dependent and an independent variable? If so, which is which?

---

iv. **Lines**

Lines may be fitted to data in two ways: either by joining the points on a graph together as a trend line or as the result of a statistical analysis, e.g. regression (Chapter 9). Trend lines are very problematic. Before using one you should ask yourself whether it is sensible to draw a line between two points on your graph and read off an intervening value. It is usually appropriate to draw trend lines on line and scatter plots but not on histograms and bar charts. In Example 12.3, only three concentrations of petrol contamination of soil were investigated. The data are measured on a continuous scale (amount of petrol added (ml)) and plotted as a line graph. A trend line that links each point on this line graph makes sense: you could read off an intervening point on the *x* axis and it would be meaningful (Fig. 12.3). If you have continuous data that are organized into classes and plotted as a histogram, it is not appropriate to link these classes together with a trend line. The imposition of classes has effectively taken away the continuous nature of the scale when visualized as a figure (e.g. Fig. 5.1).

In Example 12.2, the various behaviours of an orang-utan were recorded. When these data are plotted on a bar chart (Fig. 12.2), it is quite clear that any line joining the tops of these bars would be meaningless. This is particularly so because the data are nominal and the categories (in this case different behaviours) have no inherent order.

Beware: computer software is very keen to draw in trend lines. Do not assume this is correct. Where it is not correct, do not leave the trend line in your figure.

If a line is drawn, you must indicate clearly whether this is a trend line or one derived from statistical analysis. Some statistical tests (e.g. regression analysis) enable you to work out a mathematical equation that describes the line that best fits your data. If you have a regression equation, you should add this to your figure (e.g. Figs 9.14 and 9.15); if you do not, then you must make it clear that the line is drawn in only to indicate the possible trend.

If you are drawing a line on your figure, do not extend it beyond the range of your data. You do not know what happens outside this range and you could be suggesting a trend that is quite wrong. If you need to know what is happening outside the data set you have, you need to repeat your experiment with more observations in the area you are interested in. Chapter 9 (Q5) illustrates this graphically.

 Examine the figures in your paper. Do they fulfil all the guidance on good practice? If not, how might they be improved? If there are no figures included in the paper, do you think it would have been useful to include some? Why? To avoid overlap with data already represented in a table, would any tables need modification? In what way?

12.1.11 Results: statistics

When reporting the outcomes from your analysis, there is very little detail that is required but what is needed is critical to making sense of your analysis. In each box in Chapters 8–11, where we have illustrated worked examples for a number of statistical tests, we have included one way of reporting our results. However, there are several conventions that may be used. These conventions vary depending on the statistical test you used, the p value, and the local requirements of the journal or your department. The following, therefore, needs to be used in conjunction with these other guides.

i. Phrasing

Throughout Chapters 8–11, we have indicated the common format for reporting the results from statistical analyses. This reporting requires the inclusion of details from your hypothesis, the calculated value of your test statistic, an indication as to whether you reject or do not reject the null hypothesis, and the level of probability (p) at which this decision has been made. In each box in Chapters 8–11, we have written this information formally so that you become used to the inclusion of these various elements. For example, in Box 8.2, the outcome from the analysis is reported as 'There is no significant difference ($\chi^2_{calculated} = 4.40$, $p = 0.05$) between the observed lengths of ladybirds compared with that expected if the data are normally distributed'. This formal phrasing includes all the information required. This phrasing is helpful when you are learning to report the results from statistical analysis. However, when publishing in most scientific papers, a looser phrasing is used which has the same information but in a less structured more personal style and which often pulls together a number of different aspects of the investigation into the one sentence.

ii. Numerical values of p

p is a value that indicates the level of significance in your hypothesis testing. It can range from virtually 1.0 to virtually 0.00. In most biological research a p value of equal to or less than 0.05 is said to be significant. The smaller the p value the greater the degree of certainty that the decision we have made is correct. For example, in Box 8.1, the study examined the distribution of holly leaf miners on a single tree using a chi-squared goodness-of-fit test. It was found

Table 12.6 Common format for reporting results from statistical analysis.

p value at which null hypothesis is rejected	Phrasing	Symbol
>0.05	Not significant	NS (not significant)
≤0.05	Significant difference	*
≤0.01	Highly significant difference	**
≤0.001	Very highly significant difference	***

that $\chi^2_{calculated}$ (155.47) is greater than $\chi^2_{critical}$ (5.99) at $p = 0.05$ and therefore we reject the null hypothesis. Looking again at the chi-squared table of critical values, you can see that, in fact, at $p = 0.001$, $\chi^2_{critical}$ is 13.82. Therefore, we can reject the null hypothesis at this higher level of significance. It is important to convey this increased confidence in the decision you have made and this can be done using one of two conventions.

You may emphasize how 'significant' the difference is between your samples by using particular phrases. These are a widely adopted standard and should be adhered to (Table 12.6).

In our example from Box 8.1, we would therefore write: 'There is a very highly significant difference ($\chi^2_{calculated} = 155.47$, $p<0.001$) between the numbers of holly leaf miners found at three different heights on the tree'.

As an alternative to using these particular phrases, a system of asterisks (*) can be used to indicate the level of significance at which you rejected your null hypothesis. If you are using a table that includes the data from your investigation, then it is common practice to incorporate the * symbol within the table. The number of asterisks used is again a standardized convention that you should follow (Table 12.6).

iii. Statistical software and the exact p value

The results from the analysis of data using statistical software usually return the exact p value. Clearly, you are then able to use this value rather than the selected values for $p = 0.05$, etc. The phrasing and use of an asterisk can be adapted for the exact p values. For example, if the significance of the hypothesis testing was reported as $p = 0.0037$, you would use the phrasing relating to $p = 0.01$. This is because the significance is less than $p = 0.01$ but is not significant at $p = 0.001$.

iv. Regression analysis

If you have carried out a regression analysis (Chapter 9), the data are usually represented as a scatter plot and the regression equation and level of significance is added to the figure near the regression line (e.g. Figs 9.14 and 9.16). In 12.1.10iv, we discuss good practice in relation to drawing lines on figures including regression lines.

v. Tukey's test

The outcome from a *post hoc* test such as Tukey's test may be reported in a visual way, where lines are used to link non-significant values. Examples of this can be seen in Chapter 10.

An alternative and more frequently used method is the use of superscripts where the same letter is used to identify those means that are not significantly different from each other (10.8.4).

vi. Variation

An estimate of the probability that the population mean lies within a particular range around our sample estimate is known as a confidence limit. This statistic is usually represented in a figure as a vertical line (Fig. 12.3). Confidence limits can be added to most figure types as long as you have appropriate data to summarize (5.7).

Where parametric data are summarized in a table, you may indicate the variation around a mean value using either the standard deviation (s) from the data or the standard error of the mean (SEM). The first of these provides a valuable indication of the variation in the data in the same units as the original observations. In this case, you would report $\overline{X} \pm s$. If you use the mean and standard error of the mean, this allows a reader to calculate confidence limits for the mean value. In this case, you would be reporting $\overline{X} \pm SE$. Variation around a mean is more usually indicated using SEM. Either way, you must make it clear in row and column table headings which terms you are reporting.

 Look in the journal article you have been reading. How have the statistics been reported?

12.1.12 Discussion

In the results section, you have *reported* your results, the key trends identified in your data and any outcomes from testing hypotheses. In your discussion, you *evaluate* these findings. This is achieved by starting with the points you have identified in your data and providing an explanation and comment about each point. In this section you should determine what your findings indicate about the biological questions you posed at the outset or that have become relevant as a result of the evaluation of your data. Each part of this discussion would usually draw on other reported scientific literature and here we remind you not to rely on abstract sources only and to be a critical reader (12.1.6). You need to have an overall plan for your discussion before you start writing as this will help to ensure that you address all the trends identified in your results in a logical and coherent way.

If the structure of the results section is based on a number of objectives, you may use these as a basis for the early part of the discussion. However, you do not want to fragment your discussion unnecessarily. The discussion is where you pull aspects of your study together.

As part of the evaluation, you need to be open about any assumptions you have made or any possible bias arising from your design. You also need to demonstrate your awareness of faults and limitations within your investigation. There may be some weaknesses in the report as a result of practical requirements; however, careful planning before carrying out the research should ensure that these are minimal.

Recommendations may also form part of your discussion section. You should consider how the point reached by your research can be used to further the investigation into the biological question(s) you posed. This information needs to be integrated into your overall discussion and should not be tacked on at the end so that it appears to be an afterthought.

At the end of your discussion, you should include a conclusion. Some report formats place the conclusion in its own section; however, in most journal articles the conclusion is an integral and final part of the discussion. Conclusions are one of the key parts of your report as this is what the reader will 'take away' with them; it will remind them of the contents of your paper or report and should clearly present your conclusions about the biological questions posed in your study. It is worth thinking carefully about the conclusion and making sure that it refers only to what you found in *your* study and is supported by *your* data analysis. A conclusion should not include any new material: sometimes people wrongly embark on another discussion in the middle of a conclusion. Often there is a desire to make a great flourish at the end of a report, especially if you are writing a lengthy report for an honours project or thesis. This flourish needs to be avoided as it tends to lead you to overemphasize your findings and to make them out to be more than they are.

 Read the discussion in the journal article you have been examining. Note how the discussion is constructed, how the information from other studies is linked to the findings from this study, and the location and construction of the conclusion. How might you improve your own discussion section? Read your institution's or department's guidance. Does it differ? How?

Common errors

A discussion can suffer from the same problems that the introduction section is prey to, including plagiarism and an over-reliance on abstracts. There are further problems which we see occasionally in reports: not basing the discussion on your results.

No results For many students, having just written their results section, there is a sense that the results have been dealt with, and the discussion then plunges directly and sometimes exclusively into a consideration of other people's work with little or no reference to your own results. If you have free-standing facts derived solely from other studies, then you have probably made this error. Planning your discussion before you start writing it and making sure you cover each point identified in your results section in a balanced and systematic manner should produce a balanced discussion.

Believe in yourself You will want to give a good impression of your work and to encourage your readers to follow your thoughts to the same conclusion you have reached but there is also a need to be self-critical. However, too much negativity in your work is very off-putting. All science has its limitations, and these can be pointed out without allowing them to completely undermine your arguments. You need to find the middle way between persuasive writing and self-criticism.

12.1.13 **References: Harvard system**

Correctly referencing your work is essential. There are two parts to referencing a report or paper: you must (i) acknowledge your sources of information in the body of the report and then (ii) include full details of these sources at the end. There are several systems for referencing. In the biosciences, the most widely used system in reports is the Harvard system, which is covered in this section. However, in posters and presentations alternative systems are commonly used, particularly the British standard or numeric system, and we outline this in 12.2.1v. You should check with the journal's guide to authors or your department's regulations to confirm which system is required. Within the Harvard system, you will find some slight variations in the use of punctuation, etc., and again you should check which version is in use by referring to your local regulations.

i. Referencing in the text

The intention of referencing in the text is to acknowledge where the facts you have included in your work have come from. In the Harvard referencing system this is achieved by including the name of the author and date of publication that is the source of the fact. Where this is not possible, some other unique tag that allows unambiguous cross-referencing to the full information in the reference list may be used.

You should acknowledge the sources of information at the point at which you incorporate the facts from these sources into your report. Therefore if one sentence contains information from one source and the next sentence contains information from another source, you need to reference at the end of each sentence. If all the facts in a paragraph have been derived from a single source then the reference can be placed at the end of the paragraph.

 Examine the journal article. Find ten examples of referencing in the introduction and/or discussion. How do the format and positioning of these various reference citations differ?

Bibliography
The full details of all sources of information you have read whilst working on a particular topic.

Reference list
The full details of each source of information that has been used explicitly in the construction of your report.

Harvard referencing system
The most widely used referencing system. This is based on the author's family name and date of publication for in-text citations and the provision of full citation details in a reference list.

There are many possible scenarios for referencing. For example, the type of source may be a journal article or a video or television programme, there may be one or many authors, you may be using a table or figure taken directly from a source, or facts taken from the source but presented in your own words. It is neither feasible nor helpful to cover all possible combinations of these but the guidance in Table 12.7 is constructed so that you may use one or more rows to produce an appropriate reference tag.

ii. Reference list

Your reference citation from the text should link to a unique set of details about the reference which enables a reader to find this source of information. There are two types of lists of these details: a bibliography and a reference list. In biological sciences a bibliography is usually a record of all sources of information you have read, whereas a reference list is a list of those sources of information you have included directly in your report (i.e. you have referred to them). In most cases, you will be asked to compile a reference list at the end of your report and it is to this that the Harvard referencing system really applies, although the format can also be used when constructing a bibliography.

There are two aims for a reference list. Firstly, the tag in the text (*i*) must link uniquely to one reference in the reference list, and secondly, sufficient detail must be provided in the

Table 12.7 Summary of formats for one version of the Harvard referencing system used as a citation in the main body of text in a report. The author's name refers to their family name

Context	Referencing in the text	Example
You refer to the author in the sentence.	Only the date is added in brackets after the author's name.	In an excellent study, Herbert (2005) identified …
You refer to the facts in the sentence.	The author (or other, see below) and date of publication are given in brackets.	A rare moth was recorded in Shropshire during a study of *A. moschatellina* (Holmes, 2005).
You have integrated information from more than one source.	Both citations should be included in either alphabetical or date order depending on local practice.	Studies of the genetics of the cell cycle have demonstrated a high degree of complexity (Cook *et al.*, 2013; Gonzalez *et al.*, 2012)
You are using a table or figure taken from another source.*	You should make it clear that this is taken from another source, and include the author of that work and date of publication.	Fig. 11.1. The variation in lead concentration in a soil sample from Hartlebury Common (taken from Hiles, 2008).
You are using more than one paper by the same author from the same year.	To distinguish between several papers add a, b, etc., after the date.	(Cherry *et al.*, 2015a; Cherry *et al.*, 2015b)
You use information that is shown to have come from somewhere else but you have not read this original source.	You give details of both in the text.	(Sokal & Rohlf, 2011, as cited in Holmes *et al.*, 2016)
You need to indicate particular parts of a document.	The additional information (usually a page number) should be given after the year, but within the brackets.	(Ruxton & Colegrave, 2003, p. 64)
There are two authors for a journal article, book, or conference paper.	You include both authors' surnames and the date of the publication. The authors' names are given in the same order as that used in the source.	(Weaver & Holmes, 2013)
There are more than two authors for a journal article, book or conference paper.	Only include the first author's surname and then add *et al.* and the date of publication. *et al.* is Latin and is generally placed in italics. This is a less common practice now than formerly.	(Alma *et al.*, 2004)
Contributor to a book.	Include the name of the contributor(s) and date of publication.	(Adams-Groom, 2012)
The work is anonymous or the author is not given in the source.	Anonymous may be used in place of the author's name, or use details of the source.	(Anonymous, 2012); (Defra, 2016); (*The Independent*, 2015)
You have personally been provided with unpublished information.	Indicate that this is a personal communication, from whom, and include the date you received the information.	(J. Huffer, pers. comm., Oct 15)
e-journals	Author's name and date of publication.	(Davis, 2011)

(continued)

Table 12.7 *Continued*

Context	Referencing in the text	Example
Web material other than e-journals.	Author's surname (or other) and date of publication (if known) or date accessed.	(Titterstone Clee Heritage Trust, accessed 10.7.15)
Television programme or film.	Give the title and year of publication or date screened.	(My giant life, 3.11.15)
CD and DVDs.	Use the author's surname (or other) and year of publication.	(Joseph, 2014)

*If you are preparing a paper for publication (rather than a student essay or internal report), you need to obtain permission from the author and copyright holder to directly reproduce previously published material such as illustrations. Student work is usually covered by the institution's copyright licence.

reference list for the reader to be able to obtain the reference themselves. All the references in a reference list need to be arranged in a consistent manner to enable the reader to be able to find the full reference details quickly. The reference list can be organized in a number of ways but most often it is in alphabetical order in relation to the first author's family name. In the list of authors' names on a paper for European names this is usually the last name given for each of the authors, for Asian languages such as Chinese this is usually the first name.

 Q19   Examine the reference list in the journal article you have been scrutinizing. Are the references organized alphabetically by author? Are there any papers with very similar authorship? What criteria have been used to order these papers?

There can be many types of information and therefore many possible types of references. We give a guide that, by combining one or more rows, will provide you with the requisite detail for preparing a reference list in the Harvard style for most types of sources of information you will encounter (Table 12.8).

Common errors

Not recording the sources of information One of the most common errors in relation to referencing is a failure to record the sources of information at the time when you are taking notes. Clearly, making good this omission takes a lot of unnecessary work and can be easily avoided.

   Reference citations in the wrong place If you are discussing your ideas and your results, you need to check that any referencing in that sentence is in the right place. Often reference citations are misplaced within a sentence and appear to imply that your own ideas or your own results are in fact those of another author. For example, an undergraduate wrote 'It is clear that the organic content of the soil in this study (Holmes *et al.*, 1999) should be higher.' This implies that Holmes *et al.* carried out the study and not the student. What the undergraduate should have written was 'It is clear from other studies (Holmes *et al.*, 1999) that the organic content of the soil in this study should be higher.' Referencing in the wrong place can also lead to the reverse implication, i.e. that you carried out work that was in fact the result of someone else's efforts. Although this will have occurred in error, it is plagiarism (see 'Cheating', p.398).

Table 12.8 Format of information in a reference list following one version of the Harvard system

Context	Format in reference list	Example
Article in a journal, one author	Author's surname and initials, (the year of publication). The title of the article. The title of the journal (in italics) volume (and part number) where known: page numbers of article.	Holmes, D. S. (2005). Sexual reproduction in British populations of *Adoxa moschatellina L. Watsonia* 25(3): 265–273.
Article in a journal with two authors	As above but list both authors. Use 'and' or '&' between the two authors.	Westbury, D.B. & Dunnett, N.P. (2008). The promotion of grassland forb abundance: a chemical or biological solution? *Basic and Applied Ecology* 9: 653–662.
Article in a journal with more than two authors	As above, listing all authors' names and use 'and' or '&' before the last author.	Cherry, A.L., Dennis, C.A., Baron, A., Eisele, L.E., Thommes, P.A. & Jaeger, J. (2015). Hydrophobic and charged residues in the C-terminal arm of hepatitis C virus RNA-dependent RNA polymerase regulate initiation and elongation. *Journal of Virology* 89(4): 2052–2063.
An authored book	Author's surname(s) and initials (the year of publication). The title of the book (in italics), edition (if not a first edition). Publisher, place of publication.	Ruxton, G. D. & Colegrave, N. (2010). *Experimental design for the life sciences*, 2nd ed. Oxford University Press, Oxford.
Contribution in a book	Contributor's surname and initials (year of publication). Title of contribution. Initials and surname of the author(s) or editor(s) of the book, if the latter include (ed.) or (eds). The title of the book (in italics). The publisher and place of publication, page number(s) of contribution.	Adams-Groom, B. (2012) Forensic Palynology. In Marquez-Grant, N. & Roberts, J. (eds), *Forensic Ecology Handbook*. Chichester, Wiley-Blackwell. pp. 153–167.
Conference paper	Contributor's surname and initials (year of publication). Title of contribution. Initials and surname of editor(s) of conference proceedings. Title of conference proceedings and date and place of conference. Publisher and place of publication, page number(s) of contribution.	Joseph, J. (1992). Art and the biosciences. C. Joseph, ed. American Society of Bioscience meeting 24.3.94. New Orleans. Houston Press, Houston, p. 13.
Thesis	Author's surname and initials (the year of publication). Title of thesis. Award (Ph.D., M.Sc., etc.). Name of institution to which work submitted.	Holmes, D. S. (1986). Selection and population dynamics in *Allium schoenoprasum*. D.Phil., University of York.
Publication from a corporate body	Name of body (year of publication). Title of publication. Publisher, place of publication, report number (if any).	Botanical Society of the British Isles (2014). Code of conduct for the conservation and enjoyment of wild plants. Botanical Society of the British Isles, London.
Newspaper article	Author's surname and initials (if known) or title of newspaper (year of publication). Title of article. Title of newspaper. Day and month, page number(s) and column number.	Verkaik, R. (2005). Police investigate retaliation attacks. *The Independent*. 9 July, p. 21.

(continued)

Table 12.8 *Continued*

Context	Format in reference list	Example
Personal communications	Do not include in the reference list.	–
e-journals	Author's surname and initials (year of publication). Title of article. Title of e-journal. Volume or part if known. Publisher. URL [date accessed].	Baysal, O., Lai, D., Xu, H-H., Siragusa, M., Çalışkan, M., Carimi, F., da Silva J. A. T., and Tör, M. (2013) A proteomic approach provides new insights into the control of soil-borne plant pathogens by *Bacillus* species. *PLoS ONE* 8: e53182. doi:10.1371/journal.pone.0053182
Web material other than e-journals	Author's or editor's surname and initials if known. (Year of publication, last updated), URL [date accessed].	Titterstone Clee Heritage Trust. (Last updated 14.10.15), www.thecleehills.org.uk [2.11.15].
Television programme, film	Title (year of production). Type of material (video, TV, etc.) Director's surname and initials. Production details—place, organization. [date screened or seen].	The boy with the incredible brain (2005). TV documentary. Oxford Scientific, C4. [23.5.05].
CD s and DVDs	Author's surname and initials (year of publication). Title {type of medium}, (Edition). Publisher and place of publication. Any identifier number. [Date accessed]	Joseph, A. (1992). The wild child. {CD ROM}. Houston Press, Houston. [14.3.92].

Not a bibliography   You are usually asked to include a reference list. Therefore you should not include sources of information that have not been referred to in the text.

Only include the sources you have read   Books and reviews are a frequently used source of information. These are derived from other (primary) sources that are referenced in the book or review. If you have not read these primary sources you need to make this clear in your referencing (Table 12.8). This is particularly important, as the authors of the book or review may have misunderstood the original report and misrepresented the findings. If you have not read the original source, you will not know this, but the error will appear to be yours unless you make it clear.

Cheating   A failure to reference correctly, both in the main body of the text and by not providing a reference list when asked to, is plagiarism. This is considered to be a form of cheating. Cheating in this way is increasing with the ease of accessing electronic forms of information. Make sure you avoid this by referencing correctly throughout all written work.

 The Harvard system is used extensively in bioscience publications, although there is considerable variation between journal styles especially in relation to the type of fonts, case, and punctuation used. Examine your journal article and note how the system differs from the format used in Tables 12.7 and 12.8. Find your local regulations relating to referencing. How do these differ from the ones we have outlined?

12.1.14 Appendix

An appendix may form part of undergraduate and graduate reports, including theses or may be provided as an annex for a published paper. There is considerable variation in what may be included in an appendix, which reflects not only differences in practice between institutions and publishers, but also variation in research areas. The golden rule when considering whether to use an appendix is never to include anything that a reader needs to refer to in order for them to understand your report. Therefore, you should not include critical tables or figures in the appendix.

For many undergraduate reports, it would be appropriate to include evidence supporting your compliance with the law such as your risk assessment, ethics approval, confirmation of permission for working in an area, etc. If you have needed to use a consent form, an unsigned copy may be incorporated into the appendix. However, you should not include anything that might compromise the confidentiality of any volunteers, such as signed consent forms or completed questionnaires, if these could be used to identify individuals.

The appendix can also be useful as a repository for raw data, but only if this is not required in the method or results sections. Including this type of information in an appendix can be useful both for the institution and for yourself; however, before you commit yourself to including pages of numbers, check with your supervisor. The data on which most of the worked examples in this book and website are based come (with permission) from undergraduate projects and for many of these the raw data were included in their appendices.

If you have been carrying out an experiment that required the use of stock solutions (e.g. 0.5M hydrochloric acid), you may wish to include details on how the stock solution was made. Similarly, it is usually more appropriate to include details about how well-known solutions or compounds (e.g. Feulgen stain) were made in the appendix rather than clutter your methods with these details. Details about suppliers can be an important piece of information that will allow someone to repeat your experiment and can also be included in the appendix. If you are in doubt as to what information your institution or department prefers in an appendix you should discuss your draft report with your supervisor or tutor.

12.1.15 Your approach to writing a report

When writing a scientific report, you need to consider how to use text, figures, and tables and the scientific report format to best convey the findings from your research. The tendency is to write the report in the format order, i.e. Title, Introduction, Method, Results, Discussion and Conclusion, References, and Abstract. In fact, a much more efficient approach and one that will enable you to avoid some of the problems we have outlined in this chapter (e.g. essay-like introductions) is to write your draft report as follows: Method (and site if relevant), Results, Discussion, and Introduction, and then on reviewing the draft add the Conclusions, Abstract, and finally the Title. From our experience, this order will require the least number of revisions, give you greater mastery of the overall report, and lead to a better, more balanced report at the end.

The reason for this order is that the methods section is usually the most straightforward section and is a matter of reporting from (hopefully) a clear set of notes. Completing one section such as the methods gives you a sense of satisfaction and progress, which can give you a much-needed lift. The results section is the key to your report as it is here that you present the information you will go on to discuss. The results section usually takes the longest time to construct because all the data need to be examined, summarized, and analysed before you begin to write this section, and the communication itself can involve figures and tables that can take some time to select and construct properly. Therefore, we suggest that this is the second section to tackle. With your results in mind, it is then usually easiest to move on to the discussion. One advantage in writing the discussion before the introduction is that by drafting the discussion you will have a clearer idea as to how you need to approach the introduction to avoid unnecessary repetition. The introduction is then the last main section to write in draft. By keeping this until towards the end of the writing process you will have a much clearer idea of what background material is needed in the introduction to properly ground the rest of the report and to allow you to introduce the points you have developed throughout your report.

Writing, even using a structured format, is a creative process and you may find yourself uncomfortable with tackling one section and more comfortable about tackling another. If you feel strongly about which section you wish to work on, then you will probably be more efficient if you go with the flow.

 Make a list of the areas where your report writing could be improved. Identify three points from those on your list that you will tackle when writing your next report.

12.2 Producing a poster

In your career as a scientist or if you go on to work as a non-scientist you will need to use a range of formats to communicate information. All formats require succinct text and often include an element of the visual representation of ideas and data. The balance of these elements varies. In a research poster the written content and the visual tools used in the presentation are of equal importance.

12.2.1 Poster content

There are a number of different types of posters and a useful exercise for you to undertake is to deliberately examine the posters you see every day, from those advertising social activities or marketing products to the posters in your department. Research posters have a specific format similar to that of a research report with an Introduction, Method, Results, Discussion, and Conclusion but they are heavily influenced by how the audience engages with this format. Firstly a poster is designed to be read by one or two people at a time who are standing a few feet away. You therefore need to consider font size and layout so that these readers can easily follow the points you are making. Secondly you can expect that whilst your audience will be prepared to commit more time to your poster than they might give to an advert most will have to stand up. You therefore need to write the content carefully making sure you find the balance between brevity and detail and you can use visual tools such as using bullet points or separating sections into panels as a way of drawing attention to specific elements within the poster.

Research posters are often displayed in a dedicated poster session where your work is presented alongside that of many other scientists. It is therefore important to make your poster eye-catching and with clearly stated key points that will stay with your audience even if they read all the many posters at the symposium.

i. Title

The title will be used by your potential readers to decide whether they are going to stop and read your poster or move on. You need to consider your audience and think about who you want to read your poster, and make sure the key words appear in your title. Your title therefore needs to entirely convey the research you are presenting incorporating these key words but also be succinct. Your title is your only marketing tool so you need to make good use of it. One approach to consider is to pose a question or to make a statement followed by a question. The alternative is to use a format similar to that used when writing a report or scientific paper.

The visual impact of the title is also important. The font should be a bold stocky shape with the maximum font size that still allows the title to fit on one or two lines at most and where the letters are about 6cm in height. If the title spreads onto a second line then you should use 1.15 or 1.5 line spacing to improve the visual appearance without taking up too much space. A colour can be effective if used sensibly. A bold colour that has high contrast with the background is worth considering, or white on a dark background.

ii. Authors

It is common practice on a poster to include the names of the authors. As an undergraduate or graduate at a conference it is the convention for this list to include your supervisor's name after your own and to include the name of your institution. If you are preparing a poster for an assessment the names are restricted to the undergraduates who worked together to produce the poster. If work is to be marked anonymously you may be asked to include a student number rather than your name. The names of authors should be included in a low mid-sized font so that the final size of the letters is about 2–3cm (on this A4 page this would be about 72 point). The font should be the same type as that used elsewhere, such as in the main sections. Too many different fonts on a poster looks messy.

iii. Conference title

If you are attending a conference then the conference title can be included on the poster along with the date. This has the advantages of indicating that the work is being presented uniquely to this conference and when you come to display the poster in your own institution it indicates the conference the work was prepared for. When preparing a poster for an assessment you are usually asked to indicate some course details.

iv. Sponsors

If you have received funding or are part of a research group then you should acknowledge this on your poster. Sources of funding can be indicated by including the logos of the sponsors usually either at the top or bottom of the poster.

v. Introduction

The introduction needs to achieve the same as that for a report (12.1.6) but in far fewer words. You therefore have to carefully select and succinctly explain critical background which also conveys the rationale of the work. You need to do this in four to five paragraphs of prose written in the passive past tense. To achieve this you may wish to use tables to summarize the findings of previous key research or the definitions of key terms. The aim will appear towards the end of this introduction and should be identified explicitly.

British numeric referencing system

The citations in the text are numerical and reflect the order in which the citations appear. The in-text citations link to numerically listed reference details.

Your introduction should be fully referenced using the Harvard style or the British numeric system. In this context the British numeric system has an advantage as it is more compact, which can be helpful if space is limited. In the British system references are indicated in the text using sequential numbers. The first citation in the introduction is therefore numbered 1, the second is numbered 2, and so on. These numbers are used instead of the authors' names and date of publication. If a reference is used more than once then the original number is used every time the reference citation appears. The reference list is then presented in numerical rather than alphabetical order.

 An author submitted a review for a general science publication which required the use of the numeric British referencing system. The paper was rejected and on redrafting the author submitted it to a different publication which required the use of the Harvard system of referencing. In what order would these references now appear?

1. Treu, R. Holmes, D.S., Smith, B.M., Astley, D. Johnson, M.A.T & Truman, L. (2001). *Allium ampeloprasum*: an isoclonal plant found across a range of habitats in S.W. England. *Plant Ecology* 155: 229–235.
2. Bougourd, S.M. Plowman, A.B., Ponsford, N.R., Elias, M.L., Holmes, D.S. & Taylor, S. (1995). The case for unselfish B-chromosomes: evidence from *Allium schoenoprasum*. Kew Chromosome Conference IV. Ed. Brandham, P.E. & M.D Bennett.
3. Holmes, D.S. & Bougourd, S.M. (1989). B chromosomes and selection in *Allium schoenoprasum*. I In natural populations. *Heredity* 63: 83–87.
4. Holmes, D.S. & Bougourd, S.M. (1991). B chromosomes and selection in *Allium schoenoprasum*. II In experimental populations. *Heredity* 67: 117–122.

The introduction can be presented in sections with subheadings which can be numbered. This can be an effective visual tool for conveying a sense of the order within the introduction and the way the poster is to be navigated. The introduction will be read from a distance of several feet so you need to select a black font, at least 14 point size and with at least 1.15 line spacing.

vi. Method

The method needs to be carefully constructed. It is often not possible to include the full method details so you need to determine what you can abbreviate without losing your audience. If your audience work in similar areas with similar techniques you can use generic terms for a method such as 'PCR' whilst giving little further detail. If you are in doubt this should be discussed with your supervisor.

The method can be written as a list of bullet points as this is often a more succinct way of presenting the information. You may also find it useful to split the method into sections with subheadings. However, if you take this approach you need to number the subheadings or in some way make it very clear in which order the information should be read. The font size and type should be the same as the introduction. In your university studies you may need to follow a specific format and present the method in a more usual format as prose in paragraphs.

vii. Results

In the results section you can provide a brief overview, list the key findings using bullet points, and then carefully select the figures and tables that best illustrate these trends. Adding additional information to figure or table headings or colour coding elements can be an effective way of communicating a lot of information in few words. Unlike reports, whilst the key points in the text are still important, it is the detail in the figures and tables that will convince the readers that what you say is correct. Therefore you need to ensure that these details are clear. Your figures and tables should all be at 14 point or greater and the text in this section should match that in the method and results section for style. Other important features should be similarly sized. If you need to distinguish between lines on a figure or bars on a chart then using two methods such as colour and shading or colour and dashes with a key makes it easier to see the differences.

viii. Discussion

In the discussion you should aim at four to five paragraphs of text in which you present your evaluation of the findings and the debate about what this means to the biological question you posed. Limitations of the study should be integrated with the evaluation as should consideration for the next step in the investigation if this is relevant. The referencing style should be the same as in the introduction and may either follow the Harvard or the British numeric system. The discussion should be presented in the same font as the introduction and can also be subdivided to draw attention to the different points considered in the text. The discussion should be completed with a conclusion. Along with the title it is the conclusion that is the most important part of the poster—the take-home message. You need to keep it short (one or two sentences) and upbeat whilst accurately reflecting the findings from your study.

ix. Reference list

The inclusion of a reference list on a poster is always problematical as it takes up space. However you must properly acknowledge all the sources you have cited on the poster. (We suggest a solution to the space problem in 'additional material'.) If you are using the British numeric system for referencing then the papers are listed in the order that they appear on the poster and are numbered to reflect this order. If you are using the Harvard system then the references are listed in alphabetical order in relation to the author's family name.

12.2.2 **Poster construction**

Posters are usually produced electronically using publishing or presentation software. Your institution may have the facility to print posters but if they do not there are numbers of companies that offer this service. With this in mind, you should, as your very first task, decide if and how you are printing your poster. This is critical as it can determine the software you will use, which in turn can determine how much space you have for the content and how much flexibility you have with backgrounds and fonts. The printing of a poster incurs a cost which you should also consider at the outset. If professional printing costs are not within your means then you can produce a competent poster using card, spray glue, and a guillotine or straightedge and knife.

Conferences or assessment details will tell you the size of a poster though it is usually A1 in a landscape orientation. If you are making your poster then the hardest technical aspect to solve is how to produce a large but neat title. If you are using software the most common disappointment comes with the transfer of your design from the screen into print. There are no easy solutions to either of these challenges so we are pointing them out so you are fore-warned and prepared to give more time to these aspects than you might have anticipated was needed.

As you learn to write reports you quickly realize that writing the report is only one of three actions involved, as you also have a design element, such as selecting the most appropriate format for a figure, and a technical production aspect, such as finding a printer that is free and works. Posters have the same elements and in the same way the design and production steps take more time that you may anticipate to complete.

12.2.3 **Principles of layout on a poster**

There are four principles that inform how you design a poster: rule of three; eye line; contrast; and navigation.

i. Rule of three

Generally we are used to seeing text on A4 sheets. As a result research posters tend to be built around this basic sizing. This means that for an A1 poster you can normally expect to fit three columns of information into the space. These three columns from left to right are roughly given over to: the introduction and part of the method; the rest of the method and the results; and the discussion and reference list (Fig. 12.4).

ii. Eye line

A poster is read by an audience who are standing up, therefore the parts that are easiest to read are usually those in the middle of the poster. Placing figures and tables at or slightly above this eye line make it easier for the reader to see the detail. Text is easier to follow even if you are having to bend down or peer upwards to read it. An exception to this is the conclusion. This is such an important part of the communication role of your poster that you do not want to bury it in the bottom right-hand corner of your poster if you can avoid it.

iii. Contrast

Contrast on a poster makes the difference between having an audience and not. Therefore, although you can add in interesting and relevant 'wallpaper' and use a wide range of colours and fonts, this is not in your best interests. Use your creative skills but use them wisely. No backgrounds, one or two font types, and black with one colour, possibly with one or two shades, is more than sufficient to produce a professional-looking and effective research poster.

iv. Navigation

Posters are primarily background space on which you arrange sections of text, figures, and tables. This physical discontinuity between the sections means that you need to clearly show the order in which you want the information read. The simplest way to achieve this is to number the panels or number headings and sometimes subheadings. It is always important to cross-reference to figures and tables in the text and this applies to posters as much as reports.

12.2.4 Laying out a poster

The elements of a poster can be described as panels. For example the title would be a panel, and the names of the authors a separate panel. You have to decide what to put together onto each panel without making any one panel too long, and you then need to place each panel on the poster. We illustrate one common layout for a research poster which can be adjusted depending on the number and type of panels (Fig. 12.4).

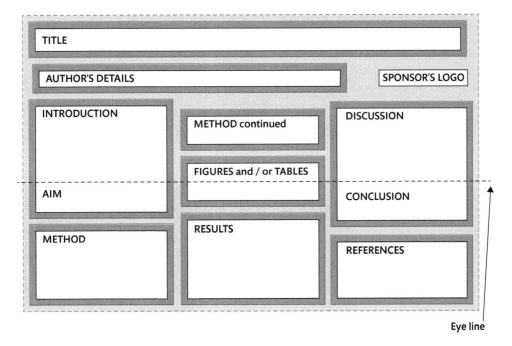

Fig. 12.4 A standard plan for a research poster illustrating the use of contrasting borders and a layout that takes into account the eyeline of the audience

One effective method for ensuring panels stand out from the background is to place a contrasting margin round the edge. If you are making your poster you do this by printing the text onto white paper and mounting the paper onto card before then mounting the whole onto the poster. The card is a contrasting colour to the background. By stiffening the paper you also find it easier to mount the panels on the poster. When using software the same principle of contrast applies (Fig. 12.4).

12.2.5 Additional material

You can maximize the value of your poster by using its potential three-dimensional element. Adding a small open envelope to the bottom of the poster can add interest to the design and provide you with a means of promoting yourself or extending the information you can otherwise cover. Depending on the circumstances, multiple copies of material placed in this folder can include your reference list, or a list of useful web addresses for further information on the topic, or business cards with your professional online and contact details.

Common errors

Many of the common errors we encounter in student posters are the same as those in reports. There are additional points, however, relating specifically to the presentation aspects of the poster.

Poor presentation We have encouraged you to consider font size, contrast, choice of colour, and positioning of the various elements in your poster. All these can be used to improve the effectiveness of the poster as a communication tool or can make it unreadable. This design and production aspect of producing a poster is therefore worth spending time on.

Cannot see in what order to read poster content One of the values of the standard format for a poster shown in Fig. 12.4 is that this mirrors the format of a report so is a construction that all your readers will be familiar with. If you select a different format then you need to take extra care to ensure that the order in which the information should be read is very clear. Headings, subheadings, and numbering sections is the simplest method you can follow to achieve this.

12.3 Presentations

The third method used to communicate your results is that of an oral presentation. In a presentation you take on the role of actor: you know your lines, you are familiar with the stage, and you know when and how to interact with the props. One way to develop your presentation skills is to critically observe your lecturers and see what works and what is annoying and off-putting. Your challenge is to improve on their performance.

12.3.1 The role of the presenter

The role of the presenter is that of an actor. You are there to inform, entertain, and persuade so that your audience are left with a memorable impression of your work and its final conclusion.

i. The explanation

Verbal communication styles differ considerably. For example, your words and tone of voice will be very different if you are talking with friends compared to giving a speech. In a presentation of research results you are talking with acquaintances and strangers. The approach is therefore semi-formal but not as formal as giving a speech. There is a really simple approach that you can take to achieve this: do not write out what you will say in full. If you do write out your text in full and then read it out word for word it sounds as though you are reading out an essay which is not a conversational style. Either use the slides in your presentation to help remind you of what you were going to say or write very brief bullet points on prompt cards as a reminder.

ii. Your interaction with the visual slide presentation

Most people use commercial presentation software to produce slides that support their verbal presentation. In our acting analogy the slides are the props and your interaction with them is critical in bringing the whole presentation together as a smooth integrated performance. The key word here is interaction. When you develop and practise your presentation think about how you will physically relate to the slides. Commonly you should point out key phrases or explain and expand information in figures, diagrams, and tables. This interaction means that you and the slides belong together as complementary elements in the presentation.

iii. Your interaction with your audience

No actor ignores the audience. They speak to them, look at them, and interact with them. If you are having a conversation with someone you do not turn your back on them. All of these are common difficulties to overcome when learning the skills of a successful presenter. There are a number of approaches you can take to help you avoid these pitfalls.

You may wish to look at your slides so that you can keep track of where you are in the presentation. You can do this to some extent by using the computer screen but where you need to turn towards the projection to point out relevant details you should make sure that as far as possible it is only side on and you then turn back to your audience.

Looking at your audience can cause some people concern. Be reassured, *en masse* your audience will no longer appear to be individuals. Nonetheless, if this is a troublesome aspect for you then the trick is to look at the far back wall, just over the heads of your audience. If you take this approach you will still appear to be talking to your audience.

Looking at your audience is important as it both engages their attention and also makes you more audible. You are more able to gauge how loudly you need to speak if you address the person in the back row. If you are nervous you tend to speak more quickly and you may need to make a conscious effort to speak slowly so that your audience can follow what you are saying.

If you present enough you will come to learn about any annoying speech habits you have. As an undergraduate one of us had a lecturer who said 'umm' very frequently in all their lectures. This became so fascinating that we would count the 'umms' rather than listen to what was being said. To be an effective communicator it is therefore of value to learn about such distractions in your speech and overcome them.

One approach teachers and lectures take is to pace about. This can keep your audience focused on you and therefore on your presentation. However, many lecture theatres are designed with a central console which has both the computer and a microphone. If this is the case and you pace about then your words will become louder and quieter depending on how close you are to the microphone. You also have to be adept at timing your pacing so that you can still interact with your slides on the screen. For anyone who needs to lip-read, a presenter who moves about can mean the presentation becomes in essence 'inaudible 'to them. So our general advice as you develop your skills would be for you to position yourself in one place where you can be seen and heard and stay there.

12.3.2 **Visual slide presentation**

Most presenters use commercial software to produce a supporting visual presentation that complements their verbal presentation. The primary role of this visual presentation is to inform the audience whilst a secondary role can be as a prompt for the presenter. However, this secondary role should not be allowed to dictate the content and format of the slides.

i. The content

The structure of the visual presentation is therefore based on the audiences needs, which are to: know where the presentation is going; be able to read the key points; be given details about key references; see figures and tables presented in a form that allows them to understand the key points that these are making; and have a summary of the main take-home message(s).

To achieve these objectives a presentation based on a research project therefore has: a title and name of presenter(s); a slide with an overview of the sections of the talk; an introduction providing the background and rationale; the aim and, if there is more than one, the objectives; the method; the results; a discussion; a conclusion and or summary or overview; and a reference list. Each of these sections has to achieve the same as that in a report but does so with very little text. Therefore, the ideas have to be distilled into one or two bullet points that you then expand on verbally. The format for the reference citations on the slides is most commonly the Harvard system. However, to save space the British system can be used. If you are presenting at a conference the Harvard system is best as some of your audience will be familiar with the literature so knowing the names of authors you refer to during the presentation has value.

ii. The design

a. Bullet points

There is a temptation to use the slides of a presentation as an alternative form of prompt. This goes against the primary role of the visual presentation which is for your audience's benefit. Your audience will not be able to read paragraphs of text quickly enough to follow what you are saying. You therefore need to translate the oral content of your presentation into a series of very short bullet points with additional diagrams, figures, and/or tables.

b. Font size and color

Simple clear fonts such as Arial and Times New Roman are most easily read, with the standard font size being 24–28 point for text and 40 for titles. The contrast between the text and background is critical. So black on white or white on a dark background is usually the most effective. The appearance of colours on the large screen does not match what you see on your computer screen, especially if there is additional lighting in the lecture theatre. It therefore helps to test your design in a similar room to see if your use of colour is effective in practice.

c. Background

There are many standard designs that the presentation software allows you to select or to develop. These can add interest but should not be used at the expense of contrast. Background often looks washed out when used in a lecture theatre so if you really wish to add colour then a very dark background colour and white text will achieve this and still provide good contrast. If you use one background for all slides you need to check that figures and diagrams are not masked by the background.

d. Navigation

Apart from providing an overview at the beginning and a summary at the end it helps the audience orient themselves within the presentation if you include numbered headings and sub-headings.

e. Animation

All commercial and free presentation software includes the facility to animate the elements in your presentation. Again this can add interest but please use this facility wisely. Animation can be used to enhance your presentation. For example, it can be used to allow you to build up step by step the content of one slide, allowing your audience to consider each point one at a time. Adding animation to figures can draw the audience's attention to specific aspects to coincide with your comments and can allow you to present two sides to an argument or an answer to a question without switching between slides. However, each time you ask the text to fly in or fade out uses up additional time in your presentation, so too much of this can interrupt the flow of your talk. One free presentation software program has made animation a major element of its basic format and the audience has been known to experience motion sickness as a result of all the comings and goings!

f. Videos and hyperlinks

If you embed videos or hyperlinks you need to anticipate two things. Firstly it will take some time for the computer to find and open these for you which interrupts the flow of your presentation. Secondly, whilst these links may work on your computer there is no guarantee they will work in another location. Therefore you need to be confident that the content is really worth the time given to it in your talk and be prepared with some additional slides that explain the material in case the links do not work.

g. Timing

There are two elements to timing you may consider. One feature of the software allows you to run the slide show automatically. We have yet to go to a presentation where this worked. There are so many factors that can change the exact timing of your delivery of a presentation. In addition, disabling this facility if it goes wrong interrupts your talk. We would therefore suggest this is not a useful tool for a research presentation.

The second aspect of timing is how many slides you should prepare. This will depend on how nervous you may get as we tend to speed up and miss out bits when nervous. It also depends on how much you wish to expand on each slide. For a complex figure you may spend several minutes discussing it, for text the time spent on a slide is likely to be much less. As a rough guide you would expect to use about 30–40 slides for a 1-hour lecture or about five to six slides for a 10-minute talk.

12.3.3 **Preparation**

As in all forms of communication, good preparation can make a lot of difference to the final output. You need to think ahead and anticipate any difficulties that might arise and plan for them. In the case of an oral presentation it is worth practising your lecture so that you become confident in the timings relating to longer explanations or the use of embedded features such as a hyperlink. If you are able to practice in a room similar to the one you will be in this will allow you to: practise loading your presentation; make sure all the features work and the contrast is effective; practise how you will physically interact with your slide show; plan how to move from one slide to the next; gauge whether you need to turn the room lights off; and, finally, check that you are audible. You may elect to take your presentation on your laptop but in this case you should take a backup. The quickest and most effective backup is on a memory stick as other remote sources are slow to access. In addition to practising, it is worth having a printed copy of the visual presentation in case for some reason you experience a technical failure.

12.3.4 **Nerves**

Nearly all presenters, even experienced ones, get nervous. The issue is not the nerves but what you do about the effect they have on your body. We have encountered numbers of ways people, including ourselves, cope with their nerves before giving a presentation. The first of these is to give yourself some encouragement. A well-known television personality used to tell herself to 'sparkle'. You can approach the task as an actor and in so doing avoid the nerves that belong to you and be someone else. If you shake then the lectern is a great place to stand behind and if necessary hold on to. Remember, your audience want you to be a success, they are your supporters.

 Identify one thing that we have suggested that you are going to try when you next give a presentation.

Summary of Chapter 12

- One aim of this chapter is to encourage you to review and improve your writing skills in relation to writing scientific reports in the format usually used in science journals (12.1).

- We examine each section of a scientific report, look at its construction, and discuss approaches towards identifying and addressing common errors.

- We discuss points to consider when preparing and presenting a poster particularly where these differ from the production and design issues that relate to a report (12.2).

- We review the points you should consider when preparing for an oral presentation in terms of yourself as the presenter and the slides you may produce using commercial presentation software (12.3).

- The Online Resource Centre includes interactive exercises that test your understanding of this chapter with other topics considered earlier in this book.

Answers to chapter questions

A2   a. 5mg of sodium chloride were added to water and stirred until it was dissolved.

    b. It is clear from the literature that ice is less dense than water.

A13   An interval scale of measurement (length of leaves (mm)) has been used but the points along this scale of measurement have not been determined by the investigator and are scattered. To impose some order on the representation of this data, a histogram with classes is more useful than a line graph (Fig. 12.5).

A14   Yes. The amount of corn given to the cows is under the investigator's control and is the independent variable. The weight of the calves at birth is the responding dependent variable.

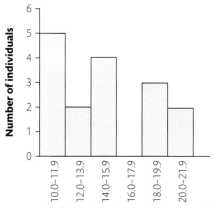

Fig. 12.5 Frequency distribution of leaf length of *Secale cereale* seedlings after 3 weeks' growth

A21 Bougourd, S.M. Plowman, A.B., Ponsford, N.R. Elias, M.L. Holmes, D.S. & Taylor, S. (1995). The case for unselfish B-chromosomes: evidence from *Allium schoenoprasum*. *Kew Chromosome Conference* IV. Ed. Brandham, P.E. & M.D Bennett.

Holmes, D.S. & Bougourd, S.M. (1989). B chromosomes and selection in *Allium schoenoprasum*. I In natural populations. *Heredity* 63: 83–87.

Holmes, D.S. & Bougourd, S.M. (1991). B chromosomes and selection in *Allium schoenoprasum*. II In experimental populations. *Heredity* 67: 117–122.

Treu, R. Holmes, D.S., Smith, B.M., Astley, D. Johnson, M.A.T & Truman, L. (2001). *Allium ampeloprasum*: an isoclonal plant found across a range of habitats in S.W. England. *Plant Ecology* 155: 229–235.

How to choose a research project

Many undergraduates are expected to carry out a piece of independent research during their final year of a degree course. For some this may be literature-based. In some institutions, the topics are listed and you indicate a preference. Some students, however, are asked to choose and design their own research project. This checklist is designed to help the last group of students in particular, but some points are also relevant for students choosing topics from a prescribed list.

Your honours project in your final year can be used to develop skills or contacts that will be valuable for obtaining employment, or the topic may be chosen because it particularly interests you. Above all, it must be a project you are enthusiastic about, as, of all your degree work, it is this research project that demonstrates your abilities as a scientist in your chosen field and as a graduate to be independent in your learning.

Most institutions do not require you, as an undergraduate, to carry out original work. You will, however, be expected to show some initiative. This means that, although you must not copy other people's work, it can usually be adapted and extended. For example, you might apply the same method to a different location or species. However, you must check your local regulations to confirm this.

The checklist that follows is a series of prompts that will help you come to a better understanding of yourself and your interests and therefore help you to focus in on a topic of research that will most suit you.

A1 How to choose a research project

Here is a checklist to help you choose a topic for your independent research.

1) Do you already have an idea?

   YES (go to 14)   NO (go to 2)

2) What do you wish to do when you complete your programme of study?

   You can often use your honours research project to strengthen links with future poten-tial employers, develop skills relevant to your planned career, or provide you with school experience.

3) When would you like to carry out the work?

   Some research is seasonal and it may not be practical for you to carry out research at a par-ticular time of year, e.g. fieldwork, observations of reproductive behaviour in vertebrates or invertebrates, or a survey of student opinion.

4) Where would you like to carry out the work?

   You may prefer to work from home or you have contacts abroad, etc.

5) Do you have any potentially useful outside contacts?

   Undergraduates may be able to set up projects with schools, sports clubs, support groups, businesses, charities, or industry. Speak to your supervisor/personal tutor about this as there will be institutional regulations that relate to such work.

6) Do you prefer fieldwork or laboratory work?

   Most people have a preference for one or other of these and this can guide you when choosing a project.

7) Which part of your course have you most enjoyed?

8) Has there been a particular subject in your course that you would like to investigate further?

9) Often lecturers and PhD students have small research projects they would like to see carried out. Ask.

10) Organizations such as English Nature and the Wildlife Trusts invariably have a list of projects they would like undergraduates to carry out. Look at their websites, or get in touch with them or ask a member of staff who is likely to have contacts with the organization that interests you. Professional bodies such as the Genetics Society or the Royal Society of Biology sometimes offer funding opportunities for honours projects. Organizations such as Operation Wallacea offer opportunities to carry out honours projects abroad and some companies such as Pharmacia can offer internships that can allow you to carry out work suitable for an honours project.

11) Go to the library and look through a journal in your area of interest, e.g. *Journal of Biological Education, Journal of Ecology, Heredity*, etc.

   See what other research has been done. Can you extend or repeat part of this work?

12) Look at projects that have been carried out by students in the past. Research reports such as these are often held in your institution's library.

13) Now do you have an idea?

   YES (go to 14)   NO (see your supervisor/personal tutor)

14) Check that your institution has the equipment you will need and that you can be supervised by someone who works in a similar area. Contact the lecturer whose research area is most similar to the topic you are interested in. Ask them: is this topic suitable?

A2 Common problems

From our experience, we are aware that students' choices of project often suffer from problems. The common ones we encounter most often are listed below. Check your ideas to make sure your project does not suffer from any of these.

A2.1 **The topic is too broad**

Often students are enthusiastic and find it really hard to narrow down their topic and to remain focused. One approach is to use a flowchart or mind map to show how different sections interlink. Then select from this a central narrow area. You can always expand later if you have time.

A2.2 **The research topic is vague**

Some of the proposals we have seen from our students are very vague, such as 'something on fish'! Vague choices usually reflect a lack of thought and planning. They can also occur when no thought is given to the justification for the research. It may well be possible to examine something, but do you have a scientific reason for doing so? To resolve this make sure you write a draft aim and objectives at the outset as this will help to guide your planning. Draw a flowchart of the investigation and the thinking behind it, as this can also help you to see whether your choice makes sense.

A2.3 **Time management**

For nearly all researchers from undergraduates in their final year to research scientists, there are time constraints that can have an impact on your research. These fall into two groups:

i. Not appreciating the seasonality of some work

We have referred to this in the checklist in A1, in that there are often time constraints on the work you may wish to carry out. For example, if you wished to examine the association between asthma and pollen, the work would have to be carried out when the pollen was present.

ii. Not thinking about how long it will take to complete the work

When carrying out research in a new area, this is always a point to remember. For example, in Chapter 2 (Example 2.2) we described an undergraduate project in which the student investigated the relative effectiveness of tea-tree oil and triclosan as antibactericides in a GP's practice. To collect the swabs from the volunteers and to inoculate the plates took 13 hours of continuous work. The student had carried out a pilot study and so was aware that this part of the procedure would take a long time and she was able to make suitable arrangements before she started.

When planning your work, it is beneficial to draw up a detailed timetable and, if you are not familiar with the techniques, to have a trial run to enable you to judge how long it will take to complete each part of the investigation. If your work is to be reported, then ensure you also leave enough time for panicking, computer downtime, drawing figures, printing, and binding. A general rule of thumb for all time management decisions is to work out carefully how long you think it should take and then double the time you have allowed.

A2.4 **Independent learners**

Undergraduate and graduate research projects are usually intended to allow you to show your abilities as an independent researcher. Therefore, they are invariably less formally structured and you will not be chased up by staff. Attending regular meetings with your supervisor and drawing up and sticking to a timetable are essential. Although research is meant to demonstrate your own abilities, this does not mean you should work in a vacuum. Students who do this nearly always perform badly. This is not what is intended by the words 'independent learners'. Although your peers are not necessarily having the same problems that you are, use their support, but also use supervisory support and that of other established researchers.

A2.5 **Trials and tribulations**

Research is almost always unpredictable and things go wrong. If you are getting upset, see your supervisor and if they do not seem to be sympathetic, go to another lecturer or a counsellor. Try to keep things in perspective.

A2.6 **Non-significant results**

Most practical work carried out by undergraduates is designed by your lecturers to illustrate differences between treatments; therefore, you get into the habit of expecting significant differences when testing hypotheses. As a result, a non-significant result is often automatically assumed to be a mistake and dismissed as such. You must be confident that your results are a true reflection of the real underlying biology. Be critical of your work but do not undervalue the importance of non-significant results.

A2.7 **Cheating**

There are many temptations in research to cheat. Perhaps you have an outlier and would like to ignore it to simplify your analysis. You may have missing data because an organism died or a treatment did not work. Failure to acknowledge all your real data or creating data is cheating and will carry significant penalties both as an undergraduate and as a professional scientist. You will not be penalized for being honest.

Maths and statistics

In this book, we come across a number of symbols and mathematical procedures. For those of you who are not familiar with these symbols or are not confident using them and the maths, we give further explanations and worked examples. There are questions at the end of each section to test your understanding. The answers are at the end of this appendix. If you work through all the questions, it will take about 2 hours. In addition, in the Online Resource Centre, we include the full step-by-step calculations for each of the worked examples that we have included in the book.

B1 Calculators

In this appendix, we encourage you to use some functions already integrated into your calculator. This requires you to own a statistical calculator. There are three types of calculators: ones that will carry out basic steps in maths such as addition and subtraction and no more; a statistical calculator; and a programmable and/or graphical calculator. The first of these does not offer you enough built-in functions to be useful and the last has considerably more than you will ever need and is unnecessarily difficult to master. For graphics and more complex statistics, use a computer package. Statistical calculators are not expensive. Look for one that calculates n, Σx, Σx^2, and the sample variance or standard deviation. Other useful functions have been highlighted below.

B2 Add, subtract, divide, multiply, equals, more than, and less than

In this book, we use a number of symbols. These are a way of giving instructions to you, telling you what type of calculation to carry out.

You will be familiar with + (add), − (subtract), × (multiply), and ÷ (divide). Some of these processes can be written in other ways.

Multiply can be indicated as ab (i.e. a × b) or if you have more complex sums you may use brackets $(1 + 2)(2 + 3)$. This tells you to multiply the result from the sum inside the first set of brackets by the result from the sum inside the second set of brackets. In this case, this would be $3 \times 5 = 15$.

Division can be written as ÷, as in 4 ÷ 2, but it is more usual to write 4/2. If the calculations are more complex, division may be written for example as $(6 \times 3)/(4–2)$.

The top part of an equation where you are asked to carry out a division is called the numerator. The lower part is called the denominator.

You will also be familiar with the symbol = (equals) as in $4 \div 2 = 2$, or $4/2 = 2$. There are a number of other symbols that can also be used here. For example, we use $\approx$, which means 'approximately equals'. Two other symbols used in place of the equals sign are $<$ and $>$. The $<$ symbol means less than, and the $>$ symbol means more than. For example, $p < 0.05$ means that p is less than 0.05. If $p > 0.05$, then p is greater than 0.05. These two symbols can be combined to set limits to a value. For example, $0.05 > p > 0.01$ means that 0.05 is greater than p which is greater than 0.01, i.e. p is a value somewhere between 0.05 and 0.01. We also use these symbols when writing out specific hypotheses, e.g. $A > B > C$ would mean that the median of sample A is greater than the median of sample B which is greater than the median of sample C.

 Write in English your understanding of the following expressions:

    a. 2/4

    b. xy

    c. a/b

    d. $p < 0.001$

    e. $12.2/2 \approx 6$

B3 Brackets and absolute values

The next group of symbols comprises those that are placed around parts of a sum. These are collectively called brackets. The first set of brackets used is usually the round brackets ( ). If you need to surround round brackets with another set of brackets, then you usually use square brackets [ ]. For example, $[(6 \times 2) - 2]/2$. You carry out the calculation inside the brackets first. So for this example you first work out $6 \times 2 = 12$ and subtract 2, i.e. $12 - 2 = 10$, before carrying out the division, which will be $10/2 = 5$. Another example of using two sets of brackets is $[(5 - 1) \times (4 - 2)] - 3$. Here, you work out the calculations for the inside brackets first, so $(5 - 1) = 4$ and $(4 - 2) = 2$. You now complete the sum inside the square brackets $[4 \times 2] = 8$ and finally complete the remaining steps in the sum, $8 - 3 = 5$.

    Another pair of symbols that can enclose parts of a calculation is the straight brackets | |. These are used to tell you that when the calculation within the brackets is completed, the sign of the answer is positive, even if the calculation produces a negative number. These are called absolute values. For the example $|3 - 1| = 2$, the sign is already positive, so being an absolute value has made no difference. But for $|5 - 6| = 1$ the actual sum of $5 - 6$ would normally equal -1. As this is an absolute value, we ignore the negative sign, so the answer is $+1$.

Absolute

If straight brackets, e.g. $|3 - 1|$, enclose a calculation or value this indicates that the absolute value is wanted. Should the result from the calculation or the value be negative you may ignore this negative sign.

 Complete the following calculations:

    a. $(10-2)/[(3\times2)-2]$

    b. $2(x-1)$ where $x=2$

    c. $|(6+3)/-3|$

Table 10.4 Height of shells in two putative species of periwinkles from the mid- and lower shore at Porthcawl, 2002

Shell height (mm)			
Periwinkles on the lower shore		Periwinkles on the mid-shore	
5.5	4.0	3.3	6.7
8.4	5.0	6.3	5.7
5.0	6.2	6.1	4.2
5.0	5.0	8.0	6.3
5.6	7.7	13.5	6.0
4.8	6.0	5.3	7.2
8.4		6.7	

B4 N, n, x, y, Σx, Σy, and Σxy

To show you how the terms n, N, x, and y can be used, we have included a table from Chapter 10 (Table 10.4). In this study, periwinkles from the lower and mid-shore at Porthcawl have been collected and the shell heights recorded.

We have two columns of data. Either to simplify our explanations of calculations using this data or when explaining about choosing and constructing figures, we need to identify one column of data. For this we use the letters x and y. In this example the periwinkles on the lower shore are the x observations and the periwinkles on the mid-shore are the y observations. There are 13 x observations, which we would record as $n_x = 13$. There are also 13 y observations, so for this sample $n_y = 13$. In many calculations, you need to work out $n - 1$, which for both of these columns of data in this example would be $n - 1 = 13 - 1 = 12$. If we needed to know the total number of observations in our data, this would be $N = n_x + n_y = 13 + 13 = 26$.

In some calculations (e.g. Chapter 10), you need to know the totals by adding all x observations together or all the y observations together. The symbol used to indicate this is Σ. In this example, $\Sigma x = 5.5 + 8.4 + 5.0 + \cdots + 5.0 + 7.7 + 6.0 = 76.6$. The same can be calculated for Σy.

Another example of the use of x and y can be seen in Example 9.4. In this example, 13 chickens have been examined and the amount of feed each has eaten and the hardness of the egg each produced has been recorded (Table 9.8, reproduced here). In this example, there are 13 items (pullets) so for each column of data $n = 13$.

In this example, instead of two separate samples as seen in Table 10.4, two observations have been recorded for each pullet: the amount of food supplement (g) and the hardness of shells. Although these are not independent samples, we still use x and y to indicate which column of data we are referring to. In this example, the amount of food eaten is x and the hardness of eggshells is y. Unlike the previous example, one of these columns of data is said to be dependent and this is always called y; the other is said to be independent and this is always called x. We explain these terms in Chapters 7 and 10.

As we showed you in relation to Table 10.4, you may need to calculate the totals for each column, i.e. Σx and Σy. The idea of a sum (Σ) can be extended to the product xy. The symbol xy means that an x observation has been multiplied by its y observation $= xy$. We show this

Table 9.8 The hardness of eggshells produced by 13 Maran pullets and their consumption of a food supplement

Pullet	Amount of food supplement (g)	Hardness of shells
1	19.5	7.1
2	11.2	3.4
3	14.0	4.5
4	15.1	5.1
5	9.5	2.1
6	7.0	1.2
7	9.8	2.1
8	11.6	3.4
9	17.5	6.1
10	11.2	3.0
11	8.2	1.7
12	12.4	3.4
13	14.2	4.2

for the pullet example in Table 9.8 (reproduced here). If all these 13 xy values are added together, this is $\Sigma xy = 661.0$. (In Table 9.10 the original data (Table 9.8) has been reorganized into numerical order to simplify other parts of a calculation which we consider in Chapter 9.)

Table 9.10 Calculating Pearson's product moment correlation for Example 9.4 between the hardness of shells and food consumption in Maran pullets

Amount of food supplement (g) (x)	x^2	Hardness of shells (y)	y^2	$x \times y$
7.00	49.00	1.20	1.44	8.40
8.20	67.24	1.70	2.89	13.94
9.50	90.25	2.10	4.41	19.95
9.80	96.04	2.10	4.41	20.58
11.20	125.44	3.00	9.00	33.60
11.20	125.44	3.40	11.56	38.08
11.60	134.56	3.40	11.56	39.44
12.40	153.76	3.40	11.56	42.16
14.00	196.00	4.50	20.25	63.00
14.20	201.64	4.20	17.64	59.64
15.10	228.01	5.10	26.01	77.01
17.50	306.25	6.10	37.21	106.75
19.50	380.25	7.10	50.41	138.46
$\Sigma x = 161.20$	$\Sigma(x)^2 = 2153.88$	$\Sigma y = 47.30$	$\Sigma(y)^2 = 208.35$	$\Sigma xy = 661.00$
$\Sigma(x)^2 = 25985.44$		$\Sigma(y)^2 = 2237.29$		
$\bar{x} = 12.40$		$\bar{y} = 3.63846$		
$n = 13$		$n = 13$		

a. What is Σy for the data in Table 10.4?

b. What is Σx for the data in Table 9.8?

c. What is Σy for the data in Table 9.8?

B5 Powers (x^n) and roots ($^n\sqrt{}$)

You will come across x and y with a superscript such as x^2 or y^2. This indicates that you must square the x or y observation by multiplying it by itself. For example, if the x observation was 2, then $x^2 = 2^2 = 2 \times 2 = 4$. If a different superscript is used, you multiply the number by itself as many times as the superscript indicates. So if $x = 2$, then $x^3 = 2 \times 2 \times 2 = 8$. This is described as x cubed. Although squaring most whole numbers is a straightforward process, which you can work out in your head, for more complex numbers, use the x^2 button on your calculator.

This process can be reversed by taking the root of a value. In this book, we only take a square root, and the symbol that denotes this is $\sqrt{}$. If you were to take the cube root, the symbol would be $^3\sqrt{}$. You may learn square roots for some numbers such as $\sqrt{4} = 2$, $\sqrt{9} = 3$, $\sqrt{81} = 9$. For other numbers you can use the $\sqrt{}$ button on your calculator.

Taking squares and square roots is an important step in the calculation of a variance and standard deviation (e.g. Box 5.1), as the variance is the square of the standard deviation and therefore the standard deviation is the square root of the variance.

When calculating the variance or standard deviation, two other important squared terms are used: $(\Sigma x)^2$ and Σx^2. These appear to be the same apart from the brackets but the brackets are especially important here. Remember, whenever you have brackets like this you first carry out whatever calculation is inside the brackets. Therefore, $(\Sigma x)^2$ is the sum of all the x values (Σx), which is then squared. This compares with Σx^2, which means first square each x observation and then sum the squared values.

These two terms can be calculated for the x observations in Table 10.4. We have already calculated $\Sigma x = 76.6$ so $(\Sigma x)^2 = (76.6)^2 = 76.6 \times 76.6 = 5867.56$. The term Σx^2 is calculated as $5.5^2 + 8.4^2 + 5.0^2 + \cdots + 5.0^2 + 7.7^2 + 6.0^2 = 475.5$. Clearly $(\Sigma x)^2$ and Σx^2 are very different calculations with very different results, so it is important to make sure you do not get them mixed up.

a. What is $(\Sigma y)^2$ for the data from Table 10.4?

b. What is Σy^2 for the data from Table 10.4?

c. What is $(\Sigma x)^2$ for the data from Table 9.8?

d. What is Σx^2 for the data from Table 9.8.?

e. These calculations can also be used for the column of $x \times y$ values in Table 9.10. What is $(\Sigma xy)^2$?

f. What is $\Sigma(xy)^2$ for the data from Table 9.10?

When you write y^2, this is sometimes referred to as y to the power 2. Here y is multiplied by itself so the calculation is $y \times y$: there are two y values in the calculation. This can be extended. For example, y could be multiplied by itself three times, $y^3 = y \times y \times y$. If $y = 2$, then $y^3 = 2 \times 2 \times 2 = 8$. The power is 3. Powers may sometimes be shown using symbols instead of numbers.

The binomial equation is written as:

$$y = \frac{n!}{x!(n-x)!} \times p^x \times q^{(n-x)}$$

Here there are two powers: p^x and $q^{(n-x)}$. If you encounter these, you first need to identify what number the power symbol is; for example, what is x or what is $n-x$? Having found these values, and if you know the numerical value for p or q, you can work out the numerical value for p^x and $q^{(n-x)}$ using the power button on your calculator. For example, if $p = 0.5$ and $x = 4$, then $p^x = 0.5^4 = 0.0625$. If $q = 6$, $n = 8$, and $x = 4$, then $q^{(n-x)} = 6^{(8-4)} = 6^4 = 1296$.

Powers can be negative, for example x^{-2}. In this case, the sign is indicating that this calculation must be inverted (turned upside down). For example, if $x = 2$, and $x^2 = 2 \times 2$, then:

$$x^{-2} = \frac{1}{2 \times 2} = \frac{1}{4} = 0.25$$

You can use the power button (x^y) and the change sign button ($^+/-$) on your calculator to work out positive and negative powers.

 Use your calculator to work out the following powers:

    a. 6^5

    b. 6^{-5}

    c. 0.5^2

    d. 0.5^{-2}

B6 Factorials ($x!$)

In the binomial equation (B5) another symbol (!) was used. Mathematics revolves around number patterns. These often occur so frequently that they are given a particular symbol. Factorials are one of these patterns that appear in this book. A factorial is the value that results from multiplying a whole number (integer) by all the integers less than and including itself. A factorial of $3! = 3 \times 2 \times 1 = 6$, a factorial of $4! = 4 \times 3 \times 2 \times 1 = 24$ and so on. In the binomial equation (5.2.2), two factorials are included: $x!$ and $(n-x)!$ To work these factorials out, you first need to know what x is, or what $n-x$ is. You could then work this sum out by hand, but it is simpler to calculate factorials using the factorial button on your calculator.

 Use a calculator to work out the following:

    a. $10!$

    b. $x!$, where $x = 6$

    c. $(n-x)!$, where $x = 6$ and $n = 13$

B7 Constants: pi (π) and exponential (e)

Patterns in mathematics can also include constant numbers or constants. They are specific numbers that have been found to be needed in many calculations for the mathematical relationships to work. The two constants that appear in this book are pi $(\pi) \approx 3.14$ and exponential $(e) \approx 2.72$. π can usually be found on your calculator. e is invariably used as a power as we see in the Gaussian equation (5.2.1). On a calculator, this button is indicated as e^x. To find the value of e itself, press the e^x button, 1, equals.

---

a. What is π to five decimal places?

b. What is e to five decimal places?

c. What is e^2?

d. What is e^{-2}?

e. What is e^x when $x = 2(6 - 4)$?

B8 Scales of measurement (%, log, ln, and $\times 10^n$)

We use many scales of measurements. The one most familiar to you is our numbering system based on 10, where we have units (numbers 1–9), tens (numbers 10–99), hundreds (numbers 100–999), etc. Very large numbers and very small numbers can be very tedious to write out in full, so there is a shorthand, which we symbolize as $\times 10^n$, where n can be any number. As examples of using this shorthand 100 can be written 10^2 (i.e. 10×10), 1000 can be written as 10^3 (i.e. $10 \times 10 \times 10$), 200 could be written as 2×10^2, 5000 could be written as 5×10^3, etc. The same idea can be used for very small numbers. We introduced the idea of negative powers in B5. Negative powers of 10 can be used as a shorthand when writing small numbers. For example, 0.1 can be written as 10^{-1} as this is telling us that the number is 1/10. So 0.01 is 10^{-2} or 1/100; 0.5 could be written as 5×10^{-1}, i.e. $5 \times 1/10$, etc. Your calculator (and computer) will report large and small numbers in this way. If you used your calculator to answer Q5(b), it would be reported as 1.28601×10^{-4}.

Apart from the base 10 system of numbering that we are very familiar with, you will encounter other scales each indicated using a particular symbol. The four you need to be able to work with are percentages (%), logarithms to the base 10 ($\log_{10}$), natural logarithms (ln = $\log_e$), and inverse sine ($\sin^{-1}$). We cannot explain here how these scales are derived, but encourage you to find the relevant buttons on your calculator and feel confident using them.

On a scientific calculator, you will probably not have a % button. This is calculated from a proportion or ratio that is multiplied by 100. For example, 3/4 as a % = $(3/4) \times 100 = 0.75 \times 100 = 75\%$. Values may need to be converted to the $\log_{10}$ scale or converted from the $\log_{10}$ scale, taking an antilog. The first is achieved by pressing the $\log_{10}$ button on your calculator (sometimes labelled log). For example, $\log_{10} 2 = 0.30103$. To take the antilog, for most calculators press shift $\log_{10}$ =. This should return the log value back to the base 10 scale. The same process can be used for ln 2 = 0.69315 and its natural antilog (antiln) using shift ln =. When using the sine functions, you again can take the sine (sin) or inverse sine ($\sin^{-1}$) usually by using the same button with or without the shift function first.

 Q8   a. What is 5/20 as a percentage?

b. Rewrite 19,080,000 as a number × 10^n.

c. What is (3 − 6)/2 as a percentage?

d. Rewrite 0.000000345 as a number × 10^n.

e. What is log 4?

f. Take the antilog of 0.05.

g. What is ln 4?

h. Take the antiln of 0.05.

i. What is the sine (sin) of 90°?

j. What is the inverse sine ($\sin^{-1}$) of 0.6?

B9 Unpacking complex equations

The four basic procedures in mathematics are addition (+), subtraction (−), multiplication (×), and division (÷). These are straightforward in their own right, but when there are several of these elements in one calculation, it is not always clear in which order to carry out the procedures and this order can be critical in obtaining the correct outcome. To help convey which order should be followed, brackets are often placed around parts of an equation.

To help unpack complex equations, you need to follow this order:

1) Brackets

2) Powers and roots

3) Division (except where considering fractions, see examples 1–3)

4) Multiplication

5) Addition

6) Subtraction

We are going to unpack some of the equations frequently encountered in this book.

Example 1 (e.g. 5.4.3i)

$$\bar{x} = \frac{3+4}{2} = \frac{7}{2} = 3.5$$

If you look at the list above, it seems that, as there are no brackets, powers, or roots, 'division' appears to be the first step. However, this is a simplification. Where you have a fraction and wish to carry out the division step, you first need to simplify the fraction so that the division step may be carried out. In this example the first step then is to add all the values in the numerator (the top of the fraction) together and then divide the total by the denominator (lower part of the fraction).

Example 2 (e.g. 5.7)

$$SE = \frac{10}{\sqrt{4}} = \frac{10}{2} = 5$$

Again, there is a fraction, only this time the denominator has a square root sign ($\sqrt{}$) in front of it. You must first work out the square root of the denominator and then use this to divide the numerator.

Example 3 (e.g. Box 8.3)

$$\chi^2_{calculated} = \frac{(12-6)^2}{4} + \frac{(12-8)^2}{2} = \frac{6^2}{4} + \frac{4^2}{2} = \frac{36}{4} + \frac{16}{2} = 9+8 = 17$$

In the numerators, there are three different procedures indicated by the symbols, subtraction $-$, a square $()^2$, and addition $+$. The subtraction is within brackets and must be carried out first. The value within the brackets can then be squared. This gives the numerator for each fraction. Each division is then carried out and finally the two numbers can be added together.

Example 4 (e.g. Box 9.1)

$$\chi^2_{calculated} = \frac{(|9-6.5|-0.5)^2}{2} + \frac{(|4-6.5|-0.5)^2}{2}$$

This example illustrates the use of another symbol $||$. A number within these two vertical lines is called an absolute number. At the end of the calculation within these brackets the sign of the number within these lines is ignored. The calculation is carried out following the same rules we have discussed above: brackets, then powers, then fractions, then addition. Absolute values also occur in t-tests (e.g. Box 10.4).

$$\chi^2_{calculated} = \frac{(|9-6.5|-0.5)^2}{2} + \frac{(|4-6.5|-0.5)^2}{4}$$

$$= \frac{(|2.5|-0.5)^2}{2} + \frac{(|-2.5|-0.5)^2}{4}$$

$$= \frac{(2.5-0.5)^2}{2} + \frac{(2.5-0.5)^2}{4} = \frac{(2.0)^2}{2} + \frac{(2.0)^2}{4} = \frac{4.0}{2} + \frac{4.0}{4}$$

$$= 2+1 = 3$$

Example 5 (e.g. 5.5.1)

In some cases we may wish to calculate two values, one at the top of the range and one at the bottom of the range. This is indicated by the symbol $\pm$. We use this in both the range for non-parametric data (5.5.1) and confidence limits (5.7).

For example:

$$\text{Confidence limits} = \bar{x} \pm (t \times \text{SEM})$$

This symbol $\pm$ indicates that you need to work out the calculation twice, once with an addition and once with a subtraction. If $\bar{x} = 15$, $t = 3$, and $\text{SEM} = 2$, then:

$$\bar{x} \pm (t \times \text{SEM}) = 15 \pm (3 \times 2) = 15 \pm 6 = \text{range is } 15 - 6 \text{ to } 15 + 6, \text{ or } 9 \text{ to } 21$$

Example 6 (e.g. Boxes 5.1 and 10.8)

A very frequently used equation is the one written here. It is used in parametric ANOVAs (Chapter 10) and is part of the calculation of variance.

$$SS_{sample} = \sum x^2 - \frac{\left(\sum x\right)^2}{n}$$

We have already looked at the terms $\sum x^2$ and $(\sum x)^2$ in B5. To complete the calculation of SS_{sample}, the fraction $(\sum x)^2/n$ is calculated first and then the subtraction. We worked out $\sum x^2$, $(\sum x)^2$, and n for the x observations in Table 9.10 in B4, where $n = 13$, $\sum x^2 = 2153.88$, and $(\sum x)^2 = 25985.44$. So for this example:

$$SS_{sample} = 2153.88 - (25985.44 / 13) = 2153.88 - 1998.88 = 155.0$$

Example 7 (e.g. 8.4.3)

In several places in the book, we have given some complex equations. These can be very off-putting, but when they are unpacked you will see that they are manageable. The Gaussian equation (5.2.1) has been unpacked for you in 8.4 using the data from Table 5.9. Examine the Gaussian equation and decide in what order you would carry out the calculation. How does this compare with the steps in 8.4.3?

Gaussian equation:

$$y = \frac{1}{\sqrt{(2\pi s^2)}} e^{-h}$$

where

$$h = \frac{(x - \bar{x})}{2s^2}$$

Answers to questions

 A1
a. Two divided by four
b. x multiplied by y
c. a divided by b
d. p is less than 0.001
e. 12.2 divided by 2 approximately equals 6.

 A2
a. First work out the values inside the round brackets. So $(10 - 2) = 8$ and $(3 \times 2) = 6$. Then work out the sum within the $[\,]$ so $[6 - 2] = 4$. Then carry out the division of $8/4 = 2$.
b. First work out the sum within the round brackets where $(x - 1) = 2 - 1 = 1$. Then multiply this by $2 = 2 \times 1 = 2$.
c. First work out the sum within the round brackets $(6 + 3) = 9$. Then carry out the division within the straight brackets $|9/{-3}|$. When a positive number is divided by a negative number the answer will be negative, so $9/{-3} = -3$. The straight brackets tell you this is to be an absolute value so you can ignore the sign. The answer to this sum is $+3$.

 A3
a. $\Sigma y = 3.3 + 6.3 + 6.1 + \cdots + 6.3 + 6.0 + 7.2 = 85.3\,\text{mm}$
b. $\Sigma x = 19.5 + 11.2 + 14.0 + \cdots + 8.2 + 12.4 + 14.2 = 161.2\,\text{g}$
c. $\Sigma y = 7.1 + 3.4 + 4.5 + \cdots + 1.7 + 3.4 + 4.2 = 47.3\,\text{units}$

 A4
a. $(\Sigma y)^2 = 85.3^2 = 85.3 \times 85.3 = 7276.09$
b. $\Sigma y^2 = 3.3^2 + 6.3^2 + 6.1^2 + \cdots + 6.3^2 + 6.0^2 + 7.2^2 = 629.57$
c. $(\Sigma x)^2 = 161.2^2 = 161.2 \times 161.2 = 25985.44$
d. $\Sigma x^2 = 19.5^2 + 11.2^2 + 14.0^2 + \cdots + 8.2^2 + 12.4^2 + 14.2^2 = 2153.88$
e. $(\Sigma xy)^2 = 661.0^2 = 661.0 \times 661.0 = 436921.0$
f. $\Sigma(xy)^2 = 8.4^2 + 13.94^2 + 19.95^2 + \cdots + 77.01^2 + 106.75^2 + 138.46^2 = 51021.652$

 A5
a. 7776
b. 0.0001286
c. 0.25
d. 4.0

 A6
a. 3628800
b. 720
c. $(13 - 6)! = 7! = 5040$

 A7
a. 3.14159
b. 2.71828
c. 7.38906
d. 0.13534
e. $x = 2 \times 2 = 4$, so $e^4 = 54.59815$

A8
a. $0.25 \times 100 = 25\%$
b. 1.908×10^7
c. $-3/2 = -1.5 \times 100 = -150\%$
d. 3.45×10^{-7}

e. 0.60206

f. 1.12202

g. 1.38629

h. 1.05127

i. 1.00 (note: set calculator to degrees not radians)

j. 36.87° (note: set calculator to degrees not radians)

Quick reference guide for choosing a statistical test

This quick guide to choosing a statistical test is supported by the more detailed information in Chapter 7.

Hypothesis	No. of treatment variables	Other attributes		Data are parametric or non-parametric	Test	Section
Do the data match an expected ratio?	One	Observations are counts or frequencies	No replicates	Non-parametric	Chi-squared goodness-of-fit	8.3
			Replicates	Non-parametric	Chi-squared goodness-of-fit with replicates	8.5
			$n < 25$, limited design	Non-parametric	Fisher's exact test	9.4
					Barnard's test	
Do you wish to test for an association?	Two treatment variables for each item	Distribution is linear	Both interval scales and neither is fixed	Parametric	Pearson's product moment correlation	9.7
		Distribution is linear; one of the treatment variables is fixed	No replicates	Both measures on interval scale	Simple linear regression	9.10
			Replicates	Parametric	Model I linear regression with replicates	9.11
		Distribution is linear; neither treatment variable is fixed	Both scales of measurement have the same units	Parametric	Principal axis regression	9.12
			The scales of measurement have different units	Parametric	Ranged principal axis regression	9.13

(continued)

Hypothesis	No. of treatment variables	Other attributes		Data are parametric or non-parametric	Test	Section
		Scatter plot indicates distribution is monotonic	Treatment variables are ordinal or interval; at least one of the treatment variables is usually not fixed	Non-parametric	Spearman's rank correlation	9.6
		Two treatment variables for each item	Observations are counts or frequencies, no replicates	Non-parametric	Chi-squared test for association	9.3
Do samples come from the same or different populations?	One	Two categories, $n < 30$	Observations are unmatched	Parametric	t-test	10.2 and 10.3
				Non-parametric	Mann–Whitney U test	11.1
			Observations are matched (repeated measures)	Parametric	t-test for matched data	10.4
				Non-parametric	Wilcoxon's matched pairs test	11.2
			One sample to compare to a single observation	Parametric	One-sample t-test	10.5
				Non-parametric	One-sample Wilcoxon's signed rank test	7.1.1
		Two categories, $n \geq 30$	Unmatched data	Data are measured on a continuous scale	z-test for unmatched data	10.1
			Matched data	Data are measured on a continuous scale	z-test for matched data	10.4
		More than two categories	Equal replicates	Parametric	Parametric one-way ANOVA for equal replicates	10.7

(continued)

Hypothesis	No. of treatment variables	Other attributes		Data are parametric or non-parametric	Test	Section
			Unequal replicates	Parametric	Parametric one-way ANOVA for unequal replicates	10.9
			Equal or unequal replicates	Non-parametric	Non-parametric one-way ANOVA (Kruskal–Wallis test)	11.3
		Post hoc tests	Equal replicates	Parametric	Tukey	10.8
		Post hoc tests	Unequal replicates	Parametric	Tukey–Kramer	10.10
		Post hoc tests	Equal or unequal replicates	Non-parametric		11.4
	Two	More than two categories	Equal replicates	Parametric	Parametric two-way ANOVA for equal replicates	10.12
				Non-parametric	Non-parametric two-way ANOVA	11.5
				Non-parametric	Scheirer–Ray–Hare test	11.7
			Unequal replicates	Parametric	Parametric two-way ANOVA for unequal replicates	10.14
			No replicates or you have matched data or repeated measures	Parametric	Parametric two-way ANOVA for no replicates	10.15
				Non-parametric	Friedman's test	7.1.1
		Nested design	No replicates	Parametric	Nested two-way ANOVA	10.16
		Post hoc tests		Parametric	Tukey	10.13
		Post hoc tests		Non-parametric		11.6
	Three treatment variables			Parametric	Three-way parametric ANOVA	10.17
				Non-parametric	Scheirer–Ray–Hare test	11.7

Tables of critical values for statistical tests

D1. Critical values for the chi-squared test (χ^2) between $p = 0.05$ and $p = 0.001$, where v is the degrees of freedom and p is the probability

D2. Critical values for Spearman's rank correlation (r_s) between $p = 0.10$ and $p = 0.001$ for a two-tailed test, where n is the number of pairs of observations

D3. Critical values for the Pearson's correlation (r) between $p = 0.10$ and $p = 0.001$ for a two-tailed test, where the degrees of freedom (v) are the number of pairs of observations (n) − 2 and p is the probability

D4. Critical values for the F-test that is carried out to confirm that the variances are homogeneous before a z- or t-test, where $p = 0.05$ and where v_1 is the degrees of freedom for the larger variance (numerator) and v_2 is the degrees of freedom for the smaller variance (denominator)

D5. Critical values for a two-tailed z-test

D6. Critical values for the t-tests between $p = 0.10$ and $p = 0.001$ for a two-tailed test, where v is the degrees of freedom and p is the probability

D7. Critical values for an F_{max} test that is carried out to confirm that the variances are homogeneous before a parametric analysis of variance, where $p = 0.05$, a is the number of samples or treatments, and v is the degrees of freedom

D8. Critical values for an F-test for a parametric ANOVA, where $p = 0.05$, v_1 is the degrees of freedom for the numerator, and v_2 is the degrees of freedom for the denominator

D9. Critical values for an F-test for a parametric ANOVA, where $p = 0.01$, v_1 is the degrees of freedom for the numerator, and v_2 is the degrees of freedom for the denominator

D10. Critical values for an F-test for a parametric ANOVA, where $p = 0.001$, v_1 is the degrees of freedom for the numerator, and v_2 is the degrees of freedom for the denominator

D11. q values for Tukey's test at $p = 0.05$, $p = 0.01$, and $p = 0.001$, where a is the number of samples and v is the degrees of freedom for MS_{within} from the ANOVA calculation

D12. Critical values of U for the Mann–Whitney U test at $p = 0.05$ for a two-tailed test, where n_1 is the number of observations in sample 1 and n_2 is the number of observations in sample 2

D13. Critical values for Wilcoxon's matched pairs test (T) between $p = 0.1$ and $p = 0.002$ for a two-tailed test, where N is the number of pairs of observations used to provide ranks for the calculation (i.e. not those where $d = 0$)

D14. Critical values for H for the Kruskal–Wallis test, between $p = 0.10$ and $p = 0.01$ for a two-tailed test, where a is the number of categories (samples) and n is the number of observations in a category

D15. Critical values for Q for the multiple comparisons test to follow a significant non-parametric ANOVA. The Q statistic is given between $p = 0.10$ and $p = 0.001$ for a two-tailed test, where a is the number of categories (samples)

Table D1 Critical values for the chi-squared test (χ^2) between $p = 0.05$ and $p = 0.001$, where v is the degrees of freedom and p is the probability

v	p			v	p		
	0.05	0.01	0.001		0.05	0.01	0.001
1	3.84	6.64	10.83	29	42.56	49.59	58.30
2	5.99	9.21	13.82	30	43.77	50.89	59.70
3	7.81	11.34	16.27	31	44.99	52.19	61.10
4	9.49	13.28	18.47	32	46.19	53.47	62.49
5	11.07	15.09	20.51	33	47.40	54.78	63.87
6	12.59	16.81	22.46	34	48.60	56.06	65.25
7	14.07	18.47	24.32	35	49.80	57.34	66.62
8	15.51	20.09	26.12	36	51.00	58.62	67.99
9	16.92	21.67	27.88	37	52.19	59.89	69.35
10	18.31	23.21	29.59	38	53.38	61.16	70.70
11	19.67	24.72	31.26	39	54.57	62.43	72.06
12	21.03	26.22	32.91	40	55.76	63.69	73.40
13	22.36	27.69	34.53	41	56.94	64.95	74.75
14	23.68	29.14	36.12	42	58.12	66.21	76.08
15	25.00	30.58	37.70	43	59.30	67.46	77.42
16	26.30	32.00	39.25	44	60.48	68.71	78.75
17	27.59	33.41	40.79	45	61.66	69.96	80.08
18	28.87	34.80	42.31	46	62.83	71.20	81.40
19	30.14	36.19	43.82	47	64.00	72.44	82.72
20	31.41	37.57	45.31	48	65.17	73.68	84.04
21	32.67	38.93	46.80	49	66.34	74.92	85.35
22	33.92	40.29	48.27	50	67.51	76.15	86.66
23	35.17	41.64	49.73	60	79.08	88.38	99.61
24	36.41	42.98	51.18	70	90.53	100.43	112.32
25	37.65	44.31	52.62	80	101.88	112.33	124.84
26	38.88	45.64	54.05	90	113.145	124.12	137.21
27	40.11	46.96	55.48	100	124.34	140.17	149.45
28	41.34	48.28	56.89				

Table D2 Critical values for Spearman's rank correlation (r_s) between $p = 0.10$ and $p = 0.001$ for a two-tailed test, where n is the number of pairs of observations and p is the probability

(To find the critical values for a one-tailed test, double the p value. For example, the critical values for a two-tailed test when p = 0.1 will be the critical values for p = 0.05 for a one-tailed test.)

n	p				
	0.10	0.05	0.02	0.01	0.001
5	0.900	1.000	1.000	–	–
6	0.829	0.886	0.943	1.000	–
7	0.714	0.786	0.893	0.929	1.000
8	0.643	0.738	0.833	0.881	0.976
9	0.600	0.683	0.783	0.833	0.933
10	0.564	0.648	0.745	0.794	0.903
11	0.523	0.623	0.709	0.818	0.873
12	0.497	0.591	0.678	0.780	0.846
13	0.475	0.566	0.648	0.745	0.824
14	0.457	0.545	0.626	0.716	0.802
15	0.441	0.525	0.604	0.689	0.779
16	0.425	0.507	0.582	0.666	0.762
17	0.412	0.490	0.564	0.645	0.748
18	0.399	0.476	0.549	0.625	0.728
19	0.388	0.462	0.534	0.608	0.712
20	0.377	0.450	0.521	0.591	0.696
21	0.368	0.438	0.508	0.576	0.681
22	0.359	0.428	0.496	0.562	0.667
23	0.351	0.418	0.485	0.549	0.654
24	0.343	0.409	0.475	0.537	0.642
25	0.336	0.400	0.465	0.526	0.630
26	0.329	0.392	0.456	0.515	0.619
27	0.323	0.385	0.448	0.505	0.608
28	0.317	0.377	0.440	0.496	0.598
29	0.311	0.370	0.432	0.487	0.589
30	0.305	0.364	0.425	0.478	0.580
35	0.283	0.335	0.394	0.433	0.539
40	0.264	0.313	0.368	0.405	0.507
45	0.248	0.294	0.347	0.382	0.479
50	0.235	0.279	0.329	0.363	0.456

Table D3 Critical values for the Pearson's correlation (r) between $p = 0.10$ and $p = 0.001$ for a two-tailed test, where the degrees of freedom (v) is the number of pairs of observations (n) − 2 and p is the probability

(To find the critical values for a one-tailed test, divide the p value in half. For example, the critical values for a two-tailed test when p = 0.10 will be the critical values for p = 0.05 in a one-tailed test.)

v	p		v	p	
	0.05	0.01		0.05	0.01
1	0.997	1.000	26	0.374	0.479
2	0.950	0.990	27	0.367	0.471
3	0.878	0.959	28	0.361	0.463
4	0.811	0.917	29	0.355	0.456
5	0.754	0.874	30	0.349	0.449
6	0.707	0.834	31	0.344	0.443
7	0.666	0.798	32	0.339	0.436
8	0.632	0.765	33	0.334	0.430
9	0.602	0.735	34	0.329	0.424
10	0.576	0.708	35	0.325	0.418
11	0.553	0.684	36	0.320	0.413
12	0.532	0.661	37	0.316	0.408
13	0.514	0.641	38	0.312	0.403
14	0.497	0.623	39	0.308	0.398
15	0.482	0.606	40	0.304	0.393
16	0.468	0.590	41	0.300	0.389
17	0.456	0.575	42	0.297	0.384
18	0.444	0.561	43	0.294	0.380
19	0.433	0.549	44	0.291	0.376
20	0.423	0.537	45	0.288	0.372
21	0.413	0.526	46	0.284	0.368
22	0.404	0.515	47	0.281	0.365
23	0.396	0.505	48	0.279	0.361
24	0.388	0.496	49	0.276	0.358
25	0.381	0.487	50	0.273	0.354

Table D4 Critical values for the F-test that is carried out to confirm that the variances are homogeneous before a z- or t-test, where $p = 0.05$ and v_1 is the degrees of freedom for the larger variance (numerator) and v_2 is the degrees of freedom for the smaller variance (denominator)

v_2	v_1																		
	1	2	3	4	5	6	7	8	9	10	12	15	20	24	30	40	60	120	∞
1	647.8	799.5	864.2	899.6	921.8	937.1	948.2	956.7	963.3	968.6	976.7	984.9	993.1	997.2	1001	1006	1010	1014	1018
2	38.51	39.00	39.17	39.25	39.30	39.33	39.36	39.57	39.39	39.40	39.41	39.43	39.45	39.46	39.46	39.47	39.48	39.49	39.50
3	17.44	16.04	15.44	15.10	14.88	14.73	14.62	14.54	14.47	14.42	14.34	14.25	14.17	14.12	14.08	14.04	13.99	13.95	13.90
4	12.22	10.65	9.98	9.60	9.36	9.20	9.07	8.98	8.90	8.84	8.75	8.66	8.56	8.51	8.46	8.41	8.36	8.31	8.26
5	10.01	8.43	7.76	7.39	7.15	6.98	6.85	6.76	6.68	6.62	6.52	6.43	6.33	6.28	6.23	6.18	6.12	6.07	6.02
6	8.81	7.26	6.60	6.23	5.99	5.82	5.70	5.60	5.52	5.46	5.37	5.27	5.17	5.12	5.07	5.01	4.96	4.90	4.85
7	8.07	6.54	5.89	5.52	5.29	5.12	4.99	4.90	4.82	4.76	4.67	4.57	4.47	4.42	4.36	4.31	4.25	4.20	4.14
8	7.57	6.06	5.42	5.05	4.82	4.65	4.53	4.43	4.36	4.30	4.20	4.10	4.00	3.95	3.89	3.84	3.78	3.73	3.67
9	7.21	5.71	5.08	4.72	4.48	4.32	4.20	4.10	4.03	3.96	3.87	3.77	3.67	3.61	3.56	3.51	3.45	3.39	3.33
10	6.94	5.46	4.83	4.47	4.24	4.07	3.95	3.85	3.78	3.72	3.62	3.52	3.42	3.37	3.31	3.26	3.20	3.14	3.08
11	6.72	5.26	4.63	4.28	4.04	3.88	3.76	3.66	3.59	3.53	3.43	3.33	3.23	3.17	3.12	3.06	3.00	2.94	2.88
12	6.55	5.10	4.47	4.12	3.89	3.73	3.61	3.51	3.44	3.37	3.28	3.18	3.07	3.02	2.96	2.91	2.85	2.79	2.72
13	6.41	4.97	4.35	4.00	3.77	3.60	3.48	3.39	3.31	3.25	3.15	3.05	2.95	2.89	2.84	2.78	2.72	2.66	2.60
14	6.30	4.86	4.24	3.89	3.66	3.50	3.38	3.29	3.21	3.15	3.05	2.95	2.84	2.79	2.73	2.67	2.61	2.55	2.49
15	6.20	4.77	4.15	3.80	3.68	3.41	3.29	3.20	3.12	3.06	2.96	2.86	2.76	2.70	2.64	2.59	2.52	2.46	2.40
16	6.12	4.69	4.08	3.73	3.50	3.34	3.22	3.12	3.05	2.99	2.89	2.79	2.68	2.63	2.57	2.51	2.45	2.38	2.32
17	6.04	4.62	4.01	3.66	3.44	3.28	3.16	3.06	2.98	2.92	2.82	2.72	2.62	2.56	2.50	2.44	2.38	2.32	2.25
18	5.98	4.56	3.95	3.61	3.38	3.22	3.10	3.01	2.93	2.87	2.77	2.67	2.56	2.50	2.44	2.38	2.32	2.26	2.19
19	5.92	4.51	3.90	3.56	3.33	3.17	3.05	2.96	2.88	2.82	2.72	2.62	2.51	2.45	2.39	2.33	2.27	2.20	2.13
20	5.87	4.46	3.86	3.51	3.29	3.13	3.01	2.91	2.84	2.77	2.68	2.57	2.46	2.41	2.35	2.29	2.22	2.16	2.09
21	5.83	4.42	3.82	3.48	3.25	3.09	2.97	2.87	2.80	2.73	2.64	2.53	2.42	2.37	2.31	2.25	2.18	2.11	2.04
22	5.79	4.38	3.78	3.44	3.22	3.05	2.93	2.84	2.76	2.70	2.60	2.50	2.39	2.33	2.27	2.21	2.14	2.08	2.00

(continued)

v_2	v_1																		
	1	2	3	4	5	6	7	8	9	10	12	15	20	24	30	40	60	120	∞
23	5.75	4.35	3.75	3.41	3.18	3.02	2.90	2.81	2.73	2.67	2.57	2.47	2.36	2.30	2.24	2.18	2.11	2.04	1.97
24	5.72	4.32	3.72	3.38	3.15	2.99	2.87	2.78	2.70	2.64	2.54	2.44	2.33	2.27	2.21	2.15	2.08	2.01	1.94
25	5.69	4.29	3.69	3.35	3.13	2.97	2.85	2.75	2.68	2.61	2.51	2.41	2.30	2.24	2.18	2.12	2.05	1.98	1.91
26	5.66	4.27	3.67	3.33	3.10	2.94	2.82	2.73	2.65	2.59	2.49	2.39	2.28	2.22	2.16	2.09	2.03	1.95	1.88
27	5.63	4.24	3.65	3.31	3.08	2.92	2.80	2.71	2.63	2.57	2.47	2.36	2.25	2.19	2.13	2.07	2.00	1.93	1.85
28	5.61	4.22	3.63	3.29	3.06	2.90	2.78	2.69	2.61	2.55	2.45	2.34	2.23	2.17	2.11	2.05	1.98	1.91	1.83
29	5.59	4.20	3.61	3.27	3.04	2.88	2.76	2.67	2.59	2.53	2.43	2.32	2.21	2.15	2.09	2.03	1.96	1.89	1.81
30	5.57	4.18	3.59	3.25	3.03	2.87	2.75	2.65	2.57	2.51	2.41	2.31	2.20	2.14	2.07	2.01	1.94	1.87	1.79
40	5.42	4.05	3.46	3.13	2.90	2.74	2.62	2.53	2.45	2.39	2.29	2.18	2.07	2.01	1.94	1.88	1.80	1.72	1.64
60	5.29	3.93	3.34	3.01	2.79	2.63	2.51	2.41	2.33	2.27	2.17	2.06	1.94	1.88	1.82	1.74	1.67	1.58	1.48
120	5.15	3.80	3.23	2.89	2.67	2.52	2.39	2.30	2.22	2.16	2.05	1.94	1.82	1.76	1.69	1.61	1.53	1.43	1.31
∞	5.02	3.69	3.12	2.79	2.57	2.41	2.29	2.19	2.11	2.05	1.94	1.83	1.71	1.64	1.57	1.48	1.39	1.27	1.00

Table D5 Critical values for a two-tailed z-test

(To find the critical values for a one-tailed test, divide the p value in half. For example, the critical values for a two-tailed test when p = 0.1 will be the critical values for p = 0.05 for a one-tailed test.)

Probability value	z
0.10	1.647
0.05	1.960
0.01	2.576
0.02	2.326
0.002	3.100
0.001	3.291

Table D6 Critical values for the t-test between $p = 0.10$ and $p = 0.001$ for a two-tailed test, where v is the degrees of freedom, and p is the probability

(To find the critical values for a one-tailed test, divide the p value in half. For example, the critical values for a two-tailed test when p = 0.1 will be the critical values for p = 0.05 for a one-tailed test.)

v	p					
	0.100	0.050	0.025	0.010	0.005	0.001
1	6.314	12.706	25.452	63.657	127.320	636.620
2	2.920	4.303	6.205	9.925	14.089	31.598
3	2.353	3.182	4.176	5.841	7.453	12.941
4	2.132	2.776	3.495	4.604	5.598	8.610
5	2.015	2.571	3.163	4.032	4.773	6.859
6	1.943	2.447	2.969	3.707	4.317	5.959
7	1.895	2.365	2.841	3.499	4.029	5.405
8	1.860	2.306	2.752	3.355	3.832	5.041
9	1.833	2.262	2.685	3.250	3.690	4.781
10	1.812	2.228	2.634	3.169	3.581	4.587
11	1.796	2.201	2.593	3.106	3.497	4.437
12	1.782	2.179	2.560	3.055	3.428	4.318
13	1.771	2.160	2.533	3.012	3.372	4.221
14	1.761	2.145	2.510	2.977	3.326	4.140
15	1.753	2.131	2.490	2.947	3.286	4.073
16	1.746	2.120	2.473	2.921	3.252	4.015
17	1.740	2.110	2.458	2.898	3.222	3.965
18	1.734	2.101	2.445	2.878	3.197	3.922
19	1.729	2.093	2.433	2.861	3.174	3.883
20	1.725	2.086	2.423	2.845	3.153	3.850
21	1.721	2.080	2.414	2.831	3.135	3.819
22	1.717	2.074	2.406	2.819	3.119	3.792
23	1.714	2.069	2.398	2.807	3.104	3.767
24	1.711	2.064	2.391	2.797	3.090	3.745

(continued)

v	p					
	0.100	0.050	0.025	0.010	0.005	0.001
25	1.708	2.060	2.385	2.787	3.078	3.725
26	1.706	2.056	2.379	2.779	3.067	3.707
27	1.703	2.052	2.373	2.771	3.056	3.690
28	1.701	2.048	2.368	2.763	3.047	3.674
29	1.699	2.045	2.364	2.756	3.038	3.659
30	1.697	2.042	2.360	2.750	3.030	3.646
50	1.676	2.008	2.310	2.678	2.937	3.496
100	1.661	1.982	2.276	2.625	2.871	3.390
∞	1.645	1.960	2.241	2.576	2.807	3.290

Table D7 Critical values for an F_{max} test that is carried out to confirm that the variances are homogeneous before a parametric ANOVA, where $p = 0.05$, a is the number of samples or treatments, and v is the degrees of freedom

N	a										
	2	3	4	5	6	7	8	9	10	11	12
2	39.0	87.5	142	202	266	333	403	475	550	626	704.0
3	15.4	27.8	39.2	50.7	62.0	72.9	83.5	93.9	104	114	124.0
4	9.60	15.5	20.6	25.2	29.5	33.6	37.5	41.1	44.6	48.0	51.4
5	7.15	10.8	13.7	16.3	18.7	20.8	22.9	24.7	26.5	28.2	29.9
6	5.82	8.38	10.4	12.1	13.7	15.0	16.3	17.5	18.6	19.7	20.7
7	4.99	6.94	8.44	9.70	10.8	11.8	12.7	13.5	14.3	15.1	15.8
8	4.43	6.00	7.18	8.12	9.03	9.78	10.5	11.1	11.7	12.2	12.7
9	4.03	5.34	6.31	7.11	7.80	8.41	8.95	9.45	9.91	10.3	10.7
10	3.72	4.85	5.67	6.34	6.92	7.42	7.87	8.28	8.66	9.01	9.34
12	3.28	4.16	4.79	5.30	5.72	6.09	6.42	6.72	7.00	7.25	7.48
15	2.86	3.54	4.01	4.37	4.68	4.95	5.19	5.40	5.59	5.77	5.93
20	2.46	2.95	3.29	3.54	3.76	3.94	4.10	4.24	4.37	4.49	4.59
30	2.07	2.40	2.61	2.78	2.91	3.02	3.12	3.21	3.29	3.36	3.39
60	1.67	1.85	1.96	2.04	2.11	2.17	2.22	2.26	2.30	2.33	2.36
∞	1.00	1.00	1.00	1.00	1.00	1.00	1.00	1.00	1.00	1.00	1.00

Table D8 Critical values for an F-test for a parametric ANOVA, where $p = 0.05$, v_1 is the degrees of freedom for the numerator, and v_2 is the degrees of freedom for the denominator

v_2	v_1									
	1	2	3	4	5	6	8	10	20	∞
1	161.4	199.5	215.7	224.6	230.2	234.0	238.9	242.0	249.0	254.3
2	18.51	19.00	19.16	19.25	19.30	19.33	19.37	19.40	19.45	19.50
3	10.13	9.55	9.28	9.12	9.01	8.94	8.84	8.79	8.66	8.53
4	7.71	6.94	6.59	6.39	6.26	6.16	6.04	5.96	5.80	5.63
5	6.61	5.79	5.41	5.19	5.05	4.95	4.82	4.74	4.56	4.36
6	5.99	5.14	4.76	4.53	4.39	4.28	4.15	4.06	3.87	3.67
7	5.59	4.74	4.35	4.12	3.97	3.87	3.73	3.64	3.44	3.23
8	5.32	4.46	4.07	3.84	3.69	3.58	3.44	3.35	3.15	2.93
9	5.12	4.26	3.86	3.63	3.48	3.37	3.23	3.14	2.94	2.71
10	4.96	4.10	3.71	3.48	3.33	3.22	3.07	2.98	2.77	2.54
11	4.85	3.98	3.59	3.36	3.20	3.09	2.95	2.85	2.65	2.40
12	4.75	3.88	3.49	3.26	3.11	3.00	2.85	2.75	2.54	2.30
13	4.67	3.80	3.41	3.18	3.02	2.92	2.77	2.67	2.46	2.21
14	4.60	3.74	3.34	3.11	2.96	2.85	2.70	2.60	2.39	2.13
15	4.54	3.68	3.29	3.06	2.90	2.79	2.64	2.54	2.33	2.07
16	4.49	3.63	3.24	3.01	2.85	2.74	2.59	2.49	2.28	2.01
17	4.45	3.59	3.20	2.96	2.81	2.70	2.55	2.45	2.23	1.96
18	4.41	3.55	3.16	2.93	2.77	2.66	2.51	2.41	2.19	1.92
19	4.38	3.52	3.13	2.90	2.74	2.63	2.48	2.38	2.16	1.88
20	4.35	3.49	3.10	2.87	2.71	2.60	2.45	2.35	2.12	1.84
21	4.32	3.47	3.07	2.84	2.68	2.57	2.42	2.32	2.10	1.81
22	4.30	3.44	3.05	2.82	2.66	2.55	2.40	2.30	2.07	1.78
23	4.28	3.42	3.03	2.80	2.64	2.53	2.38	2.27	2.05	1.76
24	4.26	3.40	3.01	2.78	2.62	2.51	2.36	2.25	2.03	1.73
25	4.24	3.38	2.99	2.76	2.60	2.49	2.34	2.24	2.01	1.71
26	4.22	3.37	2.98	2.74	2.59	2.47	2.32	2.22	1.99	1.69
27	4.21	3.35	2.96	2.73	2.57	2.46	2.30	2.20	1.97	1.67
28	4.20	3.34	2.95	2.71	2.56	2.44	2.29	2.19	1.94	1.65
29	4.18	3.33	2.93	2.70	2.54	2.43	2.28	2.18	1.93	1.64
30	4.17	3.32	2.92	2.69	2.53	2.42	2.27	2.16	1.88	1.62
40	4.08	3.23	2.84	2.61	2.45	2.34	2.18	2.08	1.84	1.51
50	4.03	3.18	2.79	2.56	2.40	2.29	2.13	2.03	1.78	1.44
60	4.00	3.15	2.76	2.52	2.37	2.25	2.10	1.99	1.75	1.39
70	3.98	3.13	2.74	2.50	2.35	2.23	2.07	1.97	1.72	1.35
80	3.96	3.11	2.72	2.49	2.33	2.21	2.06	1.95	1.70	1.31
90	3.95	3.10	2.71	2.17	2.32	2.20	2.04	1.94	1.69	1.28
100	3.94	3.09	2.70	2.46	2.30	2.19	2.03	1.93	1.68	1.26
∞	3.84	2.99	2.60	2.37	2.21	2.10	1.94	1.83	1.57	1.00

Table D9 Critical values for an F-test for a parametric ANOVA, where $p = 0.01$, v_1 is the degrees of freedom for numerator, and v_2 is the degrees of freedom for the denominator

	1	2	3	4	5	6	8	10	20	∞
1	4052	4999	5403	5625	5764	5859	5982	6106	6234	6366
2	98.50	99.00	99.17	99.25	99.30	99.33	99.37	99.42	99.40	99.50
3	34.12	30.82	29.46	28.71	28.24	27.91	27.49	27.20	26.70	26.12
4	21.20	18.00	16.69	15.98	15.52	15.21	14.80	14.50	14.0	13.46
5	16.26	13.27	12.06	11.39	10.97	10.67	10.29	10.01	9.55	9.02
6	13.74	10.92	9.78	9.15	8.75	8.47	8.10	7.87	7.40	6.88
7	12.25	9.55	8.45	7.85	7.46	7.19	6.84	6.62	6.16	5.65
8	11.26	8.65	7.59	7.01	6.63	6.37	6.03	5.81	5.36	4.86
9	10.56	8.02	6.99	6.42	6.06	5.80	5.47	5.26	4.81	4.31
10	10.04	7.56	6.55	5.99	5.64	5.39	5.06	4.85	4.41	3.91
11	9.65	7.20	6.22	5.67	5.32	5.07	4.74	4.54	4.10	3.60
12	9.33	6.93	5.95	5.41	5.06	4.82	4.50	4.30	3.86	3.36
13	9.07	6.70	5.74	5.20	4.86	4.62	4.30	4.10	3.66	3.16
14	8.86	6.51	5.56	5.03	4.69	4.46	4.14	3.94	3.51	3.00
15	8.68	6.36	5.42	4.89	4.56	4.32	4.00	3.80	3.37	2.87
16	8.53	6.23	5.29	4.77	4.44	4.20	3.89	3.69	3.26	2.75
17	8.40	6.11	5.18	4.67	4.34	4.10	3.79	3.59	3.16	2.65
18	8.28	6.01	5.09	4.58	4.25	4.01	3.71	3.51	3.08	2.57
19	8.18	5.93	5.01	4.50	4.17	3.94	3.63	3.43	3.00	2.49
20	8.10	5.85	4.94	4.43	4.10	3.87	3.56	3.37	2.94	2.42
21	8.02	5.78	4.87	4.37	4.04	3.81	3.51	3.31	2.88	2.36
22	7.94	5.72	4.82	4.31	3.99	3.76	3.45	3.26	2.83	2.31
23	7.88	5.66	4.76	4.26	3.94	3.71	3.41	3.21	2.78	2.26
24	7.82	5.61	4.72	4.22	3.90	3.67	3.36	3.17	2.74	2.21
25	7.77	5.57	4.68	4.18	3.86	3.63	3.32	3.13	2.70	2.17
26	7.72	5.53	4.64	4.14	3.82	3.59	3.29	3.09	2.66	2.13
27	7.68	5.49	4.60	4.11	3.78	3.56	3.26	3.06	2.63	2.10
28	7.64	5.45	4.57	4.07	3.75	3.53	3.23	3.03	2.60	2.06
29	7.60	5.42	4.54	4.04	3.73	3.50	3.20	3.00	2.57	2.03
30	7.56	5.39	4.51	4.02	3.70	3.47	3.17	2.98	2.55	2.01
40	7.31	5.18	4.31	3.83	3.51	3.29	2.99	2.80	2.37	1.80
50	7.17	5.06	4.20	3.72	3.41	3.19	2.89	2.70	2.27	1.68
60	7.08	4.98	4.13	3.65	3.34	3.12	2.82	2.63	2.20	1.60
70	7.01	4.92	4.07	3.60	3.29	3.07	2.78	2.59	2.15	1.53
80	6.96	4.88	4.04	3.56	3.26	3.04	2.74	2.55	2.12	1.47
90	6.92	4.85	4.01	3.53	3.23	3.01	2.72	2.52	2.09	1.43
100	6.90	4.82	3.98	3.51	3.21	2.99	2.69	2.50	2.07	1.39
∞	6.64	4.60	3.78	3.32	3.02	2.80	2.51	2.32	1.88	1.00

Table D10 Critical values for an F-test for a parametric ANOVA, where $p = 0.001$, v_1 is the degrees of freedom for the numerator, and v_2 is the degrees of freedom for the denominator

	1	2	3	4	5	6	8	10	20	∞
1										
2	998	999	999	999	999	999	999	999	999	999
3	167.0	148.5	141.1	137.1	134.6	132.8	130.6	129.00	126.00	123.5
4	74.13	61.24	56.18	53.48	51.71	50.52	49.00	48.10	46.10	44.05
5	47.04	36.61.1	33.28	31.09	29.75	28.84	27.64	26.90	25.40	23.78
6	35.51	27.00	23.70	21.90	20.81	20.03	19.03	18.40	17.10	15.75
7	29.25	21.69	18.77	17.20	16.21	15.52	14.63	14.10	12.90	11.70
8	25.42	18.49	15.83	14.39	13.49	12.86	12.05	11.50	10.50	9.34
9	22.86	16.59	13.98	12.50	11.71	11.13	10.37	9.89	8.90	7.81
10	21.04	14.91	12.55	11.28	10.48	9.93	9.20	8.75	7.80	6.76
11	19.68	13.81	11.56	10.35	9.58	9.05	8.35	7.29	7.01	6.00
12	18.64	12.97	10.81	9.63	8.89	8.38	7.71	7.29	6.40	5.42
13	17.81	12.31	10.21	9.07	8.35	7.86	7.21	6.80	5.93	4.97
14	17.14	11.78	9.73	8.62	7.92	7.43	6.80	6.40	5.56	4.60
15	16.58	11.34	9.34	8.25	7.57	7.09	6.47	6.08	5.25	4.31
16	16.12	10.97	9.00	7.94	7.27	6.81	6.19	5.81	4.99	4.06
17	15.72	10.66	8.73	7.68	7.02	6.56	5.96	5.58	4.78	3.85
18	15.38	10.39	8.49	7.46	6.81	6.35	5.76	5.39	4.59	3.67
19	15.08	10.16	8.28	7.27	6.61	6.18	5.59	5.22	4.43	3.51
20	14.82	9.95	8.10	7.10	6.46	6.02	5.44	5.08	4.29	3.38
21	14.59	9.77	7.94	6.95	6.32	5.88	5.31	4.95	4.17	3.26
22	14.38	9.61	7.80	6.81	6.19	5.76	5.19	4.83	4.06	3.15
23	14.19	9.47	7.67	6.70	6.08	5.65	5.09	4.73	3.96	3.05
24	14.03	9.34	7.55	6.59	5.98	5.55	4.99	4.64	3.87	2.97
25	13.87	9.22	7.45	6.49	5.89	5.46	4.91	4.56	3.79	2.89
26	13.74	9.12	7.36	6.41	5.80	5.38	4.83	4.48	3.72	2.82
27	13.61	9.02	7.27	6.33	5.73	5.31	4.76	4.41	3.66	2.75
28	13.50	8.93	7.19	6.25	5.66	5.24	4.69	4.35	3.60	2.69
29	13.39	8.85	7.12	6.19	5.59	5.18	4.65	4.29	3.54	2.64
30	13.29	8.77	7.05	6.12	5.53	5.12	4.58	4.24	3.49	2.59
40	12.61	8.25	6.60	5.70	5.13	4.73	4.21	3.87	3.14	2.23
50	12.22	7.96	6.34	5.46	4.90	4.51	4.00	3.67	2.95	2.02
60	11.97	7.77	6.17	5.31	4.76	4.37	3.87	3.54	2.83	1.89
70	11.80	7.64	6.06	5.20	4.66	4.28	3.77	3.45	2.74	1.78
80	11.67	7.54	5.97	5.12	4.58	4.20	3.70	3.39	2.68	1.72
90	11.57	7.47	5.91	5.06	4.53	4.15	3.87	3.34	2.63	1.66
100	11.50	7.41	5.86	5.02	4.48	4.11	3.61	3.30	2.59	1.61
∞	10.83	6.91	5.42	4.62	4.10	3.74	3.27	2.96	2.27	1.00

Table D11 q values for a Tukey's test at $p = 0.05$, $p = 0.01$, and $p = 0.001$, where a is the number of samples and v is the degrees of freedom for MS_{within} from the ANOVA calculation

v	p																			
										a										
		2	3	4	5	6	7	8	9	10	11	12	13	14	15	16	17	18	19	20
1	0.05	17.97	26.98	32.82	37.08	40.41	43.12	45.40	47.36	49.07	50.59	51.96	53.30	54.33	55.36	56.32	57.22	58.04	58.83	59.56
	0.01	90.03	135.0	164.3	185.6	202.2	215.8	227.2	237.0	245.6	253.2	260.0	266.2	271.8	277.0	281.8	286.3	290.4	294.3	298.0
	0.001	90.3	135.1	1643	1856	2022	2158	2272	2370	2455	2532	2600	2662	2718	2770	2818	2863	2904	2934	2980
2	0.05	6.09	8.33	9.80	10.90	11.74	12.44	13.03	13.54	13.99	14.39	14.75	15.08	15.38	15.65	15.91	16.14	16.37	16.57	16.77
	0.01	14.04	19.02	22.29	24.72	26.63	28.20	29.53	30.68	31.69	32.59	33.40	34.13	34.81	35.43	36.00	36.53	37.03	37.50	37.95
	0.001	44.69	60.42	70.77	78.43	84.49	89.46	93.67	97.30	100.5	103.3	105.9	108.2	110.4	112.3	114.2	115.9	117.4	118.9	120.3
3	0.05	4.50	5.91	6.83	7.50	8.04	8.48	8.85	9.18	9.46	9.72	9.95	10.15	10.35	10.53	10.69	10.84	10.98	11.11	11.24
	0.01	8.26	10.62	12.17	13.33	14.24	15.00	15.64	16.20	16.69	17.13	17.53	17.89	18.22	18.52	18.81	19.07	19.32	19.55	19.77
	0.001	18.28	23.32	26.65	29.13	31.11	32.74	34.12	35.33	36.39	37.34	38.20	38.98	39.69	40.35	40.97	41.54	42.07	42.58	43.05
4	0.05	3.93	5.04	5.76	6.29	6.71	7.05	7.35	7.60	7.83	8.03	8.21	8.37	8.53	8.66	8.79	8.91	9.03	9.13	9.23
	0.01	6.51	8.12	9.17	9.96	10.58	11.10	11.55	11.93	12.27	12.57	12.84	13.09	13.32	13.53	13.73	13.91	14.08	14.24	14.40
	0.001	12.18	14.99	16.84	18.23	19.34	20.26	21.04	21.73	22.33	22.87	23.36	23.81	24.21	24.59	24.94	25.27	25.58	25.87	26.14
5	0.05	3.64	4.60	5.22	5.67	6.03	6.33	6.58	6.80	6.99	7.17	7.32	7.47	7.60	7.72	7.83	7.93	8.03	8.12	8.21
	0.01	5.70	6.97	7.80	8.42	8.91	9.32	9.67	9.97	10.24	10.48	10.70	10.89	11.08	11.24	11.40	11.55	11.68	11.81	11.93
	0.001	9.71	11.67	12.96	13.93	14.71	15.35	15.90	16.38	16.81	17.18	17.53	17.85	18.13	18.41	18.66	18.89	19.10	19.31	19.51
6	0.05	3.46	4.34	4.90	5.31	5.63	5.89	6.12	6.32	6.49	6.65	6.79	6.92	7.03	7.14	7.24	7.34	7.43	7.51	7.59
	0.01	5.24	6.33	7.03	7.56	7.97	8.32	8.61	8.87	9.10	9.30	9.49	9.65	9.81	9.95	10.08	10.21	10.32	10.43	10.54
	0.001	8.43	9.96	10.97	11.72	12.32	12.83	13.26	13.63	13.97	14.27	14.54	14.79	15.01	15.22	15.42	15.60	15.78	15.94	17.97
7	.05	3.34	4.16	4.68	5.06	5.36	5.61	5.82	6.00	6.16	6.30	6.43	6.55	6.66	6.76	6.85	6.94	7.02	7.09	7.17
	.01	4.95	5.92	6.54	7.01	7.37	7.68	7.94	8.17	8.37	8.55	8.71	8.86	9.00	9.12	9.24	9.35	9.46	9.55	9.65
	.001	7.65	8.93	9.76	10.40	10.90	11.32	11.68	11.99	12.27	12.52	12.74	12.95	13.14	13.32	13.48	13.64	13.78	13.92	14.04
8	0.05	3.26	4.04	4.53	4.89	5.17	5.40	5.60	5.77	5.92	6.05	6.18	6.29	6.39	6.48	6.57	6.65	6.73	6.80	6.87
	0.01	4.74	5.63	6.20	6.63	6.96	7.24	7.47	7.68	7.87	8.03	8.18	8.31	8.44	8.55	8.66	8.76	8.85	8.94	9.03
	0.001	7.13	8.25	9.00	9.52	10.00	10.32	10.64	10.91	11.15	11.36	11.56	11.74	11.91	12.06	12.21	12.34	12.47	12.59	12.70

(continued)

v	p	\multicolumn{19}{c}{a}
		2	3	4	5	6	7	8	9	10	11	12	13	14	15	16	17	18	19	20
9	0.05	3.20	3.95	4.42	4.76	5.02	5.24	5.43	5.60	5.74	5.87	5.98	6.09	6.19	6.28	6.36	6.44	6.51	6.58	6.64
	0.01	4.60	5.43	5.96	6.35	6.66	6.91	7.13	7.32	7.49	7.65	7.78	7.91	8.03	8.13	8.23	8.32	8.41	8.49	8.57
	0.001	6.76	7.77	8.42	8.91	9.30	9.62	9.90	10.14	10.36	10.55	10.73	10.89	11.03	11.18	11.30	11.42	11.54	11.64	11.75
10	0.05	3.15	3.88	4.33	4.65	4.91	5.12	5.30	5.46	5.60	5.72	5.83	5.93	6.03	6.11	6.20	6.27	6.34	6.40	6.47
	0.01	4.48	5.27	5.77	6.14	6.43	6.67	6.87	7.05	7.21	7.36	7.48	7.60	7.71	7.81	7.91	7.99	8.07	8.15	8.22
	0.001	6.49	7.41	8.01	8.45	8.80	9.10	9.35	9.57	9.77	9.95	10.11	10.25	10.39	10.52	10.64	10.75	10.85	10.95	11.03
11	0.05	3.11	3.82	4.26	4.57	4.82	5.03	5.20	5.35	5.49	5.61	5.71	5.81	5.90	5.99	6.06	6.14	6.20	6.26	6.33
	0.01	4.39	5.14	5.62	5.97	6.25	6.48	6.67	6.84	6.99	7.13	7.25	7.36	7.46	7.56	7.65	7.73	7.81	7.88	7.95
	0.001	6.28	7.14	7.69	8.10	8.43	8.70	8.93	9.14	9.32	9.48	9.63	9.77	9.89	10.01	10.12	10.22	10.31	10.41	10.49
12	0.05	3.08	3.77	4.20	4.51	4.75	4.95	5.12	5.27	5.40	5.51	5.62	5.71	5.80	5.88	5.95	6.03	6.09	6.15	6.21
	0.01	4.32	5.04	5.50	5.84	6.10	6.32	6.51	6.67	6.81	6.94	7.06	7.17	7.26	7.36	7.44	7.52	7.59	7.66	7.73
	0.001	6.11	6.92	7.44	7.82	8.13	8.38	8.60	8.79	8.96	9.11	9.25	9.38	9.50	9.61	9.71	9.80	9.89	9.98	10.06
13	0.05	3.06	3.73	4.15	4.45	4.69	4.88	5.05	5.19	5.32	5.43	5.53	5.63	5.71	5.79	5.86	5.93	6.00	6.05	6.11
	0.01	4.26	4.96	5.40	5.73	5.98	6.19	6.37	6.53	6.67	6.79	6.90	7.01	7.10	7.19	7.27	7.34	7.42	7.48	7.55
	0.001	5.97	6.74	7.23	7.60	7.89	8.13	8.33	8.51	8.67	8.82	8.95	9.07	9.18	9.28	9.38	9.46	9.55	9.63	9.70
14	0.05	3.03	3.70	4.11	4.41	4.64	4.83	4.99	5.13	5.25	5.36	5.46	5.55	5.64	5.72	5.79	5.85	5.92	5.97	6.03
	0.01	4.21	4.89	5.32	5.63	5.88	6.08	6.26	6.41	6.54	6.66	6.77	6.87	6.96	7.05	7.12	7.20	7.27	7.33	7.39
	0.001	5.86	6.59	7.06	7.41	7.69	7.92	8.11	8.28	8.43	8.51	8.70	8.81	8.91	9.01	9.10	9.19	9.27	9.34	9.42
15	0.05	3.01	3.67	4.08	4.37	4.60	4.78	4.94	5.08	5.20	5.31	5.40	5.49	5.58	5.65	5.72	5.79	5.85	5.90	5.96
	0.01	4.17	4.83	5.25	5.56	5.80	5.99	6.16	6.31	6.44	6.55	6.66	6.76	6.84	6.93	7.00	7.07	7.14	7.20	7.26
	0.001	5.76	6.47	6.92	7.25	7.52	7.74	7.93	8.09	8.23	8.37	8.48	8.59	8.69	8.79	8.87	8.95	9.03	9.10	9.17
16	0.05	3.00	3.65	4.05	4.33	4.56	4.74	4.90	5.03	5.15	5.26	5.35	5.44	5.52	5.59	5.66	5.72	5.79	5.84	5.90
	0.01	4.13	4.78	5.19	5.49	5.72	5.92	6.08	6.22	6.35	6.46	6.56	6.66	6.74	6.82	6.90	6.97	7.03	7.09	7.15
	0.001	5.68	6.37	6.80	7.12	7.37	7.59	7.77	7.92	8.06	8.19	8.30	8.41	8.50	8.59	8.68	8.76	8.83	8.90	8.96
17	0.05	2.98	3.63	4.02	4.30	4.52	4.71	4.86	4.99	5.11	5.21	5.31	5.39	5.47	5.55	5.61	5.68	5.74	5.79	5.84
	0.01	4.10	4.74	5.14	5.43	5.66	5.85	6.01	6.15	6.27	6.38	6.48	6.57	6.66	6.73	6.80	6.87	6.94	7.00	7.05
	0.001	5.61	6.28	6.70	7.01	7.25	7.45	7.63	7.78	7.92	8.04	8.15	8.25	8.34	8.43	8.51	8.58	8.65	8.72	8.78

(continued)

v	p	2	3	4	5	6	7	8	9	10	11	12	13	14	15	16	17	18	19	20
18	0.05	2.97	3.61	4.00	4.28	4.49	4.67	4.82	4.96	5.07	5.17	5.27	5.35	5.43	5.50	5.57	5.63	5.69	5.74	5.79
	0.01	4.07	4.70	5.09	5.38	5.60	5.79	5.94	6.08	6.20	6.31	6.41	6.50	6.58	6.65	6.72	6.79	6.85	6.91	6.96
	0.001	5.55	6.20	6.60	6.91	7.14	7.34	7.51	7.66	7.79	7.91	8.01	8.11	8.20	8.23	8.36	8.43	8.50	8.57	8.63
19	0.05	2.96	3.59	3.98	4.25	4.47	4.65	4.79	4.92	5.04	5.14	5.23	5.32	5.39	5.46	5.53	5.59	5.65	5.70	5.75
	0.01	4.05	4.67	5.05	5.33	5.55	5.73	5.89	6.02	6.14	6.25	6.34	6.43	6.51	6.58	6.65	6.72	6.78	6.84	6.89
	0.001	5.49	6.13	6.53	6.82	7.05	7.24	7.41	7.55	7.68	7.79	7.89	8.00	8.08	8.16	8.23	8.30	8.37	8.43	8.49
20	0.05	2.95	3.58	3.96	4.23	4.45	4.62	4.77	4.90	5.01	5.11	5.20	5.28	5.36	5.43	5.49	5.55	5.61	5.66	5.71
	0.01	4.02	4.64	5.02	5.29	5.51	5.69	5.84	5.97	6.09	6.19	6.29	6.37	6.45	6.52	6.59	6.65	6.71	6.76	6.82
	0.001	5.44	6.07	6.45	6.74	7.00	7.15	7.31	7.45	7.58	7.69	7.79	7.88	8.00	8.04	8.12	8.19	8.25	8.32	8.37
24	0.05	2.92	3.53	3.90	4.17	4.37	4.54	4.68	4.81	4.92	5.01	5.10	5.18	5.25	5.32	5.38	5.44	5.50	5.54	5.59
	0.01	3.96	4.54	4.91	5.17	5.37	5.54	5.69	5.81	5.92	6.02	6.11	6.19	6.26	6.33	6.39	6.45	6.51	6.56	6.61
	0.001	5.30	5.90	6.24	6.50	6.71	6.88	7.03	7.16	7.27	7.37	7.47	7.55	7.63	7.70	7.77	7.83	7.89	7.95	8.00
30	0.05	2.89	3.49	3.84	4.10	4.30	4.46	4.60	4.72	4.83	4.92	5.00	5.08	5.15	5.21	5.27	5.33	5.38	5.43	5.48
	0.01	3.89	4.45	4.80	5.05	5.24	5.40	5.54	5.65	5.76	5.85	5.93	6.01	6.08	6.14	6.20	6.26	6.31	6.36	6.41
	0.001	5.16	5.70	6.03	6.28	6.47	6.63	6.76	6.88	6.98	7.10	7.16	7.24	7.31	7.38	7.44	7.49	7.55	7.60	7.65
40	0.05	2.86	3.44	3.79	4.04	4.23	4.39	4.52	4.63	4.74	4.82	4.91	4.98	5.05	5.11	5.16	5.22	5.27	5.31	5.36
	0.01	3.82	4.37	4.70	4.93	5.11	5.27	5.39	5.50	5.60	5.69	5.77	5.84	5.90	5.96	6.02	6.07	6.12	6.17	6.21
	0.001	5.02	5.53	5.84	6.06	6.24	6.39	6.51	6.62	6.71	6.80	6.87	6.94	7.01	7.07	7.12	7.17	7.22	7.27	7.32
60	0.05	2.83	3.40	3.74	3.98	4.16	4.31	4.44	4.55	4.65	4.73	4.81	4.88	4.94	5.00	5.06	5.11	5.16	5.20	5.24
	0.01	3.76	4.28	4.60	4.82	4.99	5.13	5.25	5.36	5.45	5.53	5.60	5.67	5.73	5.79	5.84	5.89	5.93	5.98	6.02
	0.001	4.89	5.37	5.65	5.86	6.02	6.16	6.27	6.37	6.45	6.53	6.60	6.66	6.72	6.77	6.82	6.87	6.91	6.96	7.00
120	0.05	2.80	3.36	3.69	3.92	4.10	4.24	4.36	4.48	4.56	4.64	4.72	4.78	4.84	4.90	4.95	5.00	5.05	5.09	5.13
	0.01	3.70	4.20	4.50	4.71	4.87	5.01	5.12	5.21	5.30	5.38	5.44	5.51	5.56	5.61	5.66	5.71	5.75	5.79	5.83
	0.001	4.77	5.21	5.48	5.67	5.82	5.94	6.04	6.13	6.21	6.28	6.34	6.40	6.45	6.50	6.54	6.58	6.62	6.66	6.70
∞	0.05	2.77	3.31	3.63	3.86	4.03	4.17	4.29	4.39	4.47	4.55	4.62	4.68	4.74	4.80	4.85	4.89	4.93	4.97	5.01
	0.01	3.64	4.12	4.40	4.60	4.76	4.88	4.99	5.08	5.16	5.23	5.29	5.35	5.40	5.45	5.49	5.54	5.57	5.61	5.65
	0.001	4.65	5.06	5.31	5.48	5.62	5.73	5.82	5.90	5.97	6.04	6.09	6.14	6.19	6.23	6.27	6.31	6.35	6.38	6.41

Table D12 Critical values of U for the Mann–Whitney U test at $p = 0.05$ for a two-tailed test, where n_1 is the number of observations in sample 1 and n_2 is the number of observations in sample 2

(To find the critical values for a one-tailed test, divide the p value in half. For example, the critical values for a two-tailed test when $p = 0.1$ will be the critical values for $p = 0.05$ for a one-tailed test.)

n_1 \ n_2	2	3	4	5	6	7	8	9	10	11	12	13	14	15	16	17	18	19	20
2							0	0	0	0	1	1	1	1	1	2	2	2	2
3				0	1	1	2	2	3	3	4	4	5	5	6	6	7	7	8
4			0	1	2	3	4	4	5	6	7	8	9	10	11	11	12	13	13
5		0	1	2	3	5	6	7	8	9	11	12	13	14	15	17	18	19	20
6		1	2	3	5	6	8	10	11	13	14	16	17	19	21	22	24	25	27
7		1	3	5	6	8	10	12	14	16	18	20	22	24	26	28	30	32	34
8	0	2	4	6	8	10	13	15	17	19	22	24	26	29	31	34	36	38	41
9	0	2	4	7	10	12	15	17	20	23	26	28	31	34	37	39	42	45	48
10	0	3	5	8	11	14	17	20	23	26	29	33	36	39	42	45	48	52	55
11	0	3	6	9	13	16	19	23	26	30	33	37	40	44	47	51	55	58	62
12	1	4	7	11	14	18	22	26	29	33	37	41	45	49	53	57	61	65	69
13	1	4	8	12	16	20	24	28	33	37	41	45	50	54	59	63	67	72	76
14	1	5	9	13	17	22	26	31	36	40	45	50	55	59	64	67	74	78	83
15	1	5	10	14	19	24	29	34	39	44	49	54	59	64	70	75	80	85	90
16	1	6	11	15	21	26	31	37	42	47	53	59	64	70	75	81	86	92	98
17	2	6	11	17	22	28	34	39	45	51	57	63	67	75	81	87	93	99	105
18	2	7	12	18	24	30	36	42	48	55	61	67	74	80	86	93	99	106	112
19	2	7	13	19	25	32	38	45	52	58	65	72	78	85	92	99	106	113	119
20	2	8	13	20	27	34	41	48	55	62	69	76	83	90	98	105	112	119	127

Table D13 Critical values for T for Wilcoxon's *post hoc* test between $p = 0.1$ and $p = 0.001$ for a two-tailed test, where N is the number of pairs of observations used to provide ranks for the calculation (i.e. not those where $d = 0$)

(To find the critical values for a one-tailed test, divide the p value in half. For example, the critical values for a two-tailed test when p = 0.1 will be the critical values for p = 0.05 for a one-tailed test.)

N	p			
	0.1	0.05	0.01	0.001
5	0			
6	2	0		
7	3	2		
8	5	3	0	
9	8	5	1	
10	10	8	3	
11	13	10	5	0
12	17	13	7	1
13	21	17	9	2
14	25	21	12	4
15	30	25	15	6
16	35	29	19	8
17	41	34	23	11
18	47	40	27	14
19	53	46	32	18
20	60	52	37	21
21	67	58	42	25
22	75	65	48	30
23	83	73	54	35
24	91	81	61	40
25	100	89	68	45
26	110	98	75	51
27	119	107	83	57
28	130	116	91	64
29	140	126	100	71
30	151	137	109	78
31	163	147	118	86
32	175	159	128	94
33	187	170	138	102
34	200	182	148	111
35	213	195	159	120

Table D14 Critical values for H for the Kruskal–Wallis test, between $p = 0.10$ and $p = 0.01$ for a two-tailed test, where a is the number of categories (samples) and n is the number of observations in a category

| | a | | | | | | | | | | | | | | | | | |
| | 3 | | | 4 | | | 5 | | | 6 | | | | | | | | |
n	p 0.10	0.05	0.01	0.10	0.05	0.01	0.10	0.05	0.01	0.10	0.05	0.01
2	4.571	–	–	5.667	6.167	6.667	6.982	7.418	8.291	8.154	8.846	9.846
3	4.622	5.600	7.200	6.026	7.000	8.538	7.333	8.333	10.200	8.620	9.789	11.820
4	4.654	5.692	7.654	6.088	7.235	9.287	7.457	8.685	11.070	8.800	10.140	12.720
5	4.560	5.780	8.00	6.120	7.377	9.789	7.532	8.876	11.570	8.902	10.360	13.260
6	4.643	5.801	8.222	6.127	7.453	10.090	7.557	9.002	11.910	8.958	10.500	13.600
7	4.594	5.819	8.378	6.141	7.501	10.250	7.600	9.080	12.140	8.992	10.590	13.840
8	4.595	5.805	8.465	6.148	7.534	10.420	7.624	9.126	12.290	9.037	10.660	13.990
9	4.586	5.831	8.529	6.161	7.557	10.530	7.637	9.166	12.410	9.057	10.710	14.130
10	4.581	5.853	8.607	6.167	7.586	10.620	7.650	9.200	12.500	9.078	10.750	14.240
11	4.587	5.885	8.648	6.163	7.623	10.690	7.660	9.242	12.580	9.093	10.760	14.320
12	4.578	5.872	8.712	6.185	7.629	10.750	7.675	9.274	12.630	9.105	10.790	14.380
13	4.601	5.901	8.735	6.191	7.645	10.800	7.685	9.303	12.690	9.115	10.830	14.440
14	4.592	5.896	8.754	6.198	7.658	10.840	7.695	9.307	12.740	9.125	10.840	14.490
15	4.591	5.902	8.821	6.201	7.676	10.870	7.701	9.302	12.770	9.133	10.860	14.530
16	4.595	5.909	8.822	6.205	7.678	10.900	7.705	9.313	12.790	9.140	10.880	14.560
17	4.593	5.915	8.856	6.206	7.682	10.920	7.709	9.325	12.830	9.144	10.880	14.600
18	4.596	5.932	8.865	6.212	7.698	10.950	7.714	9.334	12.850	9.149	10.890	14.630
19	4.598	5.923	8.887	6.212	7.701	10.980	7.717	9.342	12.870	9.156	10.900	14.640
20	4.594	5.926	8.905	6.216	7.703	10.980	7.719	9.353	12.910	9.159	10.920	14.670
21	4.597	5.930	8.918	6.218	7.709	11.010	7.723	9.356	12.920	9.164	10.930	14.700
22	4.597	5.932	8.928	6.215	7.714	11.030	7.724	9.362	12.920	9.168	10.940	14.720
23	4.598	5.937	8.947	6.220	7.719	11.030	7.727	9.368	12.940	9.171	10.930	14.740
24	4.598	5.936	8.964	6.221	7.724	11.060	7.729	9.375	12.960	9.170	10.930	14.740
25	4.605	5.942	8.975	6.222	7.727	11.070	7.730	9.377	12.960	9.177	10.940	14.770

Table D15 Critical values for Q for the multiple comparisons test to follow a significant non-parametric ANOVA. The Q statistic is given between $p = 0.10$ and $p = 0.001$ for a two-tailed test, where a is the number of categories (samples).

a	p			
	0.10	0.05	0.01	0.001
2	1.645	1.960	2.576	3.291
3	2.128	2.394	2.936	3.588
4	2.394	2.639	3.144	3.765
5	2.576	2.807	3.291	3.891
6	2.713	2.936	3.403	3.988
7	2.823	3.038	3.494	4.067
8	2.914	3.124	3.570	4.134
9	2.992	3.197	3.635	4.191
10	3.059	3.261	3.692	4.241
11	3.119	3.317	3.743	4.286
12	3.172	3.368	3.789	4.326
13	3.220	3.414	3.830	4.363
14	3.264	3.456	3.878	4.397
15	3.304	3.494	3.902	4.428
16	3.342	3.529	3.935	4.456
17	3.376	3.562	3.965	4.483
18	3.409	3.593	3.993	4.508
19	3.439	3.662	4.019	4.532
20	3.467	3.649	4.044	4.554
21	3.494	3.675	4.067	4.575
22	3.519	3.699	4.089	4.595
23	3.543	3.722	4.110	4.614
24	3.566	3.744	4.130	4.632
25	3.588	3.765	4.149	4.649

Glossary

a priori Before the event. Usually refers to a theoretical expectation of the outcome of an experiment, which can therefore be stated before any data are gathered.

Absolute You may find that having calculated the difference between two values you have a negative value. For some parts of some calculations, any negative sign must be removed. In these instances, the number without its sign is called the absolute value and is indicated as |number|. For example |–10| would be used in the calculation as + 10.

Aim A generalized statement about the topic you are investigating.

Alternate hypotheses (H_1) In hypothesis testing we choose between the null hypothesis and an alternate hypothesis. There are three possible generic alternate hypotheses which can be illustrated as models relating to the mean: the samples/observations do not come from the same populations ($\mu \neq \mu_s$); the samples/observations come from a population whose mean is smaller than the original population ($\mu > \mu_s$); the samples/observations come from a population whose mean is larger than the original population ($\mu < \mu_s$).

Association Where one variable changes in a similar manner to another variable. The reasons for this are not necessarily known.

Autonomy Being able to make choices and decisions for oneself and by oneself.

Bar chart A figure for observations from an investigation with one treatment variable. The data are measured on a nominal or ordinal scale. The x axis is the treatment variable. The y axis is the number of observations (or frequency or percentage). Each category from the nominal or ordinal scale is represented as a discrete bar on the figure. There must be a space between each bar to indicate that the data are not measured on a continuous scale. The tops of these bars should not be joined together by a trend line. If the top of the bar is an average of many observations, then confidence intervals can be used to indicate the variation around the mean.

Beneficence To do good.

Bibliography The full details of all sources of information you have read whilst working on a particular topic.

Bimodal When a distribution has two high points (peaks).

Binomial distribution May be obtained if each item examined can have only one of two possible states. For example, a seed can either germinate or not germinate.

British numeric referencing system The citations in the text are numerical and reflect the order in which the citations appear. The in-text citations link to numerically listed reference details.

Case See Item.

Categories When each observation may fall into only one of two or more mutually exclusive groups (e.g. blood groups A, B, AB, or O). These groups are known as categories. The categories may be either qualitative (e.g. flower colours described as red, pink, blue, green, yellow, purple) or quantitative (e.g. the number of piglets in a litter: 0, 1, 2, 3, 4, 5, 6, etc.). The data in one category may be a sample but not inevitably (see definition of Sample for explanation).

Closed questions Where you give the participant a limited number of answers and they have to choose one or more of these options. Often used in questionnaires. See Open questions.

Codes of conduct Desired actions outlined by a public body where compliance is usually statutory.

Coefficient of variation The ratio between the mean and the standard deviation expressed as a percentage. Used if you wish to compare variation in one set of data with that of another set of data.

Confidence limits for a mean The values derived from a sample between which the population mean probably falls. These limits are calculated using the standard error of the mean.

Confounding variables Variables that are outside the scope of the investigation but may affect the results. Also called non-treatment variables. See Variables, Sampling error, and Treatment variables.

Confounding variation Variation in your data that is caused by the combined effect of confounding variables and sampling error. You need to explicitly work to minimize the confounding variation in your experimental design.

Consent form A signed statement by the participant confirming their understanding of the research requirements and indicating their willingness to take part.

Continuous Observations that do not fall into a series of distinct categories and may take any value in the scale of measurement, e.g. height (cm).

Control An experimental baseline against which any effects of the treatment(s) may be compared. It must be an integral part of an experiment.

COSHH assessment Where the use of hazardous substances is reviewed to ensure compliance with current legislation.

Critical reader Making a balanced assessment of the quality of the experimental design, analysis, and reported evaluation for each item of published work and reflecting this evaluation in your own work.

Critical value The point on the x axis where the extreme values are divided from the common values in a distribution using mathematical relationships such as the mean ± standard deviation.

Curvilinear When data are plotted and where part or all of the distribution is found to curve rather than follow a straight line.

Data A number of observations or measurements on the subject you are investigating.

Degrees of freedom (ν) A measure that reflects the number of observations in your calculation and the experimental design.

Denominator In an equation where one term is divided by another, the value that is being used to divide is the denominator. For example, in the fraction 2/4, 4 is the denominator.

Dependent variable Variables that you measure to assess whether they respond to the treatment variables. Not all investigations include dependent variables. This term does not automatically imply that there is a significant biological relationship between the dependent variables and treatment variables.

Derived variables The unit of measure for the observations is the result of a calculation, e.g. ratio, proportion, percentage, or rate.

Discrete Observations measured on a discontinuous scale or a variable that falls into a series of distinct categories and where the number of categories is limited.

Distribution The shape seen on a graph when data are plotted. See also Normal distribution, Binomial distribution, and Poisson distribution.

EU Directive A binding EU agreement that must then be passed as national law by the national government.

EU Regulation Legislation that is drawn up by members of the EU and takes immediate effect as law.

Expectation A mathematical (e.g. Gaussian equation) or biological (e.g. genetic segregation ratio) model of the outcome of an experiment against which the observations can be compared. The model can be consistently applied. Your personal prediction of an outcome is not a statistical expectation.

Expected values Values calculated using a mathematical or biological model against which the observations can be compared.

Fisher's exact test An alternative to any chi-squared test particularly when sample sizes are small or you have a 2×2 contingency table. This is best calculated using statistical software.

Fixed variable When each treatment is consistent every time, there is no sampling error. Most fixed variables are under the control of the investigator.

Focus group A discussion-based interview involving more than two people. A theme or focus is provided by the researcher, who also directs the discussion. These discussions are usually recorded as audio- or videotapes and evaluated later.

General hypothesis This is phrased in a general way; for example, there is no difference between the samples.

General linear and general linearized models Statistical approaches that may be taken when you have three or more treatment variables with data that are measured on a continuous scale and some of the apparent associations appear to be linear.

Genetically modified organism An organism where any of the genes or other genetic material have been artificially modified by a process which does not occur naturally in mating or natural recombination or if the organism has inherited or otherwise derived, through any number of replications, genes, or other genetic material (from any source) which were so modified.

Harvard referencing system The most widely used referencing system. This is based on the author's family name and date of publication for in-text citations and the provision of full citation details in a reference list.

Histogram A type of figure for observations from an investigation with one treatment variable. The data are measured on a continuous (interval) scale. There are many data points along the x axis. Usually because the treatment variable is not under the investigator's control. Therefore, these observations are organized into classes in a frequency table and these classes and the number of observations in each class are used for the x and y axes respectively. Each class is represented as a bar and each bar abuts the next, indicating that the scale of measurement is continuous not discrete. The tops of these bars should not be joined together with a trend line. If many observations are recorded within each class, the variation in each class may be represented by confidence intervals.

Homogeneous stands Where a number of areas that are similar to each other are selected for investigation.

Hypothesis The formal phrasing of each objective; includes details relating to the experiment and the way in which the data will be tested statistically.

Independent observations Where only one observation is measured from each item for one treatment variable.

Independent variable Is a term usually used when the data are plotted on a scatter plot and describes the treatment variable plotted on the x axis.

Information-theoretic models When data are collected and analysed to see whether there are significant groupings. Only then may hypotheses about the factors causing these groupings or the nature of these groupings be tested, i.e. models are constructed from the observations.

Interaction The response to a treatment shown by one group of items is reversed in a second sample in such a way that it is clear that, for example, two treatment variables are not having the same effect in two groups. For example, the number of rabbits was recorded in two locations A and B at three points in the year. In location 1, the number of rabbits increased during the year, but in location 2, the number of rabbits decreased during the year. The effect of 'time' on the number of rabbits is different for the two samples, so the effect of time is dependent on location. Time and location would be said to be interacting.

Interpolation The estimation of a value when only nearby values are known, one greater than the required value and the other less than the required value.

Interquartile range See Range.

Interval data These data are measured on a continuous scale; the data are rankable, and it is possible to measure the difference between each observation. Interval scales include temperature (°C), distance (cm), and time (mins).

Interview A meeting between the researcher and a participant. Interviews can be structured, in that the interviewer asks the same series of questions to all the interviewees, or unstructured, where the interviewee is left to make their own comments about a topic.

Item One representative of the statistical population you wish to measure. In other texts, an item may be referred to as an 'experimental unit', subject, or case.

Keywords A word or short phrase used by potential readers when searching for relevant research publications. A list of keywords may be requested by publishers.

Kurtosis A measurement that indicates how sharp the peak of the central point of a data set is.

Likert scale An ordinal scale commonly used to assess attitudes to a number of different statements.

Line graph A figure for observations from an investigation with one treatment variable. The data are measured on a continuous (interval) scale. There are few data points along the x axis (treatment variable), usually because the treatment variable is under the investigator's control. Each observation or mean from a group of observations for a given x value is plotted as a point on the figure. Variation around a mean can be indicated using confidence intervals. These points may be joined together with a trend line.

Literature review To identify all items of literature in the area you are considering followed by a review and selection process which identifies the research that best reflects the current thinking in the area.

Matched When two or more observations are recorded for one variable for each item; for example, in a study of heart rate, a resting rate was taken for ten people first, and then following exercise another heart-rate reading was taken for the same ten people. For each person, there is a before and after reading, so these data are matched. An alternative approach would be to assign ten people at random and record a resting heart rate, and then assign a different group of ten people to the exercise treatment and take an 'after exercise' heart rate from this second, independent group. These data would not be matched. Matched data can extend to more than two observations. For example, if a mouse was run through a maze once a day for a week and the time taken was recorded, there would be five 'matched' observations. These are often called repeated measures.

Mean One measure of central tendency for interval data where the distribution is known. For a normal distribution, the sample mean ($\bar{x}$) and statistical population mean (μ) are calculated in the same way. The sample mean is used as an estimate of the population mean and a value called the confidence interval (5.7) can be calculated to demonstrate the area around a sample mean in which the population mean will probably fall. The sample mean for normally distributed data is calculated as the sum (Σ) of all observations (x) divided by the number of observations in the data set (n).

Median One measure of central tendency, calculated as the middle value of an ordered data set.

Mode One measure of central tendency. When data are organized into categories, the mode is the category that contains the greatest number of observations. The categories may be inherent in the data (e.g. the number of blossoms on a rose plant) or imposed on continuous data (e.g. the heights of trees placed in ranges of 0.00–0.99m, 1.00–1.99m, etc.).

Monotonic Is an association between two variables such that when you plot a graph of one of the variables against the other, the graph will either go up from left to right (monotonic increasing) or go down from left to right (monotonic decreasing). The association does not have to be a straight line, but it must either always increase or always decrease. An example of a monotonic distribution can be seen in Fig 9.5d.

National law Law developed in the UK from common law, legislation, institutional writers, canon law, and custom.

Nominal Observations or categories of a variable that are discrete but where the values or categories cannot be ordered, e.g. colours (red, green, or blue).

Non-maleficence To do no harm.

Non-parametric data Data that, when plotted, have a non-normal distribution or, more usually, where the distribution is unknown. When testing hypotheses and you have non-parametric data, you may either transform the data or use non-parametric statistics. Parametric data may be analysed using non-parametric statistics but not the other way round.

Non-treatment variable See Confounding variables.

Normal distribution A symmetrical, 'bell-shaped' distribution with a single central peak (unimodal), which can be described by the Gaussian equation. The data are measured on an interval scale.

Null hypothesis (H_0) An assumption that two or more samples or observations are derived from the same population. In a population with a normal distribution this model may be described in terms of the means, $\mu = \mu_s$.

Numerator In an equation where one term is divided by another term, the value that is being divided into is the numerator. For example, in the fraction 2/4, 2 is the numerator.

Objective Provides succinct and specific details about an experiment that is being carried out in order to examine all or part of an aim. Objectives can be 'clumped' or 'split' and can be experimental or personal.

Observation A single measurement taken from one item.

One-tailed test When the hypothesis testing is such that only one end of a distribution is being considered. One-tailed tests are most often encountered when testing specific hypotheses. The alternate hypothesis can be illustrated as a model relating to the mean and will be either $\mu > \mu_s$ or $\mu < \mu_s$.

Open question Used in questionnaires where the answers are not prescribed. See also Closed questions.

Ordinal Observations or categories of a variable that are discrete and where the values or categories can be consistently ordered. The interval between each observation or category has no meaning, e.g. the ACFOR scale (abundant, common, frequent, occasional, rare).

Orthogonal A particular experimental design where every category for one treatment is found in combination with every category for every other treatment.

Outlier An observation that is noticeably different from all other observations, one that does not follow the apparent trend.

Parametric data Data measured on an interval scale and so quantitative, continuous, and rankable. The data have a normal distribution, e.g. height (cm). When testing hypotheses and you have parametric data, you should use parametric statistics. See also Non-parametric data.

Participant information sheet Information given to potential volunteers which outlines what the participant will be asked to do and why.

Pie chart A figure for observations from an investigation with one treatment variable. The data are measured on a nominal or ordinal scale. A circle is divided up proportionately to reflect the relative proportions (or percentages) of the observations in each category or sample. If many observations are recorded within each category, you should consider using a bar chart so that the variation in each category may be represented by confidence intervals.

Poisson distribution A unimodal distribution, which is often very asymmetrical with a protracted tail either to the right (a positive skew) or to the left (a negative skew). A distribution often found where events are randomly distributed in time or space, such as the distribution of cells within a liquid culture or the dispersal of pollen or seed by wind.

(Statistical) Population All the individual items that are the subject of your research. A statistical population may be the same as an ecological or genetic population but not necessarily.

Power of a test An indicator of how effective a statistical test is in not detecting a difference when one is not present as a result of the action of the factors under investigation.

Qualitative Categories of a treatment variable or observations that are descriptive and non-numerical. If categories are qualitative they are mutually exclusive and therefore discrete.

Quantitative Categories of a variable or observations that are numerical, e.g. number of eggs in a clutch, or height (cm). Data may be either discrete or continuous.

Questionnaire A collection of questions given to all participants in an investigation. The participant writes down their answer and returns the questionnaire to the researcher. Questionnaires are usually anonymous and confidential. They require very little contact between the researcher and volunteers and usually there is no interaction between participants.

Random sampling Each item must have an equal chance of being sampled each time.

Random variable When the variable is not under the control of the investigator and is subject to sampling error.

Range The range is the difference between the highest and lowest observations. A reduced range may be used (e.g. from the 25th observation to the 75th observation in a data set of 100 observations). This avoids any undue influence by extreme values on the range. These are known as percentiles and the interquartile range.

Rankable Observations consist of named categories or values that have an order to them; for example, finishing positions in a race, or height (cm).

Reference list The full details of each source of information that has been used explicitly in the construction of your report.

Regions of rejection and acceptance A distribution can be divided into regions using mathematical relationships such as between the mean and standard deviation. These regions separate rare or extreme values from more common values. In hypothesis testing the regions with the rare values are called the regions of rejection and the region with the common values is called the region of acceptance.

Relationship Where one variable is directly responsible for causing the change in another variable. See also Association.

Repeated measures See Matched.

Replication Where more than one item is exposed to a treatment, more than one group of items is exposed to one treatment, or more than one item is included in a sample. All replicates should be an integral part of one experiment and should be independent of each other.

Representative sample of items A sample that generates sample statistics that closely match the population parameters through the use of a clearly defined and consistently applied method.

Risk assessment A process whereby risks associated with work are identified, minimized, and finally evaluated to determine if the work can proceed.

Sample A number of observations recorded from a subset of items from the statistical population.

Sampling error Variation between samples collected from a single population that has occurred by chance and the variation between the statistical population value and sample value that has arisen by chance.

Scatter plot A figure that can be drawn when observations for two or more variables are made for each item. Each of these observations relates to one treatment variable. If you have an independent variable, this is plotted on the x axis. The second variable, which may be a dependent variable, is plotted on the y axis. If more than two variables are recorded, the plot will be

three (or more) dimensions. The two observations from each item are marked as a point on the figure. If both measurements are interval, these points may be joined together with a trend line. You may carry out regression analysis, and if significant, you may draw a regression line. If more than one y value has been recorded for each x value, you may use confidence intervals to reflect the variation around the mean y value.

Significant difference If a sample or observation is found to fall into the region of rejection then the sample or observation is said to be significantly different from the original population. This is usually reported along with the test statistic and the probability.

Skew When observations are plotted, the distribution may not be symmetrical. Where there is a tail of observations to the right, the distribution is said to have a positive skew. Where there is a tail of observations to the left, the distribution is said to have a negative skew.

Special Areas of Conservation (SACs) Areas designated within the EU with the intention of improving the conservation of specific habitats.

Specific hypothesis Some statistical tests allow you to ask more discerning (specific) questions. Specific tests of hypotheses are usually one-tailed tests.

Standard deviation This is a measure of the variation in data where the distribution is known. For data with a normal distribution, it is calculated as the square root of the variance and is a value with the same units as the original observations.

Standard error of the mean For a given statistical population, you may take many samples and these can all be described in terms of a mean. If you then used these means as your data, you could calculate a standard deviation of these mean values. This standard deviation of the mean is more commonly known as the standard error of the mean.

Statistical population See Population.

Stratified random sampling Where random sampling occurs at regular intervals.

Sum of squares of x or y SS(x) or SS(y). The sum of squared deviations from the mean. This is the numerator in a calculation of the variance for normally distributed data.

Systematic/periodic sampling A regularized sampling method where an item is chosen for observation at regular intervals.

Test statistic The value calculated during a test of hypothesis using the data in your experiment and based on a specific distribution. The test statistic is named after the distribution such as z, t, or χ^2.

Transforming data A mathematical calculation that has the effect of squeezing and/or stretching the scale you used when making your measurements so that the distribution takes on the approximate shape of a bell-shaped curve. The process of transforming data can be carried out when you wish to normalize non-parametric data.

Treatment (definition 1) When carrying out an experiment, your items are exposed to a particular environment that is manipulated by the investigator.

Treatment (definition 2) A statistical term for any samples that are being compared.

Treatment variables The factor(s) under investigation where you may examine your data for a difference, or the effect of the treatment variables, or you are looking to see if there is an association.

Two-tailed test Where the hypothesis testing is against both ends of a distribution. Most general tests of hypotheses are usually two-tailed. The alternate hypothesis can be illustrated as a model relating to the mean as $\mu \neq \mu_s$.

Type I error If we reject the null hypothesis even when it is really true and the factors we are testing are not affecting the biology of the system. The probability of a Type I error occurring is α ($= p$, level of significance (%)).

Type II error Occurs when a test of hypothesis does not reject a null hypothesis when the factor being tested actually is having an effect on the biological system. The probability of this happening is called β.

Unimodal The description of a distribution that has a single high point or peak. See Bimodal.

Unmatched See Matched.

Variable See Treatment variable, Dependent variable, and Confounding variable.

Variance A measure of the variation in the data. The calculation differs depending on the underlying distribution and whether it is for a population variance (σ^2) or a sample variance (s^2). For normally distributed data the sample variance is calculated as:

$$s^2 = \frac{\Sigma(x - \bar{x})^2}{n-1}$$

References

Barnard, C., Gilbert, F., & McGregor, P. (2001). *Asking Questions in Biology*. Longman Scientific and Technical, Harlow.

Bullock, J. M., Clear Hill, B., & Silvertown, J. (1994). Demography of *Cirsium vulgare* in a grazing experiment. *Journal of Ecology* **82**: 101-11.

Fielding, A. H. (2007). *Cluster and Classification Techniques for the Biosciences*. Cambridge University Press, Cambridge.

Foa, E. B., Huppert, J. D., Leiberg, S., Langner, R., Kichic, R., Hajcak, G., & Salkovskis P. M. (2002). *The Obsessive-Compulsive Inventory: development and validation of a short version*. [Online] 14(4): 485-96. Available from: http://www.ncbi.nlm.nih.gov/pubmed/12501574

Fowler, J., Cohen, L., & Jarvis, P. (1998). *Practical Statistics for Field Biologists*. John Wiley & Sons, Chichester.

Fresnillo, B. & Ehlers, B. K. (2008). Variation in dispersability of mainland and island populations of three wind dispersed plant species. *Plant Systematics and Evolution* **270**: 243-55.

Hawkins, D. (2009). *Biomeasurement: Understanding, Analyzing and Communicating Data in the Biosciences*, 2nd edn. Oxford University Press, Oxford.

Johnson, S. & Scott, J. (2014). *Study and Communication Skills for the Biosciences*. Oxford University Press, Oxford.

Klinkhamer, P. & de Jong, T. (1993). Biological flora of the British Isles: *Cirsium vulgare. Journal of Ecology* **81**: 177-91.

Kruskal, W. H. & Wallis, W. A. (1952). Use of ranks in one-criterion analysis of variance. *Journal of the American Statistical Association* **47**: 583-621.

Legendre, P. & Legendre, L. (2012). *Numerical Ecology*, 3rd English edn. Elsevier Science, Amsterdam.

Meddis, R. (1984). *Statistics Using Ranks: a Unified Approach*. Blackwell Publishers, Oxford.

Quinn, G. P. & Keough, M. J. (2002). *Experimental Design and Data Analysis for Biologists*. Cambridge University Press, Cambridge.

Robson, C. (2011). *Real World Research*. 3rd edn. John Wiley & Sons, Chichester.

Rodwell, J. S. (ed.) (1991). *British Plant Communities*, Vol. 1: *Woodlands and Scrub*. Cambridge University Press, Cambridge.

Ruxton, G. D. (2006). The unequal variance *t* test is an underused alternative to Student's *t* test and the Mann Whitney U test. *Behavioural Ecology* **17**: 688-90.

Scheirer, C. J., Ray, W. S., & Hare, N. (1976). The analysis of ranked data derived from completely randomised factorial designs. *Biometrics* **32**: 429-34.

Shearer, P. R. (1973). Missing data in quantitative designs. *Journal of the Royal Statistical Society Ser. C. Applied Statistics* **22**: 135-40.

Skarpaas, O., Stabbetorp, O. E., Ronning, I., & Svennungsen, T. O. (2004). How far can a hawk's beard fly? Measuring and modelling the dispersal of *Crepis praemorsa. Journal of Ecology* **92**: 747-57.

Sokal, R. R. & Rohlf, F. J. (2011). *Biometry*, 4th edn. W. H. Freeman & Co., New York.

Student (1908). The probable error of a mean. *Biometrika* **6**: 1-25.

Thompson, K., Gaston, K. J., & Band, S. (1999). Range size, dispersal and niche breadth in herbaceous flora of central England. *Journal of Ecology* **87**: 150-5.

Westbury, D. B., Woodcock, B. A., Harris, S. J., Brown, V. K., & Potts, S. G. (2008). The effects of seed mix and management on the abundance of desirable and pernicious unsown species in arable buffer strip communities. *Weed Research* **48**: 113-23.

Wright, N. R. (1997). Breath alcohol concentrations in men 7-8 hours after prolonged, heavy drinking: influence of habitual alcohol intake. *Lancet* **349**: 182.

Zar, J. H. (2009) *Biostatistical Analysis*, 5th edn (international edition). Pearson Higher Education, New York.

Index

Note: If a definition of a term is required, readers are directed to the glossary (pages 450–5).